AF560387

QUICK REVIEW IN ORGANIC CHEMISTRY

QUICK REVIEW IN ORGANIC CHEMISTRY

M.S. Yadav

ANMOL PUBLICATIONS PVT. LTD.
NEW DELHI - 110 002 (iNDIA)

ANMOL PUBLICATIONS PVT. LTD.
4374/4B, Ansari Road, Daryaganj
New Delhi - 110 002
Ph.: 23261597, 23278000
Visit us at: www.anmolpublications.com

Quick Review in Organic Chemistry

First Published, 2004

ISBN 81-261-1895-4

PRINTED IN INDIA

Published by J.L. Kumar for Anmol Publications Pvt. Ltd., New Delhi - 110 002 and Printed at Mehra Offset Press, Delhi.

Contents

Preface

In this book the important concepts, rules, mechanisms, reactions involved in the organic compounds have been given in a very concise manner. It gives an over view of the whole of the organic chemistry studied by degree students. Some of the more difficult to understand concepts have been simplified to such an extent that they can be easily understood by an average student.

The book has been divided into 15 chapters and each chapter deals with a particular aspect. Each topic is self sufficient. To keep the volume in limited size the duplication of reactions has been farsided and at places the reactions given in text have not been translated into chemical equations.

While dealing with chemical reactions the emphasis has been put to explain not only the mechanism of the reaction but also as to why a particular reaction occurs or does not occur? Why the reactivity has changed in the series of compounds? etc.

It is hoped that the book would be liked by readers as it provides all the important facts of organic chemistry and a comprehensive form.

Author

1

Structure and Bonding

1.1 Atomic Structure of Carbon

Atomic Orbitals

Carbon has six electrons i.e. $1s^2 2s^2 p_x^1 p_y^1$ and is placed in second period of the periodic table. In it there are two shells of atomic orbitals available for these electrons. The first shell closest to the nucleus has a single *s* orbital–the 1*s* orbital. The second shell has a single *s* orbital (the 2*s* orbital) and three *p* orbitals (3 × 2*p*). Therefore, there are a total of five atomic orbitals into which these six electrons can fit. The *s* orbitals are spherical in shape with the 2*s* orbital being much larger than the 1*s* orbital. The *p* orbitals are dumbbell-shaped and are aligned along the *x, y* and *z* axes. Therefore, they are assigned the $2p_x$, $2p_y$ and $2p_z$ atomic orbitals (Fig. 1.1).

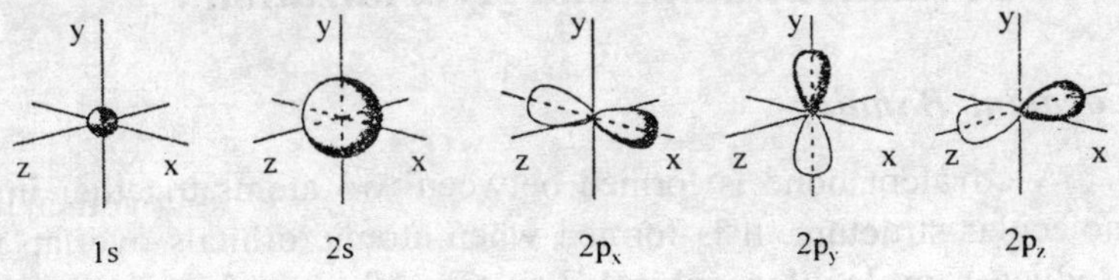

Fig. 1.1. Atomic orbitals.

Energy Levels

Various atomic orbitals of carbon atom are not of equal energy (Fig. 1.2). The 1*s* orbital has the lowest energy. The 2*s* orbital is next in energy and the 2*p* orbitals have the highest energies. The three 2*p* orbitals have the same energy, i.e. they are **degenerate.**

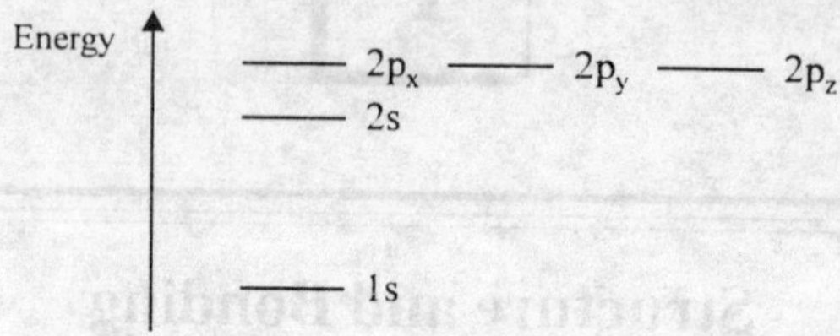

Fig. 1.2. Energy levels of atomic orbitals.

Electronic Configuration

According to the **aufbau principle, Pauli exclusion principle** and **Hund's rule,** the electronic configuration of carbon is $1s^2 2s^2 2p_x^{\,1} 2p_y^{\,1}$ (see Fig. 1.3).

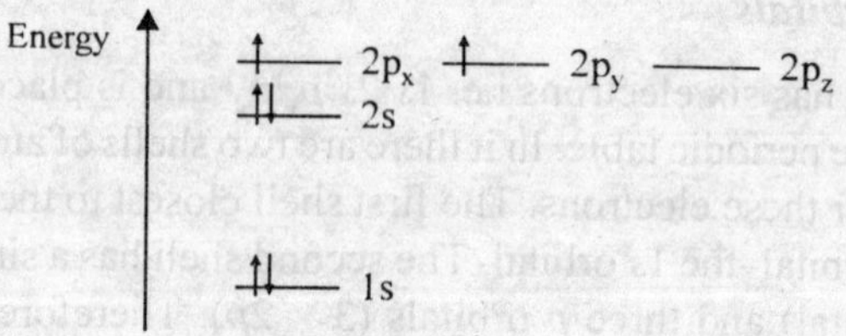

Fig. 1.3. Electronic configuration for carbon.

The electronic configuration for carbon is $1s^2\ 2s^2\ 2p_x^{\,1}\ 2p_y^{\,1}$. The numbers in superscript refer to the numbers of electrons in each orbital.

1.2 Covalent Bonding and Hybridization

Covalent Bonding

A covalent bond is formed between two atoms together in a molecular structure. It is formed when atomic orbitals overlap to produce a **molecular orbital**. For example, the formation of a hydrogen molecule (H_2) from two hydrogen atoms. Each hydrogen atom has a half-filled 1*s* atomic orbital and when the atoms approach each other, the atomic orbitals interact to produce two MOs (the number of resulting MOs must equal the number of original atomic orbitals, Fig. 1.4).

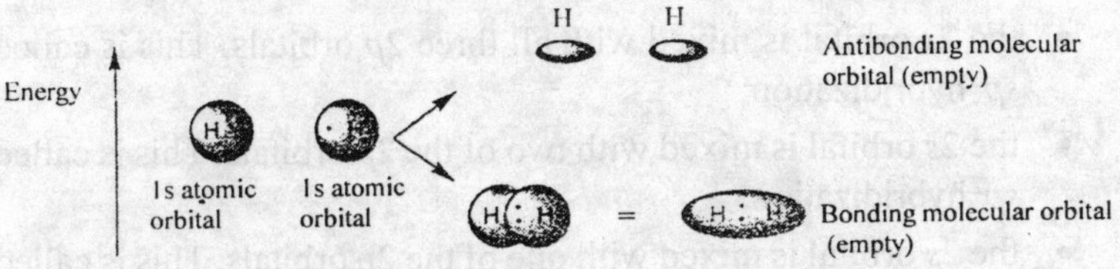

Fig. 1.4. Molecular orbitals for hydrogen (H_2).

The MOs are of different energies. One is more stable than the combining atomic orbitals and is known as the **bonding** MO. The other is less stable and is known as **antibonding** MO. The bonding MO is like a rugby ball and is formed by the combination of the 1*s* atomic orbitals. As this is the more stable MO, the valence electrons (one from each hydrogen) enter this orbital and pair up. The antibonding MO is of higher energy and consists of two deformed spheres. *This remains empty*. As the electrons end up in a bonding MO that is more stable than the original atomic orbitals, energy is released and bond formation is favoured.

Sigma Bonds

σ-bonds have a circular cross-section and are formed by the head-on overlap of two atomic orbitals. This involves a strong interaction and thus sigma bonds are strong bonds.

Hybridization

Atoms can form bonds with each other by sharing unpaired electrons such that each bond contains two electrons. A carbon atom has two unpaired electrons and so we expect carbon to form two bonds. However, carbon forms four bonds! How does a carbon atom form four bonds with only two unpaired electrons?

When a carbon atom forms bonds and is part of a molecular structure, it can 'mix' the *s* and *p* orbitals of its second shell (the valence shell). This is called **hybridization** and it allows carbon to form the four bonds which are observed in its compounds.

There are three important ways in which the mixing process can occurs are:-

- the 2*s* orbital is mixed with all three 2*p* orbitals. This is called sp^3 hybridization;
- the 2*s* orbital is mixed with two of the 2*p* orbitals. This is called sp^2 hybridization;
- the 2*s* orbital is mixed with one of the 2*p* orbitals. This is called as *sp* hybridization.

The other types of hybridization observed in some other compounds are

(i) dsp^2 – in it one $(n - 1)d$ orbital and one *ns* orbital and 2 *np* orbitals combine to form 4 hybridized orbitals.

(ii) d^2sp^3 or sp^3d^2: The six hybridized orbitals are formed by mixing up of two $\lambda(n - 1)d$ orbitals, or two *nd* orbitals with one *ns* and three *np* orbitals.

1.3 sp^3 Hybridization

Definition

The sp^3 hybridization of carbon involves mixing up of the 2*s* orbitals with all three of the 2*p* orbitals to give a set of four sp^3 hybrid orbitals. The hybrid orbitals will each have the same energy but will be different in energy from the original atomic orbitals. That energy difference will reflect the mixing of the respective atomic orbitals. The energy of each hybrid orbital is greater than the original *s* orbital but less than the original *p* orbitals (Fig. 1.5).

Electronic Configuration

The valence electrons for carbon can now be fitted into the sp^3 hybridized orbitals (Fig. 1.5). There are a total of four electrons in the 2*s* and 2*p* orbitals of carbon. The *s* orbital was filled and two of the *p* orbitals were half filled. After hybridization there is a total of four hybridized sp^3 orbitals all of equal energy. According to Hund's rule, all the four hybridized orbitals are half filled with electrons thus there are four unpaired electrons. Four bonds are now-possible.

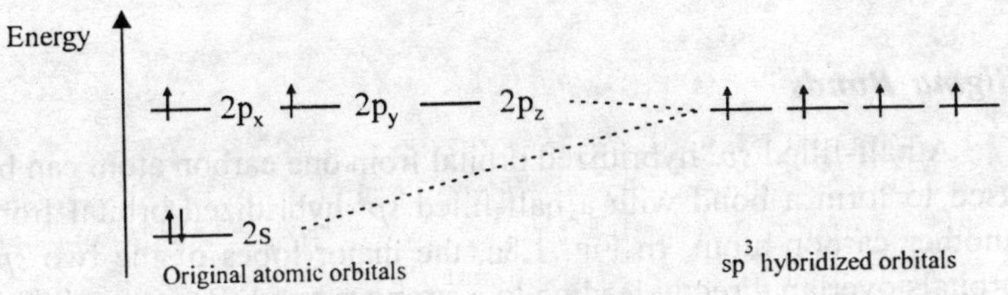

Fig. 1.5. sp^3 Hybridization.

Geometry

Each of the sp^3 hybridized orbitals has the same shape (Fig. 1.6). This deformed dumbbell looks more like a *p* orbital than an *s* orbital since more *p* orbitals were involved in the mixing process, i.e. one lobe is larger as compared to other lobe.

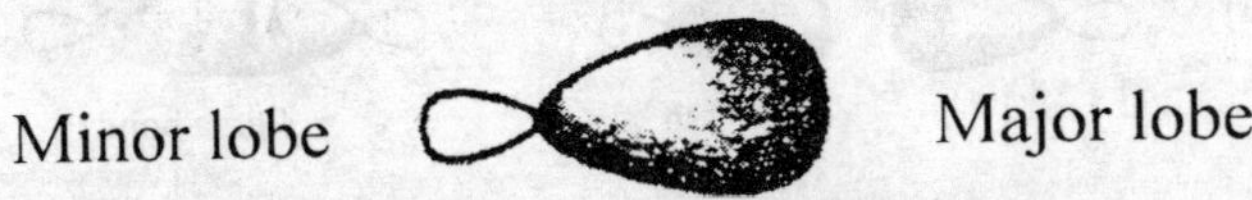

Fig. 1.6. sp^3 Hybridized orbital.

Each sp^3 orbital occupies a space as far apart from each other as possible by pointing to the concerns of tetrahedron (Fig. 1.7). The angle between each of these lobes is about 109.5°. This is what is meant by the expression **tetrahedral carbon.** The three-dimensional shape of the tetrahedral carbon can be represented by drawing a normal line for bonds in the plane of the page. Bonds going behind the page are represented by hatched wedge, and bonds coming out the page are represented by a solid wedge.

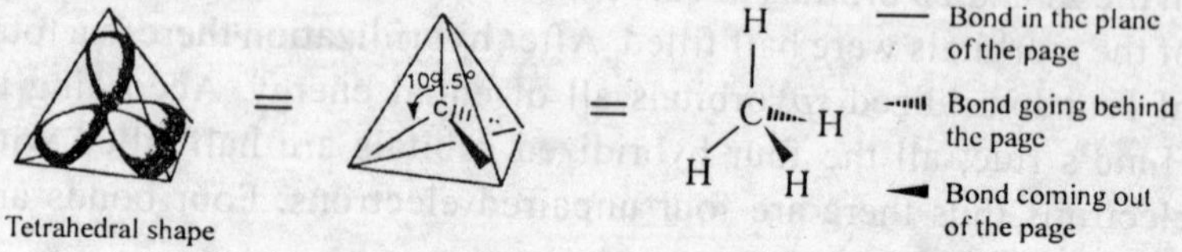

Fig. 1.7. Tetrahedral shape of an sp^3 hybridized carbon.

Sigma Bonds

A half-filled sp^3 hybridized orbital from one carbon atom can be used to form a bond with a half-filled sp^3 hybridized orbital from another carbon atom. In Fig. 1.8a, the major lobes of the two sp^3 orbitals overlap directly leading to a strong σ bond. Because of their ability to form strong σ bonds hybridization takes place. The deformed dumbbell shapes permit a better orbital overlap than would be obtained from a pure *s* orbital or a pure *p* orbital. A σ bond between an sp^3 hybridized carbon atom and a hydrogen atom involves the carbon atom using one of its half-filled sp^3 orbitals and the hydrogen atom using its half-filled 1*s* orbital (Fig. 1.8b).

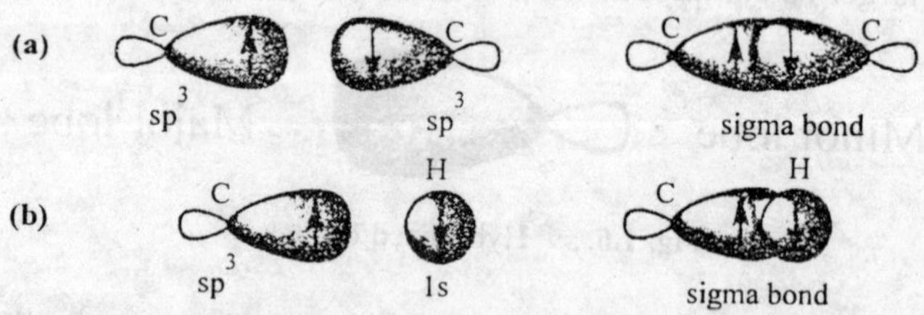

Fig. 1.8. (a) σ Bond between two sp^3 hybridized carbons; (b) σ bond between an sp^3 hybridized carbon and hydrogen.

Nitrogen, Oxygen and Chloride

The sp^3 type of hybridization is also observed for nitrogen, oxygen and chlorine atoms in organic structures. Nirtogen has five valence electrons in its second shell (i.e. $2s^22p^3$). After hybridization, it will have three half-filled sp^3 orbitals and thus can form three bonds. Oxygen has six valence electrons (i.e. $2s^22p^4$). After hybridization, it will have two half-filled sp^3 orbitals and will form two bonds. Chlorine has seven valence electrons (i.e. $2s^22p^5$). After hybridization, it will have one half-filled sp^3 orbital and will form one bond.

The four sp^3 orbitals for these three atoms (i.e. N, O and Cl) form a tetrahedral arrangement having one or more of the hybridized orbitals occupied by a lone pair of electrons. For an isolated atom, nitrogen forms a pyramidal shape where the bond angles are slightly less than 109.5° (c. 107°) (Fig. 1.9a). This compression of the bond angles is because of the orbital containing the lone pair of electrons, which requires a slightly greater amount of space than a bond. Oxygen forms an angled or bent shape where two lone pairs of electrons compress the bond angle from 109.5° to c. 104° (Fig. 1.9b).

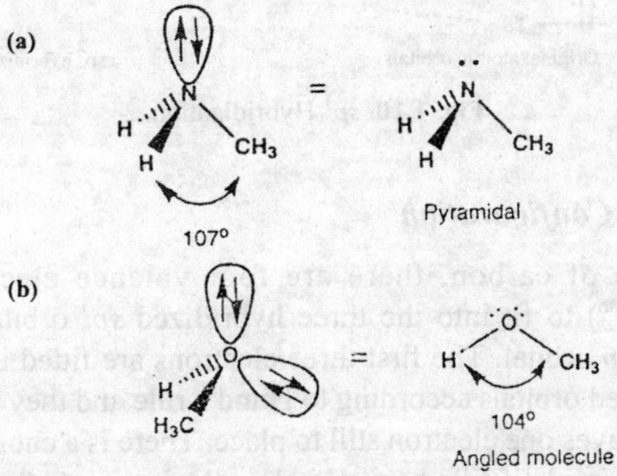

(a)

N, H, H, CH_3 = N, H, H, CH_3

107°

Pyramidal

(b)

O, H, H_3C = O, H, CH_3

104°

Angled molecule

Fig. 1.9. (a) Geometry of sp^3 hybridized nitrogen; (b) geometry of sp^3 hybridized oxygen.

Alcohols, amines, alkyl halides, and ethers all contain sigma bonds that involve nitrogen, oxygen, or chlorine. Bonds between these atoms and carbon are formed by the overlap of half-filled sp^3 hybridized orbitals from each atom. Bonds involving hydrogen atoms (e.g. O–H and N–H) are formed by the overlap of the half-filled 1*s* orbital from hydrogen and a half-filled sp^3 orbital from oxygen or nitrogen.

1.4 sp^2 Hybridization

Definition

In sp^2 hybridization, the 2*s* orbital is mixed with two of the 2*p* orbitals (e.g. $2p_x$ and $2p_z$) to form three sp^2 hybridized orbitals of equivalent energy. The remaining $2p_y$ orbital remains unaffected. The energy of each hybridized orbital is greater than that of the original *s* orbital but less than that of the original *p* orbitals. The remaining 2*p* orbital (in this case the $2p_y$ orbital) remains at its original energy level (Fig. 1.10).

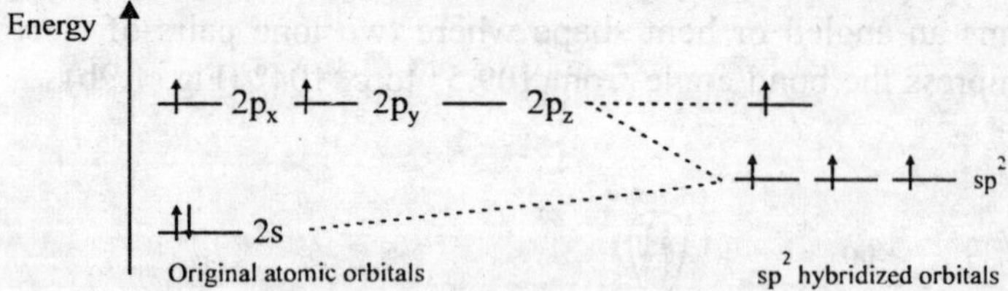

Fig. 1.10. sp^2 Hybridization.

Electronic Configuration

In case of carbon, there are four valence electrons (i.e. $2s^2 2p_x^{\ 1} 2p_y^{\ 1} p_z^{\ 0}$) to fit into the three hybridized sp^2 orbitals and the remaining 2*p* orbital. The first three electrons are fitted into each of the hybridized orbitals according to Hund's rule and they are all half-filled this leaves one electron still to place. There is a choice between pairing it up in a half-filled sp^2 orbital or placing it into the vacant $2p_y$ orbital. Generally we fill up orbitals of equal energy before moving to an orbital of higher energy. However, if the energy difference between orbitals is small (as here) it is easier for the electron to go into the higher energy $2p_y$ orbital resulting in three half-filled sp^2 orbitals and one half-filled orbital (Fig. 10). Four bonds are now possible.

Geometry

The unhybridized $2p_y$ orbital has the usual dumbbell shape. Each of the sp^2 hybridized orbitals has a deformed dumbbell shape similar to an sp^3 hybridized orbital. The difference between the sizes of the major and minor lobes is larger for the sp^2 hybridized orbital than that in case of sp^3 hybridized orbitals.

The hybridized orbitals and the $2p_y$ orbital occupy spaces as far apart from each other as possible. The lobes of the $2p_y$ orbital occupy the space above and below the plane of the *x* and *z* axes (Fig. 1.11a). The three hybridized sp^2 orbitals (major loves shown only) will then occupy the remaining space such that they are as far apart from the $2p_y$ orbital and from each other as possible. Thus, they are all placed in the *x-z* plane pointing toward the corner of a triangle (trigonal planar shape; Fig. 1.11b). The angle between each of these lobes is 120°. We are now ready to look at the bonding of an sp^2 hybridized carbon.

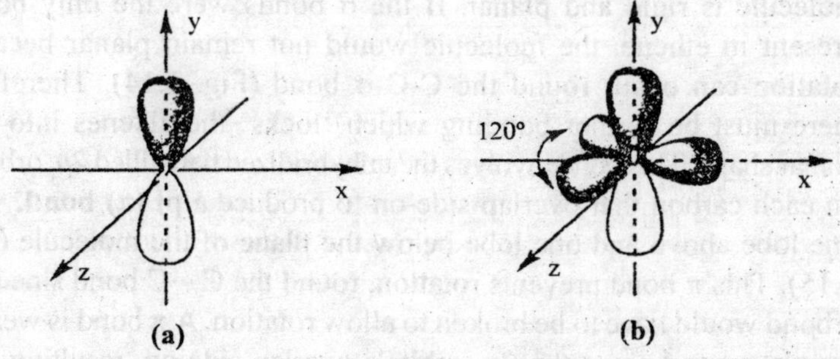

Fig. 1.11. (a) Geometry of the $2p_y$ orbital; (b) geometry of the $2p_y$ orbital and the sp^2 hybridized orbitals.

Alkenes

In alkenes all the four bonds formed by carbon are not σ-bonds. In this case of sp^2 Hybridization the three half-filled sp^2 hybridized orbitals form a trigonal planar shape. The use of these three orbitals in bonding explains the shape of an alkene, for example ethene ($H_2C{=}CH_2$). As far as the C-H bonds are concerned, the hydrogen atoms uses a half-filled 1*s* orbital to form a strong σ bond with a half filled sp^2 orbital from carbon (Fig. 1.12a). A strong σ bond is also possible between the two carbon atoms of ethene due to the overlap of sp^2 hybridized orbitals from each carbon (Fig. 1.12b).

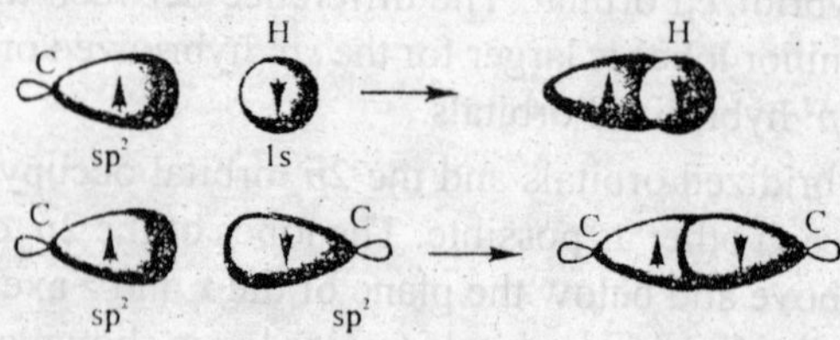

Fig. 1.12. (a) Formation of a C-H s bond; (b) formation of a C-C s bond.

The full σ bonding diagram for ethene is shown in Fig. 1.13a and it can be simplified as shown in Fig. 1.13b. *Ethene is a flat, rigid molecule where each carbon is trigonal planar.* sp^2 hybridization explains the trigonal planar carbons but can not explained why the molecule is rigid and planar. If the σ bonds were the only bonds present in ethene, the molecule would not remain planar because rotation can occur round the C-C σ bond (Fig. 1.14). Therefore, there must be further bonding which 'locks' the alkenes into this planar shape. This bond involves the unhybridized half-filled $2p_y$ orbitals on each carbon that overlap side-on to produce a **pi (π) bond**, with one lobe above and one lobe below the plane of the molecule (Fig. 1.15). This π bond prevents rotation. round the C – C bond since the π bond would have to be broken to allow rotation. A π bond is weaker than a σ bond since the $2p_y$ orbitals overlap side-on, resulting in a weaker overlap. The presence of a π bond can also explains the greater reactivity of alkenes than alkanes, since a π bond is more easily broken and is more likely to take part in reactions.

Fig. 1.13. (a) s Bonding diagram for ethene; (b) simple representation of s bonds for ethene.

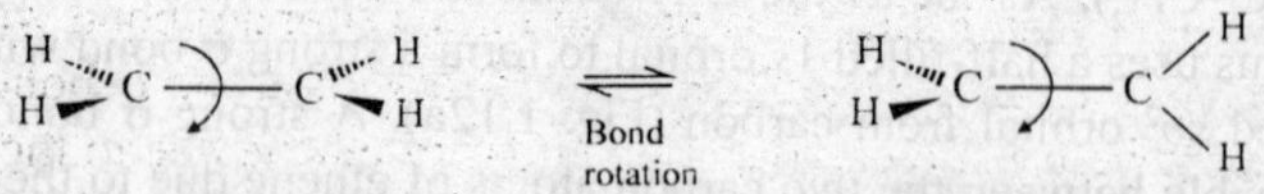

Bond rotation

Fig. 1.14. Bond rotation around a s bond.

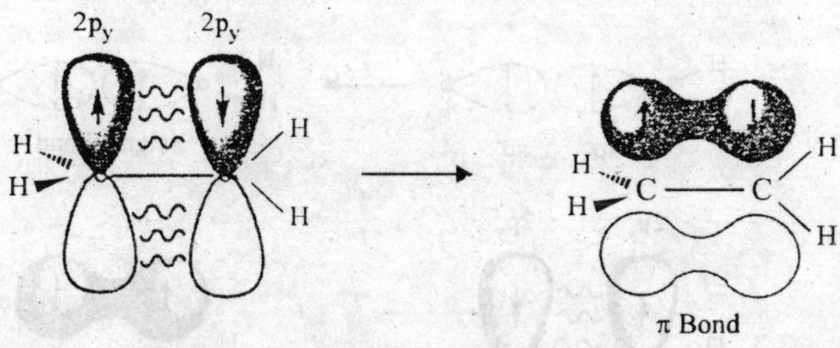

Fig. 1.15. Formation of a p bond.

Carbonyl Groups

In a carbonyl group (C=O) we find that both the carbon and oxygen atoms are sp^2 hybridized. The energy level diagram (Fig. 1.16) shows the arrangement of valence electrons of oxygen after sp^2 hybridization. Two of the sp^2 hybridized orbitals are filled with lone pairs of electrons, that leaves two half-filled orbitals available for bonding. The sp^2 orbital can be used to form a strong σ bond, while the $2p_y$ orbital can be used for the weaker π bond. Figure 1.17 shows how the σ and π bonds are formed in the carbonyl group and it also explains why carbonyl groups are planar with the carbon atom having a trigonal planar space. It also explains the reactivity of carbonyl groups since the π bond is weaker than the σ bond and is more likely to be involved in reactions.

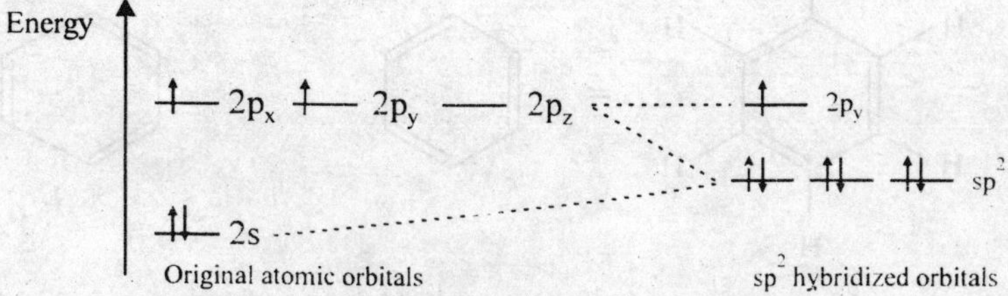

Fig. 1.16. Energy level diagram for sp^2 hybridized oxygen.

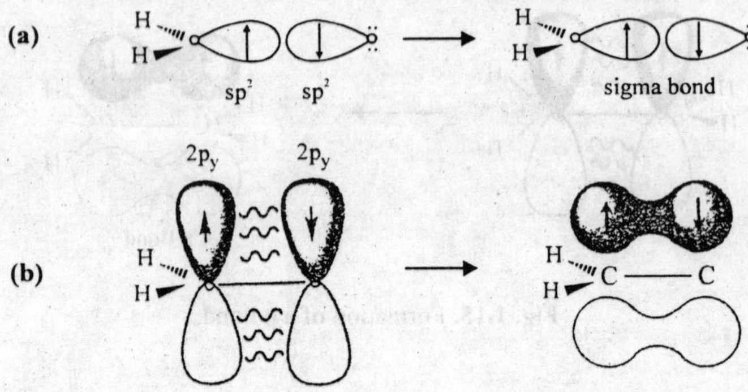

Fig. 1.17. (a) Formation of the carbonyl s bond; (b) formation of the carbonyl p bond.

Aromatic Rings

All the carbon atoms in an aromatic ring are sp^2 hybridized that means that each carbon can form three σ bonds and one π bond. All the single bonds are σ while a double bond consists of one σ bond and one π bond (Fig. 1.18a). For example, double bonds are shorter than single bonds and if benzene had this exact structure, the ring would be deformed with longer single bonds than double bonds (Fig. 1.18b).

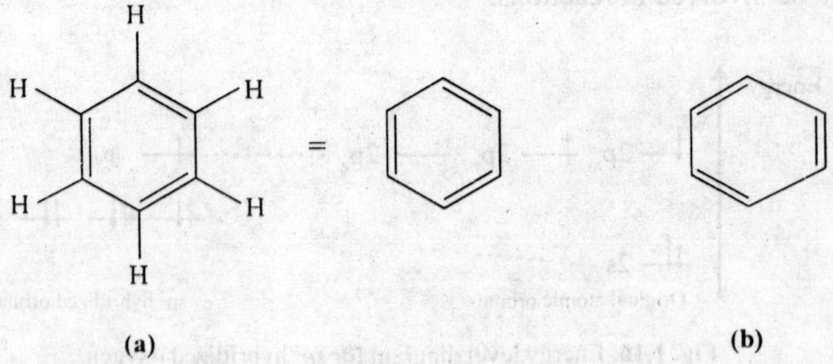

(a) (b)

Fig. 1.18. (a) Representation of the aromatic ring; (b) 'deformed' structure resulting from fixed bonds.

Actually all the C-C bonds in benzene are of the same length. To understand this, we must look more closely at the bonding which occurs. Figure 1.19a shows benzene with all its σ bonds and is drawn such that we are looking into the plane of the benzene ring. Since all the carbons are sp^2 hybridized, there is a $2p_y$ orbital left over on each carbon which can overlap with a $2p_y$ orbital on either side of it (Fig. 1.19b). From this, it is clear that each $2p_y$ orbital can overlap with its neighbours right round the ring. This leads to a molecular orbital that involves all the $2p_y$ orbitals where the upper and lower lobes merge to give two doughnut-like lobes above and below the plane of the ring (Fig. 1.20a), The molecular orbital is symmetrical and the six π electrons are said to be delocalized around the aromatic ring since they are not localized between any two particular carbon atoms. The aromatic ring is generally represented as shown in Fig. 1.20b to represent this delocalization of the π electrons. *Delocalization increases the stability of aromatic rings* and so they are less reactive than alkenes (i.e. it requires more energy to disrupt the delocalized π system of an aromatic ring than it does to break the isolated π bond of an alkene).

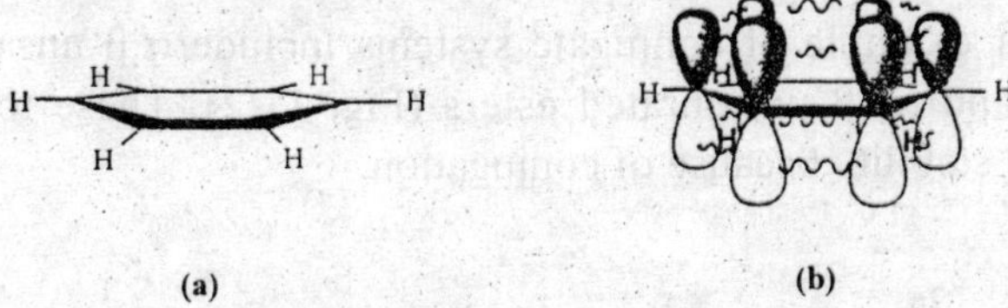

Fig. 1.19. (a) s Bonding diagram for benzene, (b) p Bonding diagram for benzene.

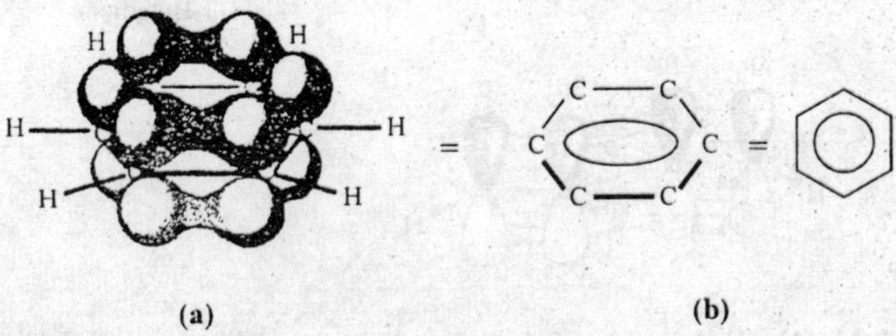

Fig. 1.20. Bonding molecular orbital for benzene; (b) representation of benzene to illustrate delocalization.

Conjugated Systems

Aromatic rings are not the only structures where delocalization of π electrons can occur. Delocalization can also occur in conjugated systems where there are alternating single and double bonds (e.g. 1,3-butadiene). All four carbons in 1,3-butadiene and sp^2 hybridized and so each of these carbons has a half-filled p orbital that can interact to give two π bonds (Fig. 1.21a). However, a certain amount of overlap is also possible between the p orbitals of the middle two carbon atoms and so the bond connecting the two alkenes has some double bond character (Fig. 1.21b). The delocalization also increases stability. However, the conjugation in a conjugated alkene is not as great as in the aromatic system. In the latter system, the π electrons are completely delocalized round the ring and all the bonds are equal in length. In 1,3-butadiene, the π electrons are not fully delocalized and are more likely to be found in the terminal C–C bonds. Although there is a certain amount of π character in the middle bond, the latter is more like a single bond than a double bond.

Other example of conjugate systems include α,β-unsaturated ketones and α, β-unsaturated esters (Fig. 1.22). These too have increased stability because of conjugation.

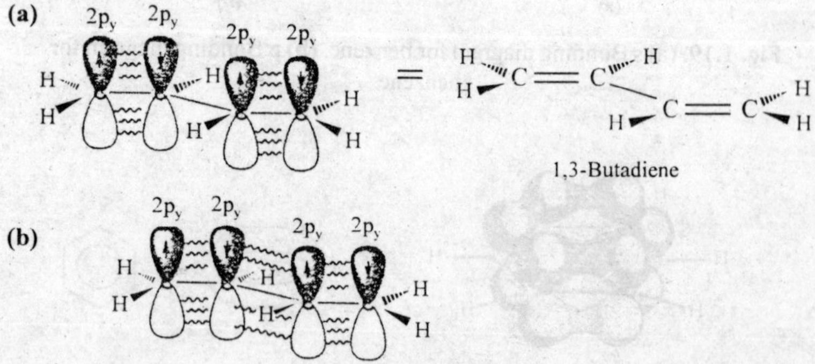

Fig. 1.21. (a) p Bonding in 1,3-butadiene; (b) delocalization in 1,3-butadiene.

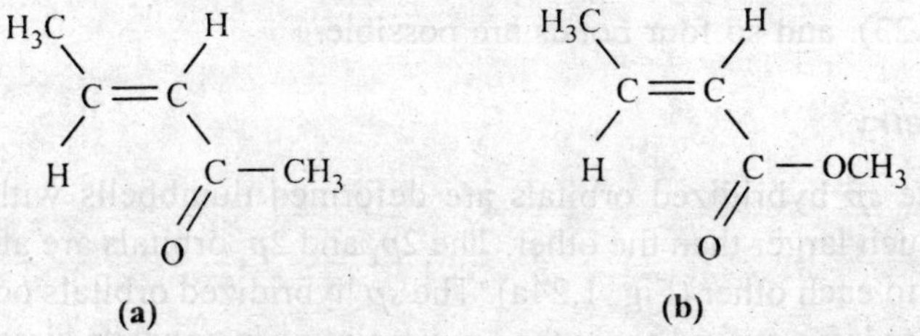

(a) (b)

Fig. 1.22. (a) a,b-unsaturated ketone; (b) a,b-unsaturated ester.

1.5 *sp*-Hybridization

Definition

In *sp* hybridization of carbon the 2*s* orbital is mixed with one of the 2*p* orbitals (e.g. $2p_x$) to form two *sp* hybrid orbitals or equal energy. This leaves two *2p* orbitals unaffected (i.e. unhybridized) ($2p_y$ and $2p_z$) with slightly higher energy than the hybridized orbitals (Fig. 1.23).

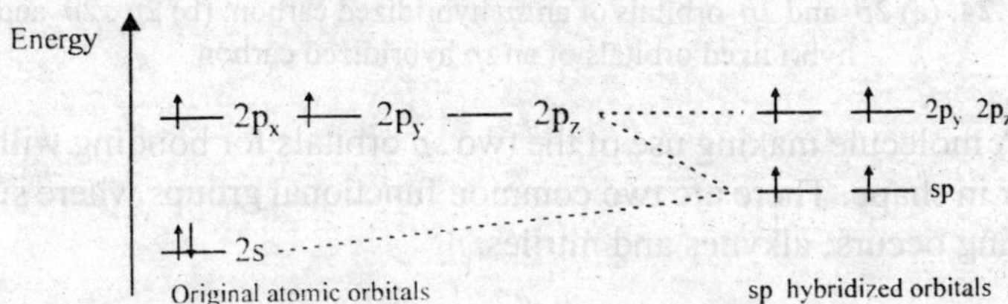

Fig. 1.23. *sp*-Hybridization of carbon.

Electronic Configuration

In a carbon atom the first two electrons fit into each *sp* orbital according to Hund's rule such that each orbital has a single unpaired electron. This leaves two electrons that can be paired up in the half-filled *sp* orbitals or placed in the vacant $2p_y$ and $2p_z$ orbitals. The energy difference between the orbitals is small and so it is easier for the electrons to fit into the higher energy orbitals than to pair up. This results in two half-filled *sp* orbitals and two half-filled 2*p* orbitals (Fig. 1.23), and so four bonds are possible.

Geometry

The *sp* hybridized orbitals are deformed dumbbells with one lobe much larger than the other. The $2p_y$ and $2p_z$ orbitals are at right angles to each other (Fig. 1.24a). The *sp* hybridized orbitals occupy the space left over and are in the *x* axis pointing in opposite directions (only the major lobe of the *sp* orbitals are shown in black; Fig. 1.24b).

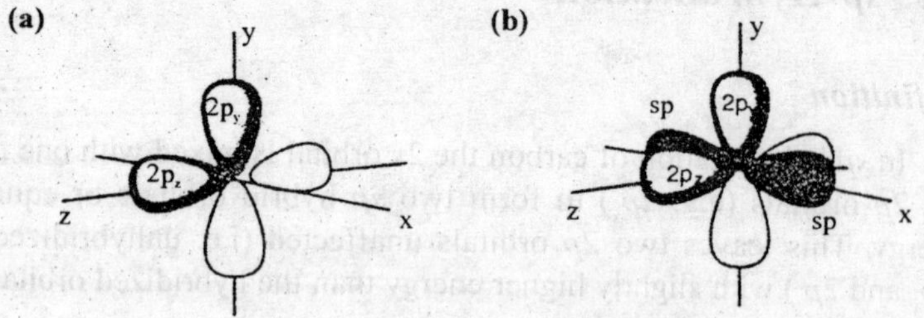

Fig. 1.24. (a) $2p_y$ and $2p_z$ orbitals of an *sp* hybridized carbon; (b) $2p_y$, $2p_z$ and *sp* hybridized orbitals of an *sp* hybridized carbon.

A molecule making use of the two *sp* orbitals for bonding will be linear in shape. There are two common functional groups where such bonding occurs; alkynes and nitriles.

Alkynes

In ethyne (Fig. 1.25) each carbon is *sp* hybridized. The C-H bonds are strong σ bonds where each hydrogen atom uses its half-filled 1*s* orbital to bond with a half-filled *sp* orbital on carbon. The remaining *sp* orbital on each carbon is used to form a strong σ carbon-carbon bond. The full σ bonding diagram for ethyne is linear (Fig. 1.26a) and can be simplified as shown (Fig. 1.26b).

$$H{-}C{\equiv}C{-}H$$

Fig. 1.25. Ethyne.

Fig. 1.26. (a) s Bonding for ethyne; (b) representation of s bonding.

Further bonding is possible because each carbon has half-filled *p* orbitals. Hence, the $2p_y$ and $2p_z$ orbitals of each atom can overlap side-on to form two π bonds (Fig. 1.27). The π bond formed by the overlap of the $2p_y$ orbitals is represented in dark gray. The π bond resulting from the overlap of the $2p_z$ orbitals is represented in light gray. Alkynes are linear molecules and are reactive due to the relatively weak π bonds.

Nitrile Groups

To explain the bonding within a nitrile group (C≡N) where both the carbon and the nitrogen are *sp* hybridized. Consider the energy level diagram in Fig. 1.28 shows how the valence electrons of nitrogen are arranged after *sp* hybridization. A lone pair of electrons occupies one of the *sp* orbitals, but the other *sp* orbital can be used for a strong σ bond. The $2p_y$ and $2p_z$ orbitals can be used for two π bonds. Figure 1.29 represents the σ bonds of HCN as lines and how the remaining 2*p* orbitals are used to form two π bonds.

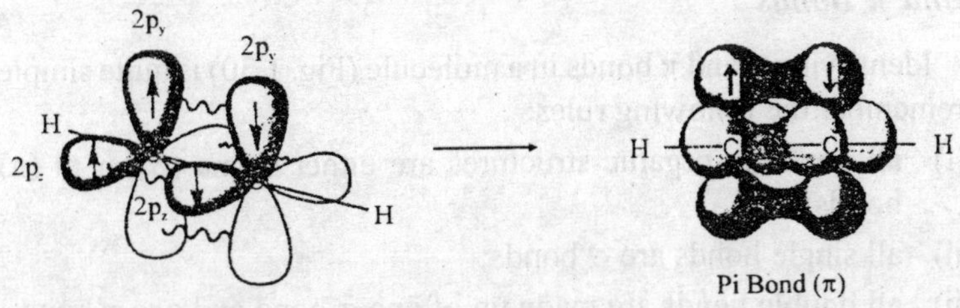

Fig. 1.27. p-Bonding in ethyne.

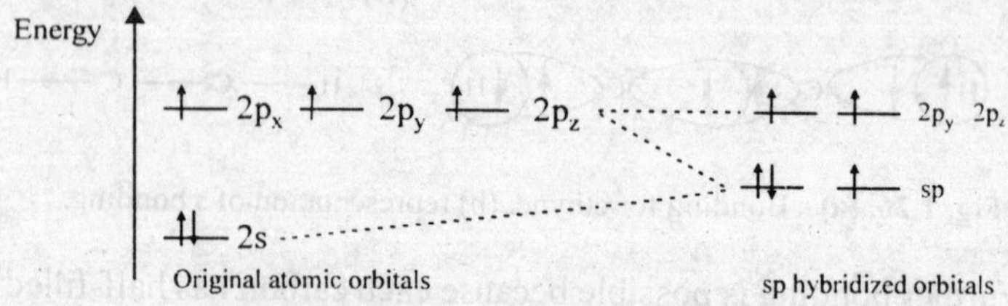

Fig. 1.28. *sp* Hybridization of nitrogen.

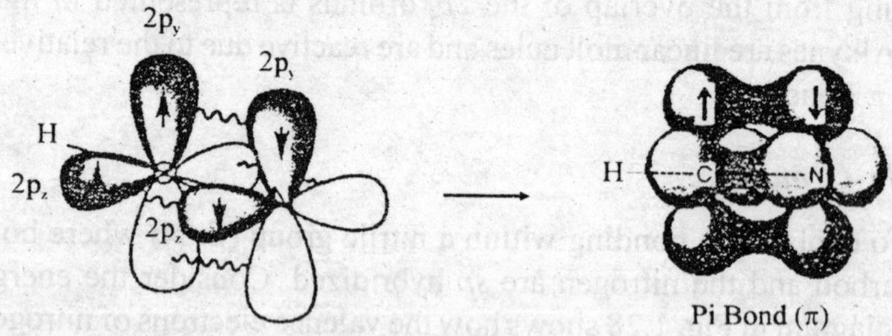

Fig. 1.29. p-Bonding in HCN.

1.6 Bonds and Hybridized Centers

σ *and* π *Bonds*

Identifying σ and π bonds in a molecule (Fig. 1.30) is quite simple if remember the following rules:

(i) all bonds in organic structures are either sigma (σ) or pi (π) bonds;

(ii) all single bonds are σ bonds;

(iii) all double bonds are made up of one σ bond and one π bond;

(iv) all triple bonds are made up of one σ bond and two π bonds.

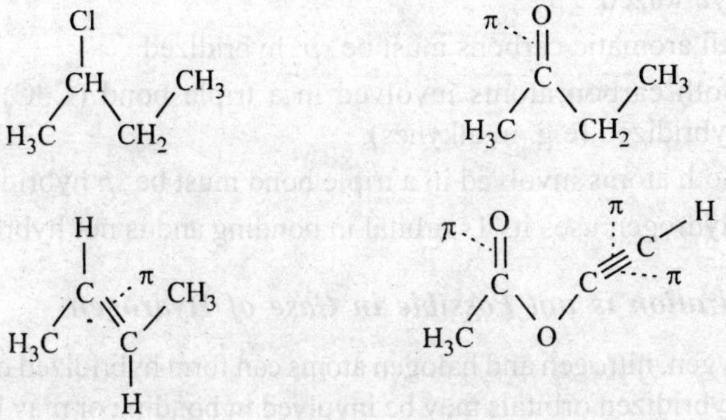

Fig. 1.30. Examples–all the bonds shown are s bonds except those labelled as p.

Hybridized Centers

All the atoms in an organic structure (except hydrogen) are either *sp*, sp^2 *or* sp^3 hybridized (Fig. 1.31).

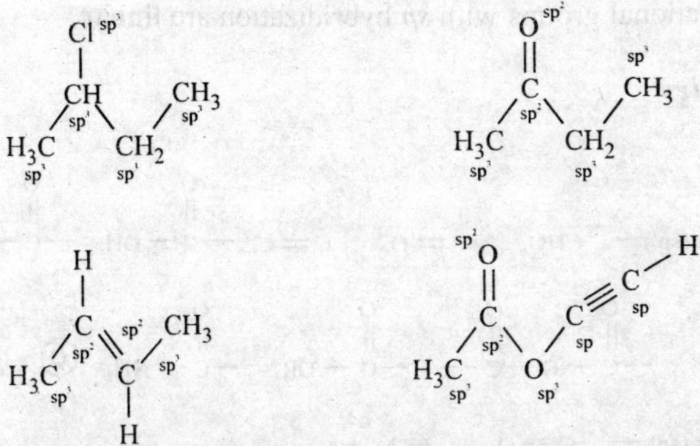

Fig. 1.31. Example of *sp*, sp^2 and sp^3 hybridized centers.

To identify the type of hybridization remember the following rules:

(i) All atoms joined by single bonds are sp^3 hybridized (except hydrogen).

(ii) Both the carbons involved in C=C (alkene) are sp^2 hybridized.

(iii) Both carbon and oxygen in carboxyl group (>C=0) must be sp^2 hybridized.

(iv) All aromatic carbons must be sp^2 hybridized.

(v) Both carbon atoms involved in a triple bond (C≡C) are sp hybridized (e.g. in alkynes).

(vi) Both atoms involved in a triple bond must be sp hybridized.

(vii) Hydrogen uses its 1s orbital in bonding and is not hybridized.

Hybridization is not Possible in Case of Hydrogen

Oxygen, nitrogen and halogen atoms can form hybridized orbitals. These hybridized orbitals may be involved in bonding or may be used to accommodate a lone pair of electrons.

Shape

The shape of a molecule and the functional group depends on the type of hybridization of atoms involved in its formation e.g. Functional groups with sp^3 hybridization are **tetrahedral**.

Functional groups having sp^2 hybridization are **planar**.

Functional groups with sp hybridization are **linear**.

Examples

Planar (sp^2)— CHO, >C=O, C=C, —C(=O)—OH, —C(=O)—Cl,

—C(=O)—O—C(=O)—, —C(=O)—OR, —C(=O)—NH_2, (benzene ring) etc.

Linear (sp) — C≡O, C≡N, etc.

Tetrahedral (sp^3)— Carbon R — OH, R — O — R R — X

where R is an alkyl group.

Reactivity

Functional groups containing π-bonds are more reactive because π-bond is weaker and can be easily broken, e.g. aromatic rings, alkenes, alkynes, aldehydes and ketones, carboxylic acids, esters, amides, acid chlorides, acid anhydrides, nitriles etc.

2

Alkanes and Cycloalkanes

2.1 Definitions

Alkanes

These are the open chain organic compounds having the general formula C_n h_{2n+2}. In them all the bonds are σ-bouds and so they are also called **saturated hydrocarbons**. All the carbon atoms in any alkane are sp^3 hybridised and so their shape in tetrahederal. Since C–C and C–H σ-bonds are strong. So the alkanes are unreactive to most of the chemical reagents.

Alkanes are also referred to as **straight chain** or **acyclic** compounds (Hydrocarbons).

Cycloalkanes

These are cyclic alkanes (hydrocarbons) and are also called **alicyclic compounds.** Their general formula in C_n H_{2n}. In these the carbon atoms are linked to form a rings of various sizes. The six membered carbon ring being most commonly found. Most of there cycloalkanes are unreactive to chemical reagents.

The cycloalkanes with three or four carbon atom rings behave like alkanes (i.e., they are reactive). Their reactivity is due to the fact that such cyclic structures are highly strained.

2.2 Drawing Structures

C-H Bond Omission

There are various ways to drawing structures of organic molecules. A molecule like ethane can be drawn showing every C–C and C–H bond. However, this becomes difficult particularly with more complex molecules, and it is much easier to miss out the C–H bonds (Fig. 2.1).

$$\mathrm{H_3C{-}CH_3}$$

Fig. 2.1. Ethane.

Skeletal Drawings

A further simplification is generally used when only the carbon-carbon bonds are shown. This is a skeletal drawing of the molecule (Fig. 2.2). In such drawings, it is understood that a carbon atom is present at every bond junction and that every carbon has sufficient hydrogens attached to make up four bonds.

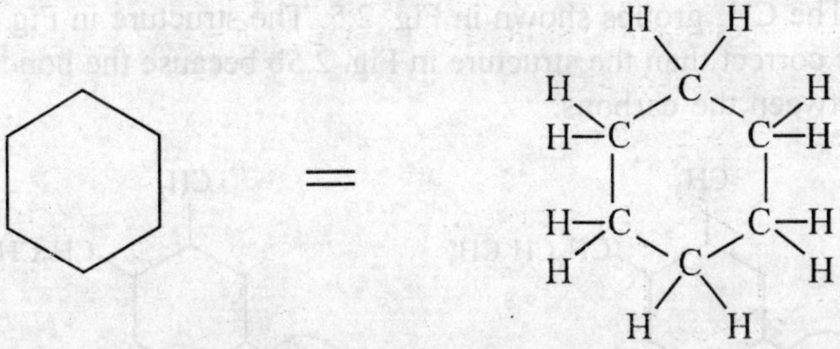

Fig. 2.2. Skeletal drawing of cyclohexane.

Straight chain alkanes can also be represented by drawing the C–C bonds in a zigzag fashion (Fig. 2.3).

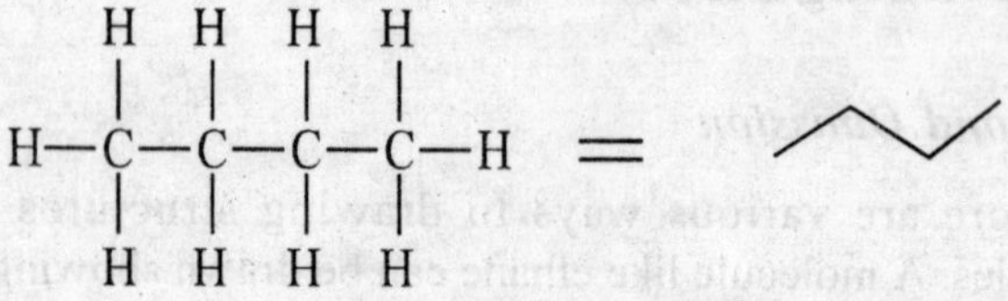

Fig. 2.3. Skeletal drawing of butane.

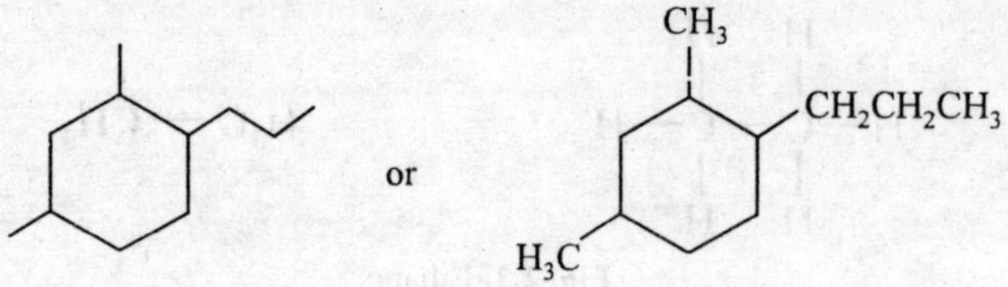

or

(a) (b)

Fig. 2.4. Drawings of an alkyl substituted cyclohexane.

Alkyl Groups

Alkyl groups (C_nH_{2n+1}) are the substituents of a complex molecule. Simple alkyl groups can be indicated in skeletal form (Fig. 2.4a), or as CH_3, CH_2CH_3, $CH_2CH_2CH_3$, *et cetera.* (Fig. 2.4b).

The CH_3 groups shown in Fig. 2.5. The structure in Fig. 2.5a is more correct than the structure in Fig. 2.5b because the bond shown is between the carbons.

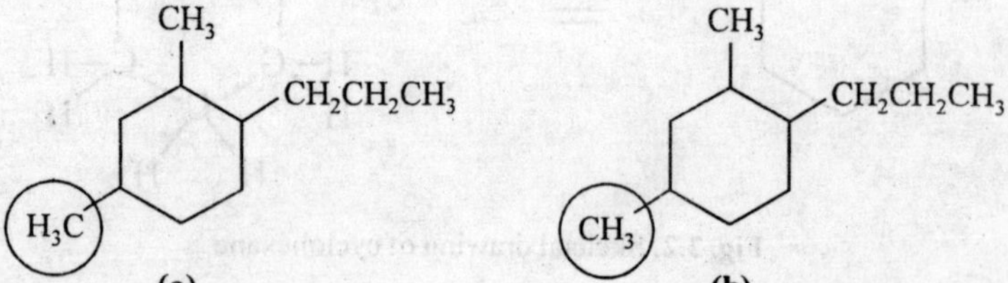

(a) (b)

Fig. 2.5. (a) Correct depiction of methyl group; (b) wrong depiction of methyl group.

2.3 Nomenclature

Simple Alkanes

To names sample alkanes the longest carbon chain is selected. The names of the simplest straight chain alkanes are shown in Fig. 2.6.

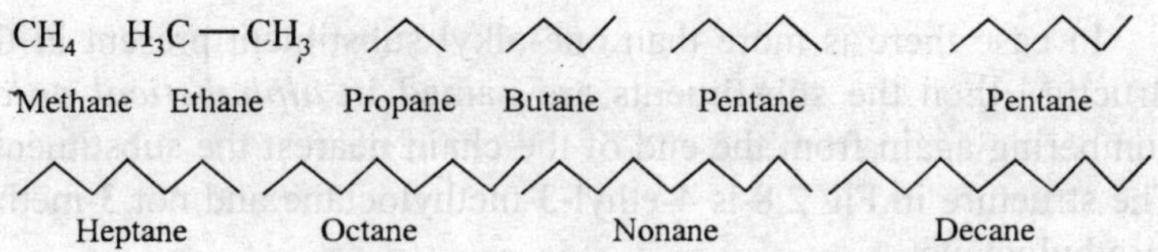

Fig. 2.6. Nomenclature of simple alkanes.

Branched Alkanes

Branched alkanes are named by the following procedure:

(i) identify the longest chain of carbon atoms. In the example shown (Fig. 2.7a), the longest chain consists of five carbon atoms and a pentane chain;

Fig. 2.7. (a) identify the longest chain; (b) number the longest chain.

(ii) number the longest chain of carbons, starting from the end nearest the branch point (Fig. 2.7b);

(iii) identify the carbon with the branching group (number 2 in Fig. 2.7b);

(iv) identify and name the branching group. (In this example it is CH_3. Branching groups (or **substituents**) are referred to as **alkyl groups** (C_nH_{2n+1}) rather than alkanes (C_nH_{2n+2}). Therefore, CH_3 is called methyl and not methane.)

(v) name the structure by first identifying the substituent and its position in the chain, then naming the longest chain. The structure in Fig. 2.6 is called 2-methylpentane. Notice that the substituent and the main chain is one complete word, that is, 2-methylpentane rather than 2-methly pentane.

Multi-Branched Alkanes

In case there is more than one alkyl substituent present in the structure then the substituents are named in *alphabetical order*, numbering again from the end of the chain nearest the substituents. The structure in Fig 2.8 is 4-ethyl-3-methyloctane and not 3-methyl -4-ethyloctane.

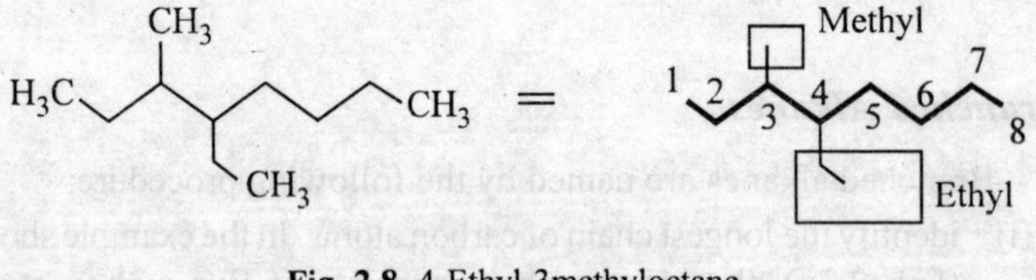

CH_3 H_3C CH_3 = CH_3 Methyl 1 2 3 4 5 6 7 8 Ethyl

Fig. 2.8. 4-Ethyl-3methyloctane.

If a structure has identical substituents, then the prefixes di-,tri-, tetra-, *et cetera* are used to represent the number of substituents. For example, the structure in Fig. 2.9 is called 2,2-dimethylpentane and not 2-methyl-2-methylpentane.

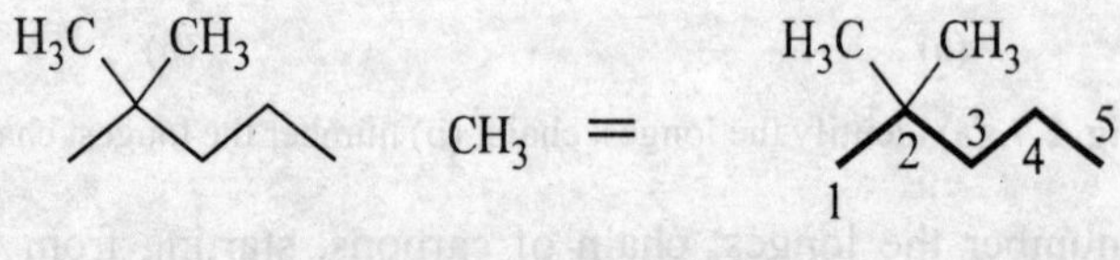

H_3C CH_3 CH_3 = H_3C CH_3 1 2 3 4 5

Fig. 2.9. 2.2-Dimethylpentane.

The prefixes di-, tri-, tetra-etc., are used for identical substituents, but the order in which they are written is still dependent on the alphabetical order of the substituents themselves (i.e. ignore the di-, tri-, tetra-, *et cetera*). For example, the structure in Fig. 2.10 is called 5-ethyl-2, 2-dimethyldecane and not 2, 2-dimethy1-5-ehtyldecane.

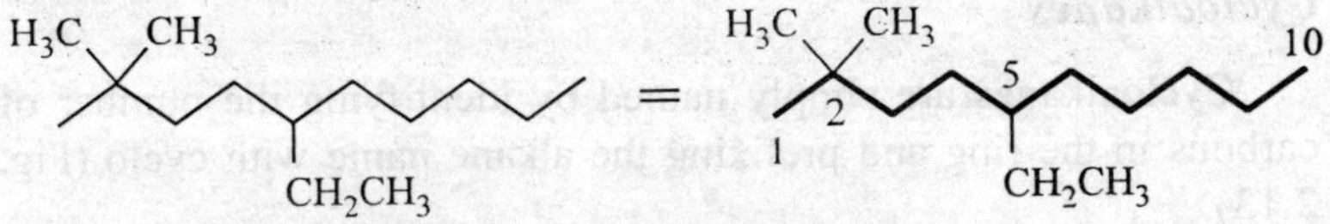

H$_3$C CH$_3$ CH$_2$CH$_3$ = H$_3$C CH$_3$ 1 2 5 10 CH$_2$CH$_3$

Fig. 2.10. 5-Ethyl-2,2-dimethyldecane.

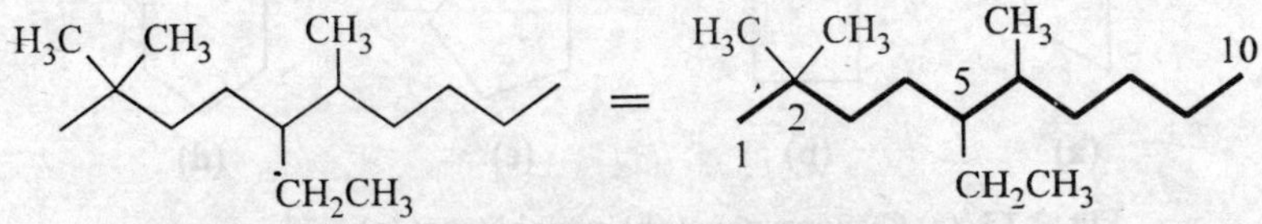

H$_3$C CH$_3$ CH$_3$ CH$_2$CH$_3$ = H$_3$C CH$_3$ CH$_3$ 1 2 5 10 CH$_2$CH$_3$

Fig. 2.11. 5-Ethyl-2,2,6-trimethyldecane.

Identical substituents can be in different positions on the chain, but the same rules apply. For example, the structure in Fig. 2.11 is called 5-ethyl-2,2,6-trimethyldecane.

In some structures, it is difficult to decide which end of the chain to number from. For example, two different substituents might be placed at equal distances from either end of the chain. In such a case, the group with alphabetical priority should be given the lowest numbering. For example, the structure in Fig. 2.12a is 3-ethyl-5-methylheptane and not 5-ethyl-3-methylheptane.

There is another rule that might take precedence over the above rule. The structure (Fig. 2.12c) has ethyl and methyl groups equally placed from each end of the chain, but there are two methyl groups to one ethyl group. Numbering should be chosen such that the smallest total is obtained. In this example, the structure is called 5-ethyl-3,3-dimethylheptane (Fig. 2.12c) rather than 3-ethyl-5,5 dimethylheptane (Fig. 2.12b) since 5+3+3 = 11 is less than 3+ 5+5 = 13.

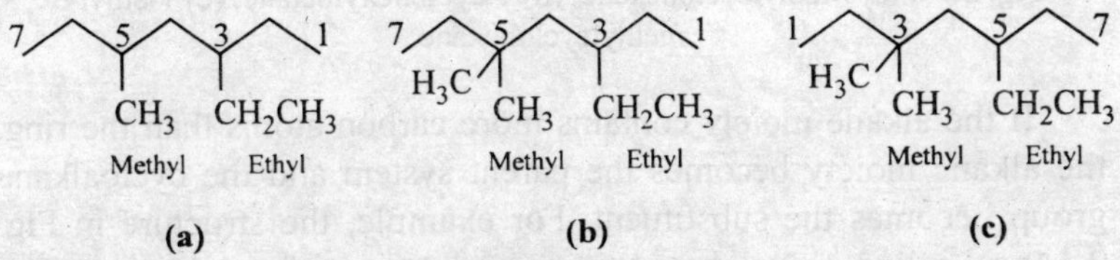

(a) 7 5 3 1 CH$_3$ CH$_2$CH$_3$ Methyl Ethyl

(b) 7 5 3 1 H$_3$C CH$_3$ CH$_2$CH$_3$ Methyl Ethyl

(c) 1 3 5 7 H$_3$C CH$_3$ CH$_2$CH$_3$ Methyl Ethyl

Fig. 2.12. (a) 3-Ethyl-5-methylheptane; (b) incorrect numbering; (c) 5-ethyl-3,3-dimethylheptane.

Cycloalkanes

Cycloalkanes are simply named by identifying the number of carbons in the ring and prefixing the alkane name with cyclo (Fig. 2.13).

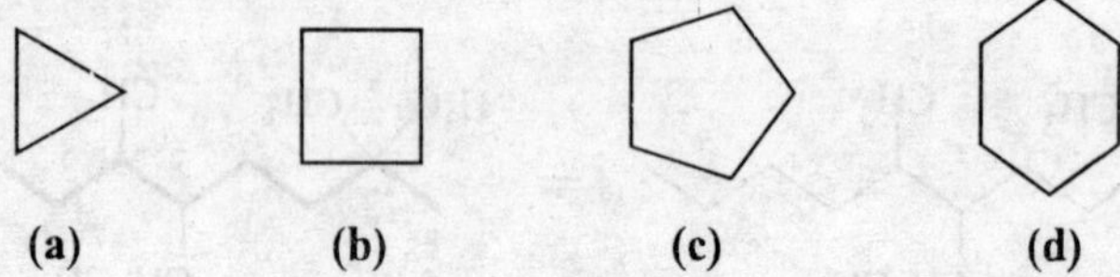

Fig. 2.13. (a) Cyclopropane; (b) cyclobutane; (c) cyclopentane; (d) cyclohexane.

Branched Cyclohexanes

Cycloalkanes made up of a cycloalkane moiety linked to an alkane moiety are generally named in such a way that the cycloalkane is the parent system and the alkane moiety is considered to be an alkyl substituent. Threfore, the structure in Fig. 2.14a is methylcyclohexane and not cyclohexylmethane. In it there is no need to number the cycloalkane ring when only one substituent is present.

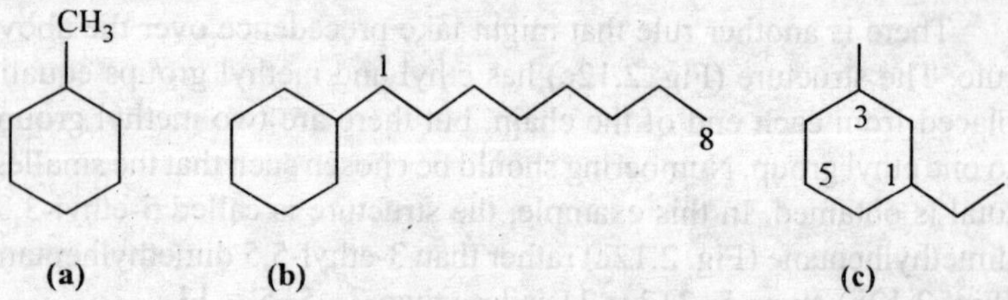

Fig. 2.14. (a) Methylcyclohexane; (b) 1-cyclohexyloctane; (c) 1-ethyl-3 methylcyclohexane.

If the alkane moiety contains more carbon atoms than the ring, the alkane moiety becomes the parent system and the cycloalkane group becomes the substituent. For example, the structure in Fig. 2.14b is called 1-cyclohexyloctane and not octylcyclohexane. The numbering is necessary to identify the position of the cycloalkane on the alkane chain.

Multi-Branched Cycloalkanes

Branched cycloalkanes having different substituents are numbered such that the alkyl substituent having alphabetical priority is at position 1. The numbering of the rest of the ring is then done such that the substituent positions add up to a minimum. For example, the structure in Fig. 2.14c is called 1-ethyl-3-methylcyclohexane rather than 1-methyl-3-ethylcyclohexane or 1-ethyl-5-methylcyclohexane. The last name is incorrect since the total obtained by adding the substituent positions together is 5+1 = 6 which is higher than the total obtained from the correct name (i.e. 1 + 3 =4).

3

Functional Groups

3.1 Recognition of Functional Groups

Definition

A functional group refers to that portion of an organic molecule that is made up of atoms other than carbon and hydrogen, or which contains bonds other than C–C and C–H bonds. For example, ethane (Fig. 3.1a) is an **alkane** and has no functional group. All the atoms are carbon and hydrogen and all the bonds are C–C and C–H. **Ethanoic acid** on the other hand (Fig. 3.1b), has a portion of the molecule (boxed portion), which contains atoms other than carbon and hydrogen, and bonds other than C–H and C–C. This portion of the molecule is called a **functional group**–in this case a **carboxylic acid.**

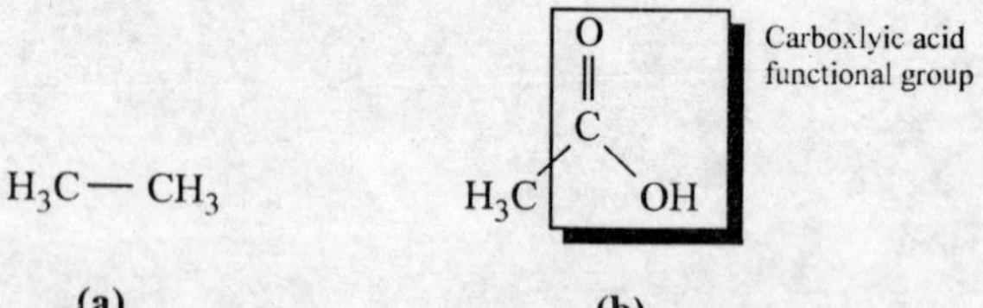

$H_3C—CH_3$

(a)

$H_3C—C(=O)—OH$

Carboxlyic acid functional group

(b)

Fig. 3.1. (a) Ethane; (b) ethanoic acid.

Common Functional Groups

Some of the more common functional groups in organic chemistry are as follows:

(i) ***Functional groups containing carbon and hydrogen only*** (Fig. 3.2);

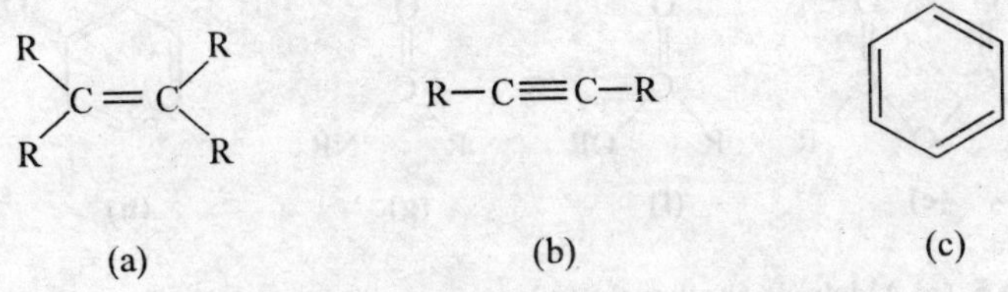

Fig. 3.2. (a) Alkene; (b) alkyne; (c) aromatic.

(ii) ***Functional groups containing nitrogen*** (Fig. 3.3);

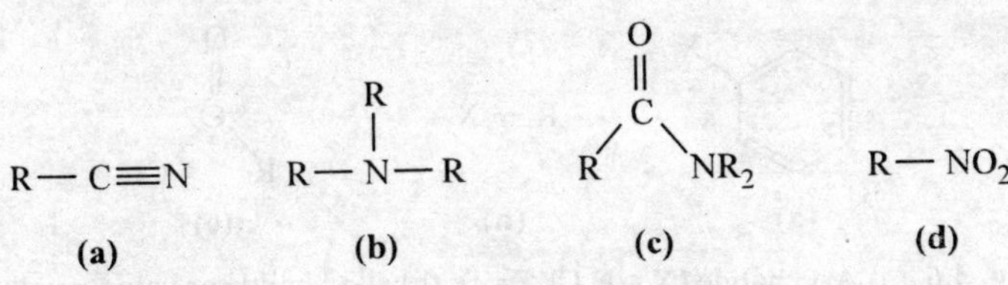

Fig. 3.3. (a) Nitrile; (b) amine; (c) amide; (d) nitro.

(iii) ***Functional groups involving single bonds and containing oxygen*** (Fig. 3.4);

Fig. 3.4. (a) Alcohol or alkanol; (b) ether.

(iv) ***Functional groups involving double bonds and containing oxygen*** (Fig. 3.5);

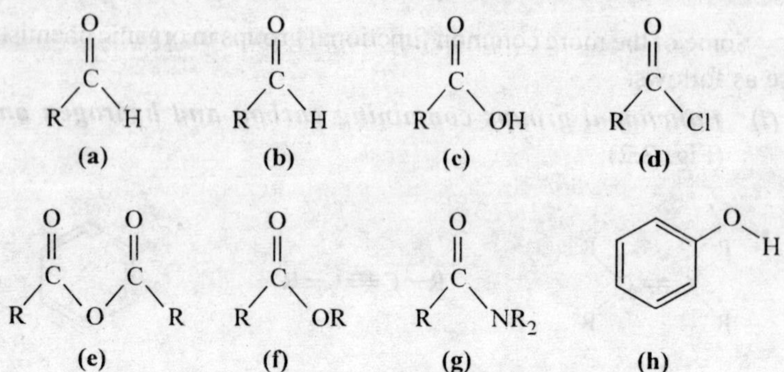

Fig. 3.5. (a) Aldehyde or alkanal; (b) ketone or alkanone; (c) carboxylic acid; (d) carboxylic acid chloride; (e) carboxylic acid anhydride; (f) ester; (g) amide; (h) phenol.

(v) Functional groups containing a halogen atom (Fig. 3.6);

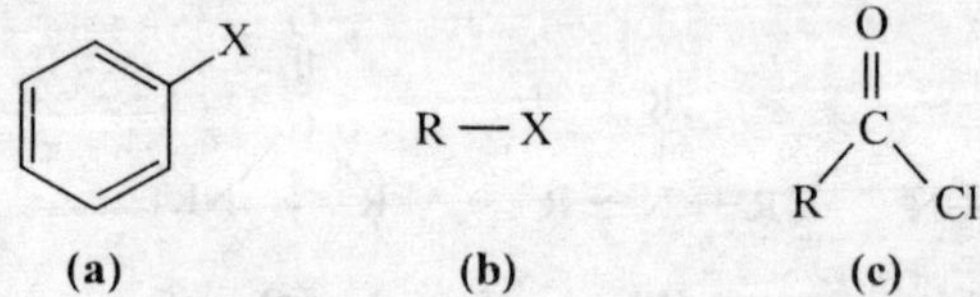

Fig. 3.6. (a) Aryl halide (X = F, Cl, Br, I); (b) alkyl halide or halogenoalkane (X = F, Cl, Br, I); (c) acid chloride.

(vi) Functional groups containing sulfur (Fig. 3.7);

R—SH R—S—R

(a) (b)

Fig. 3.7. (a) Thiol; (b) thioether.

3.2 Aliphatic and Aromatic Functional Groups

Aliphatic Functional Groups

An **aliphatic functional** group is one in which there is no aromatic ring directly attached to the functional group (Fig. 3.8a and b).

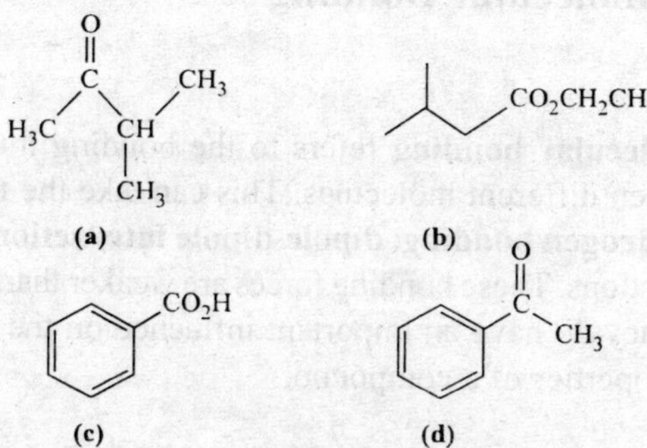

Fig. 3.8. (a) Aliphatic ketone; (b) aliphatic ester; (c) aromatic carboxylic acid; (d) aromatic ketone.

Aromatic Functional Groups

An **aromatic functional** group is one in which an aromatic ring is directly attached to the functional group (Fig. 3.8c and d).

In case of esters and amides, the functional groups are defined as aromatic or aliphatic depending on whether the aryl group is directly attached to the **carbonyl** end of the functional group, i.e., Ar-CO-X. If the aromatic ring is attached to the heteroatom instead, then the ester or amide is classified as an aliphatic amide (Fig. 3.9).

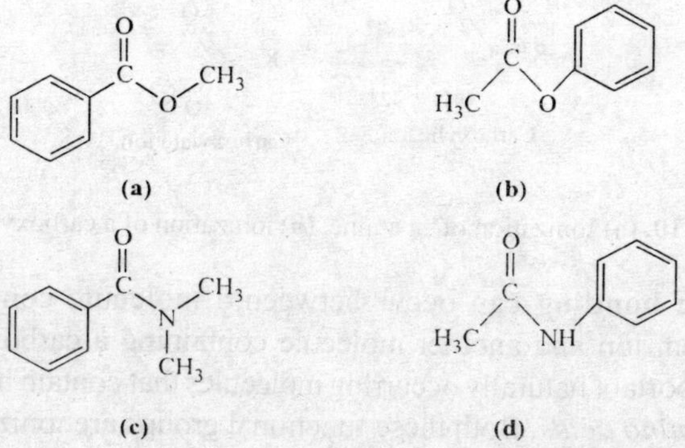

Fig. 3.9. (a) Aromatic ester; (b) aliphatic ester; (c) aromatic amide; (d) aliphatic amide.

3.3 Intermolecular Bonding

Definition

Intermolecular bonding refers to the bonding interaction that occurs between different molecules. This can take the form of **ionic bonding, hydrogen bonding, dipole-dipole interactions** or **van der Waals** interactions. These bonding forces are weaker than the **covalent bonds**, but they do have an important influence on the physical and biological properties of a compound.

Ionic Bonding

Ionic bonding occurs between molecules which have opposite charges and it involves an **electrostatic** interaction between the two opposite charges. the functional groups that most easily ionize are **amines** and **carboxylic acids** (Fig. 3.10).

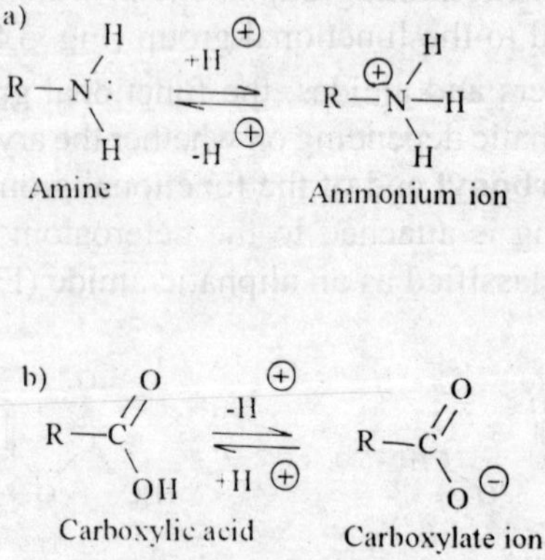

a)

$$R-\ddot{N}H_2 \underset{-H^{\oplus}}{\overset{+H^{\oplus}}{\rightleftharpoons}} R-\overset{\oplus}{N}H_3$$

Amine — Ammonium ion

b)

$$R-C(=O)-OH \underset{+H^{\oplus}}{\overset{-H^{\oplus}}{\rightleftharpoons}} R-C(=O)-O^{\ominus}$$

Carboxylic acid — Carboxylate ion

Fig. 3.10. (a) Ionization of an amine; (b) ionization of a carboxylic acid.

Ionic bonding can occur between a molecule containing an ammonium ion and another molecule containing a carboxylate ion. Some important naturally occurring molecules that contain both groups are the *amino acids*. Both these functional groups are ionized to form a structure called **zwitterion** (a neutral molecule bearing both a positive and a negative charge) and intermolecular ionic bonding can occur (Fig. 3.11).

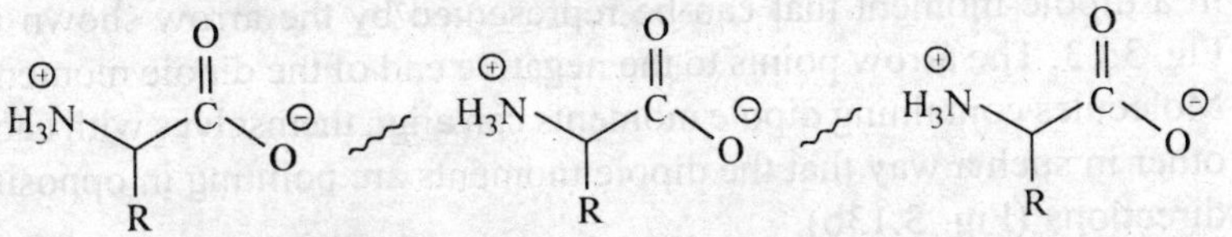

Fig. 3.11. Intermolecular ionic bonding of amino acids.

Hydrogen Bonding

Hydrogen bonding can occur if molecules have a hydrogen atom attached to a heteroatom like nitrogen or oxygen. The common functional groups that can participate in hydrogen bonding are *alcohols, phenols, carboxylic acids, amides,* and *amines*. Hydrogen bonding is possible because of the polar nature of the N–H or O–H bond. Nitrogen and oxygen are more electronegative than hydrogen. Thus, the heteroatom gains a slightly negative charge and the hydrogen gains a slightly positive charge. Hydrogen bonding involves the partially charged hydrogen of one molecule (the H bond donor) interacting with the partially charged heteroatom of another molecule (the H bond acceptor) (Fig. 3.12).

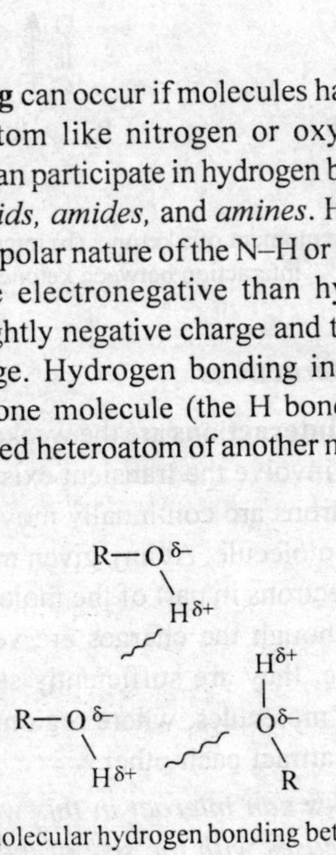

Fig. 3.12. Intermolecular hydrogen bonding between alcohols.

Dipole-dipole Interactions

Dipole-dipole interactions can occur between polarized bonds other than N–H or O–H bonds. The most likely functional groups that can interact in this way are those containing a carbonyl group (C = O). the electrons in the carbonyl bond are polarized towards the more electronegative oxygen such that the oxygen acquires a partial negative charge and the carbon acquires a partial positive charge. This results in a dipole moment that can be represented by the arrow shown in Fig. 3.13. The arrow points to the negative end of the dipole moment. Molecules containing dipole moments can align themselves with each other in such a way that the dipole moments are pointing in opposite directions (Fig. 3.13b).

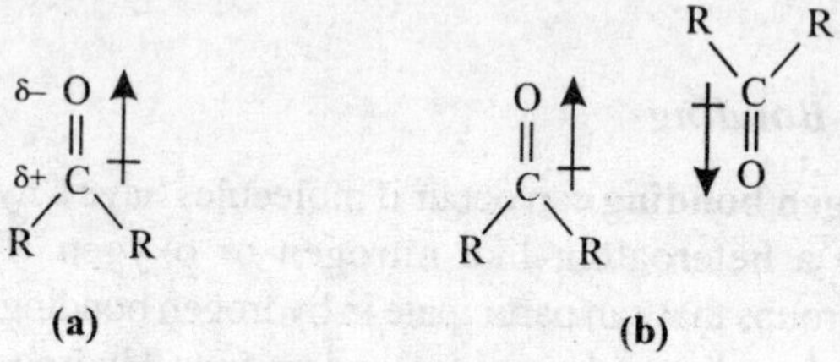

Fig. 3.13. (a) Dipole moment of a ketone; (b) intermolecular dipole-dipole interaction between ketones.

van der Waals Interactions

van der Waals interactions are the weakest of the intermolecular bonding forces and involve the transient existence of partial charges in a molecule. Electrons are continually moving in an unpredictable fashion around any molecule. At any given moment of time, there is a slight excess of electrons in part of the molecule and a slight deficit in another part. Although the charges are very weak and fluctuate around the molecule, they are sufficiently strong to permit a weak interaction between molecules, where regions of opposite charge in different molecules attract each other.

Alkane molecules can interact in this way and the strength of the interaction increases with the size of the alkane molecule. van der Waals interactions are also important for alkanes, alkynes and aromatic rings. The types of molecules involved in this form of intermolecular bonding are 'fatty' molecules that do not dissolve easily in water and such molecules are called **hydrophobic** (water-hating). Hydrophobic molecules can dissolve in nonpolar, hydrophobic solvents because of van der Waals interactions and so this form of intermolecular bonding is also called a **hydrophobic interaction**.

3.4 Properties and Reactions

Properties

The chemical and physical properties of an organic compound depend on the sort of intermolecular bonding forces present, which in turn depends on the functional group present. A molecule like methane has a low boiling point and is a gas at room temperature because its molecules are bound together by weak van der Waals forces (Fig. 3.14a). In contrast, methanol is a liquid at room temperature because hydrogen bonding is possible between the alcoholic functional groups (Fig. 3.14b).

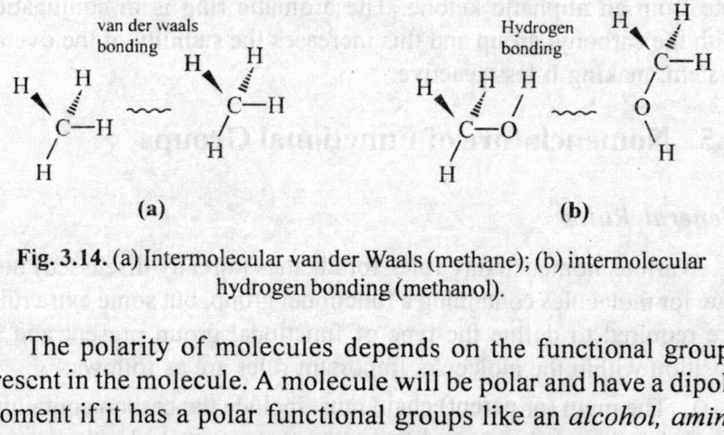

van der waals bonding

Hydrogen bonding

(a) (b)

Fig. 3.14. (a) Intermolecular van der Waals (methane); (b) intermolecular hydrogen bonding (methanol).

The polarity of molecules depends on the functional groups present in the molecule. A molecule will be polar and have a dipole moment if it has a polar functional groups like an *alcohol, amine* or *ketone.* Polarity also determines solubility in different solvents. Polar molecules dissolve in polar solvents like water or alcohols, whereas nonpolar molecules dissolve in nonpolar solvents like ether and chloroform. Polar molecules that can dissolve in water are called **hydrophilic** (water-loving) while nonpolar molecules are called **hydrophobic** (water-hating).

Generally, the presence of a polar functional group determines the physical properties of the molecule. But this is not always true. If a molecule has a polar group like a carboxylic acid, but has a long hydrophobic alkane chain, then the molecule will be **hydrophobic**.

Reactions

Most of the organic reactions occur at functional groups and are characteristic of that functional group. However, the reactivity of the functional group is affected by **stereoelectronic** effects. For example, a functional group may be surrounded by bulky groups that hinder the approach of a reagent and slow down the rate of reaction. This is called as **steric shielding. Electronic effects** can also influence the rate of a reaction. Neighboring groups can influence the reactivity of a functional group if they are *electron-withdrawing* or *electron-donating* and influence the electronic density within the functional group. *Conjugation* and *aromaticity* also effect the reactivity of functional groups. For example, an aromatic ketone reacts at a different rate from an aliphatic ketone. The aromatic ring is in conjugation with the carbonyl group and this increases the stability of the overall system, making it less reactive.

3.5 Nomenclature of Functional Groups

General Rules

Various nomenclature rules for alkanes (already discussed) hold true for molecules containing a functional group, but some extra rules are required to define the type of functional group present and its position within the molecule. Important rules are as follows:-

(i) The main (or parent) chain must include the carbon containing functional group, and so may not necessarily be the longest chain (Fig. 3.15);

Main chain = 4C
Correct

Main chain = 5 C
Wrong

Fig. 3.15. Identification of the main chain.

(ii) The presence of some functional groups is indicated by replacing -ane for the parent alkane chain with a suffixes depending on the functional group present. e.g.,

functional group	*suffix*	*functional group*	*suffix*
alkene	-ene	alkyne	-yne
alcohol	-anol	aldehyde	-anal
ketone	-anone	carboxylic acid	-anoic acid
acid chloride	-anoyl chloride	amine	-ylamine.

The example given in Fig. 3.15 is a **butanol**.

(iii) Numbering of carbon atoms must start from the end of the main chain nearest to the functional group. Therefore, the numbering should place the alcohol (Fig. 3.16) at position 1 and not position 4. (Lowest position number to the carbon containing the functional group).

Fig. 3.16. Numbering of the longest chain.

(iv) The position of the functional group must be defined in the name of the compound. Therefore, the alcohol (Fig. 3.16) is a 1-butanol.

(v) Other substituents if present are named and ordered in the same way as for alkanes. The alcohol (Fig. 3.16) has an ethyl group at position 3 and so the full name for the structure is 3-ethyl-1-butanol.

Some other rules are needed to deal with a specific situation. For example, if the functional group is at equal distance from either end of the main chain, the numbering starts from the end of the chain nearest to any substituents. For example, the alcohol (Fig. 3.17) is 2-methyl-3-pentanol and not 4-methyl-3-pentanol. (Lowest number rule).

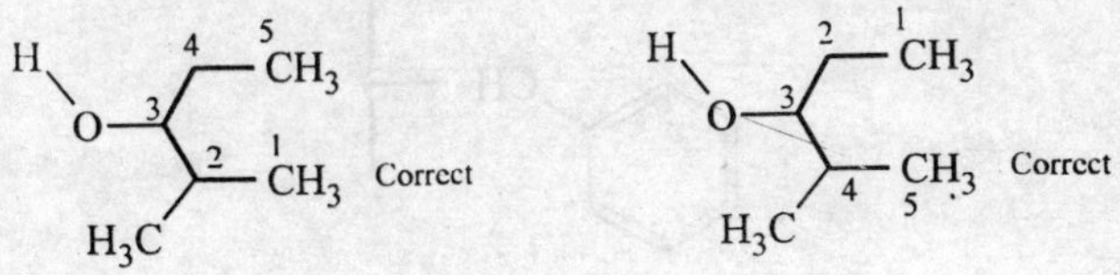

Fig. 3.17. 2-Methyl-3-pentanol.

Alkenes and Alkynes

The names of alkenes and alkynes contain the suffixes -ene and -yne respectively (Fig. 3.18). With some alkenes it is necessary to define the stereochemistry of the double bond.

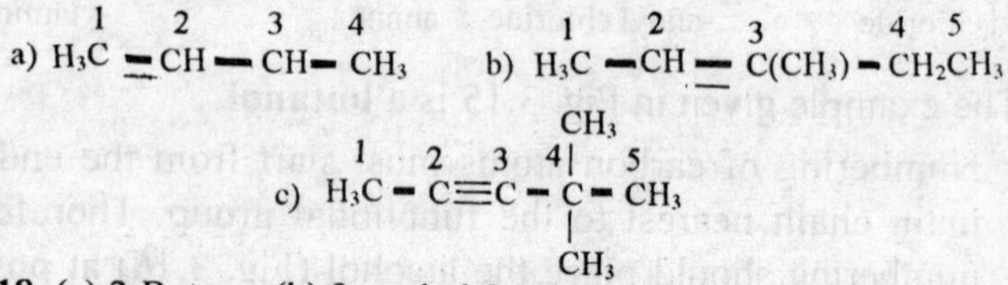

a) $\overset{1}{H_3C}=\overset{2}{CH}-\overset{3}{CH}-\overset{4}{CH_3}$ b) $\overset{1}{H_3C}-\overset{2}{CH}=\overset{3}{C}(CH_3)-\overset{4}{CH_2}\overset{5}{CH_3}$

c) $\overset{1}{H_3C}-\overset{2}{C}\equiv\overset{3}{C}-\overset{4}{C}(CH_3)_2-\overset{5}{CH_3}$

Fig. 3.18. (a) 2-Butene; (b) 3-methyl-2-pentene; (c) 4,4-dimethyl-2-pentyne.

Aromatics

The well-known aromatic structure is **benzene**. If an alkane chain is linked to a benzene molecule, then the alkane chain is generally considered to be an alkyl substituent of the benzene ring. However, if the alkane chain contains more than six carbons, then the benzene molecule is considered to be a **phenyl** substituent of the alkane chain (Fig. 3.19).

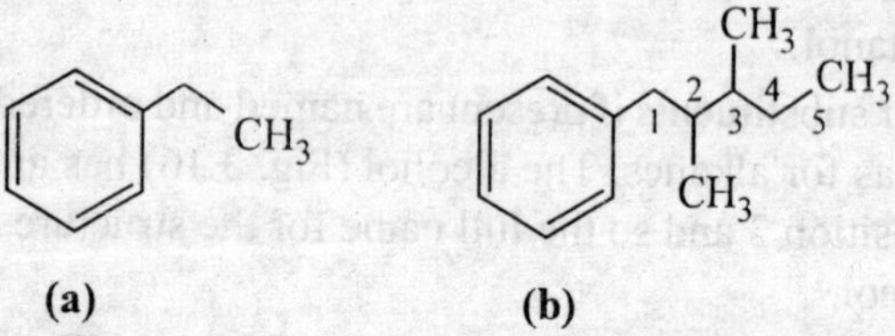

Fig. 3.19. (a) Ethylbenzene; (b) 3-phenyl-2,3-dimethylpentane.

A **benzyl** group is made up of an *aromatic ring* and a *methylene group* (Fig. 3.20).

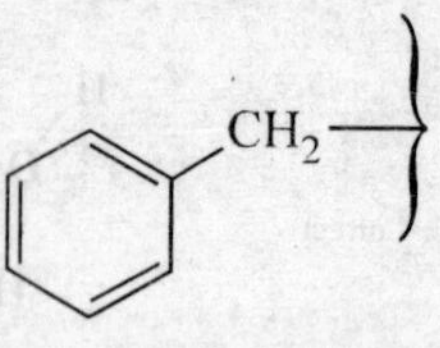

Fig. 3.20. Benzyl group.

Benzene is not the only parent name that can be used for aromatic compounds (Fig. 3.21).

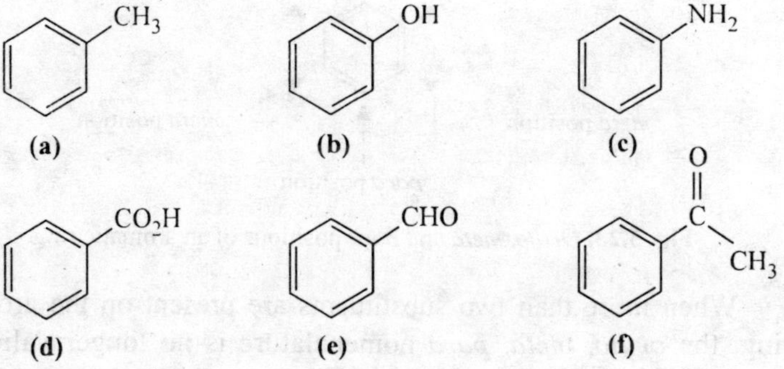

Fig. 3.21. (a) Toluene; (b) phenol; (c) aniline; (d) benzoic acid; (e) benzaldehyde; (f) acetophenone.

In case of distributed aromatic rings, the position of substituents has to be defined by numbering around the ring, in such a way that the substituents are positioned at the lowest numbers possible, for example, the structure (Fig. 3.22) is 1,3-dichlorobenzene and not 1,5-dichlorobenzene.

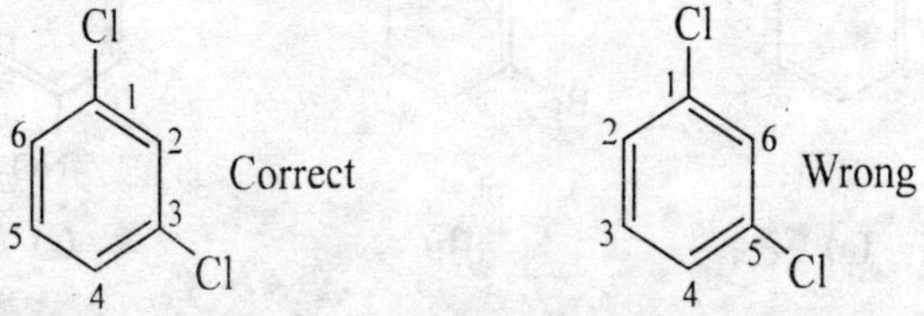

Fig. 3.22. 1,3-Dichlorobenzene.

Alternatively, the terms *ortho, meta,* and *para* can be used. These terms define the relative position of one substituent to another (Fig. 3.23). Thus, 1,3-dichlorobenzene can also be called *meta*-dichlorobenzene. This can be shortened to *m*-dichlorobenzene. The examples in Fig. 3.24 shows how different parent names can be used. The substituent which defines the parent name is given as position 1. For example, if the parent name is toluene, the methyl group must be at position 1.

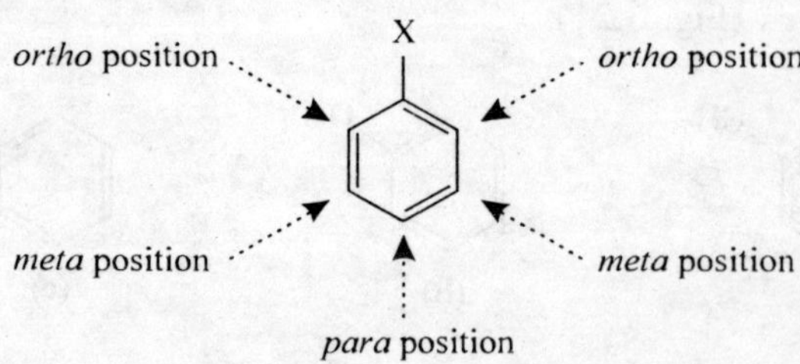

Fig. 3.23. *Ortho, meta* and *para* positions of an aromatic ring.

When more than two substituents are present on the aromatic ring, the *ortho, meta, para* nomenclature is no longer valid and numbering has to be used (Fig. 3.25). In such a case the relevant substituent has to be placed at position 1 if the parent name is toluene, aniline, *et cetera*. If the parent name is benzene, the numbering is done in such a way that the lowest possible numbers are used. In the example shown, any other numbering would result in the substituents having higher numbers (Fig. 3.26).

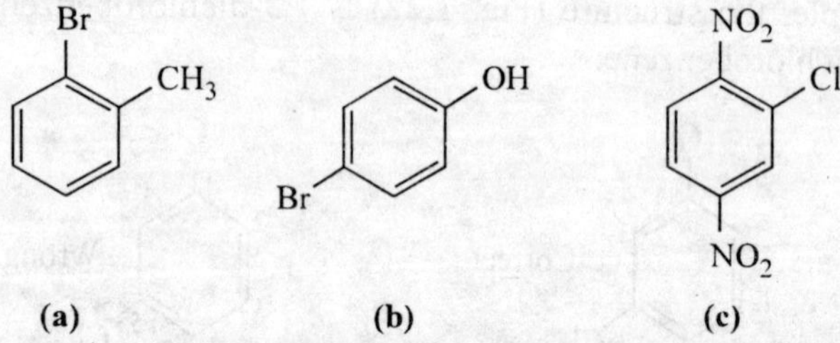

Fig. 3.24. (a) 2-Bromotoleune or o-bromotoluene; (b) 4-bromophenol or p-bromophenol; (c) 3-chloroaniline or m-chloroaniline.

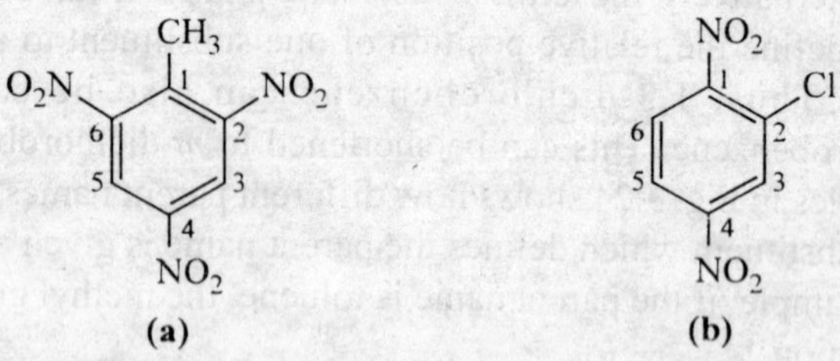

Fig. 3.25. (a) 2,4,6-Trinitrotoluene; (b) 2-chloro-1,4-dinitrobenzene.

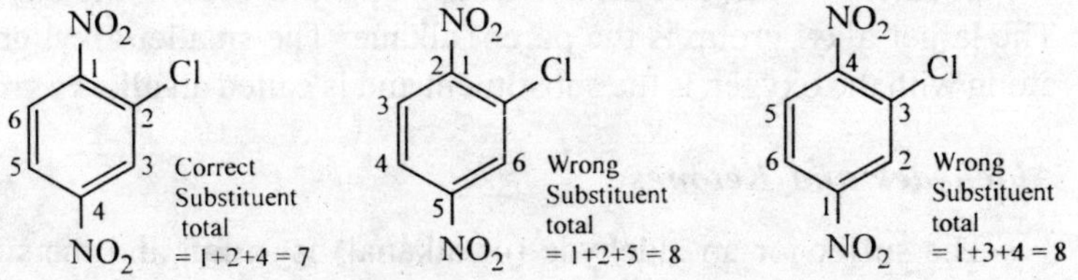

Correct Substituent total = 1+2+4 = 7

Wrong Substituent total = 1+2+5 = 8

Wrong Substituent total = 1+3+4 = 8

Fig. 3.26. Possible numbering systems if tri-substituted aromatic ring.

Alcohols

Alcohols or **alkanols** are named by using the suffix -anol. The general rules discussed earlier are used to name alcohols (Fig. 3.27).

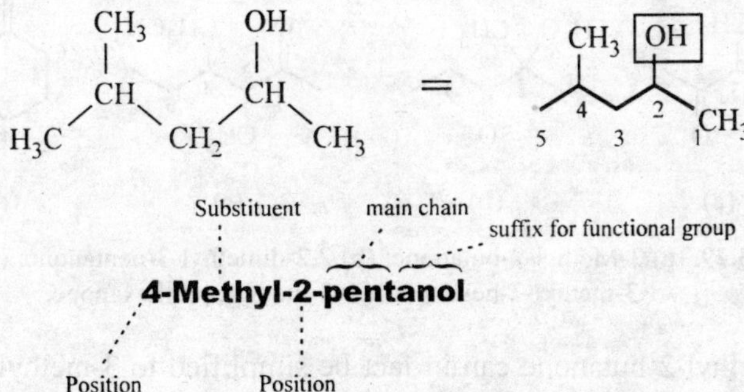

Fig. 3.27. 4-Methyl-2-pentanol.

Ethers and Alkyl Halides

For the nomenclature for ethers and alkyl halides the functional group is considered to be a substituent of the main alkane chain. The functional group is numbered and named as a substituent (Fig. 3.28).

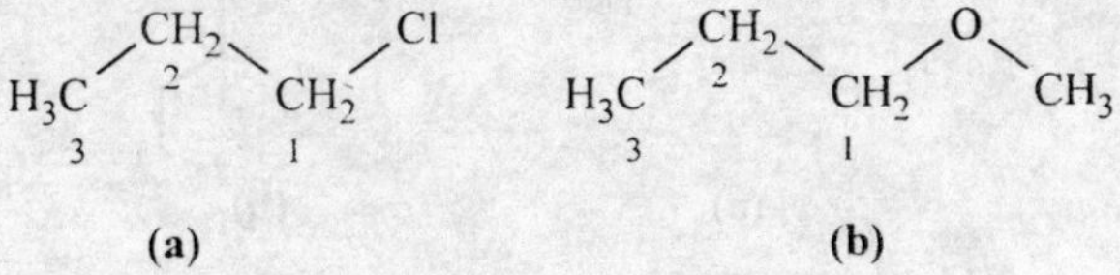

(a) **(b)**

Fig. 3.28. (a) 1-Chloropropane; (b) 1-methoxypropane.

In ethers we have two alkyl groups on either side of the oxygen. The larger alkyl group is the parent alkane. The smaller alkyl group along with the oxygen is the substituent and is called an **alkoxy** group.

Aldehydes and Ketones

The suffix for an aldehyde (or alkanal) is **-anal**, and the suffix for a ketone (or alkanone) is **-anone**. The main chain must include the functional group and the numbering is such that the functional group is at the lowest number possible. If the functional group is in the center of the main chain, the numbering is done in such a way that other substituents have the lowest numbers possible (*e.g.*, 2,2-dimethyl-3-pentanone and not 4,4-dimethyl-3-pentanone; Fig. 3.29).

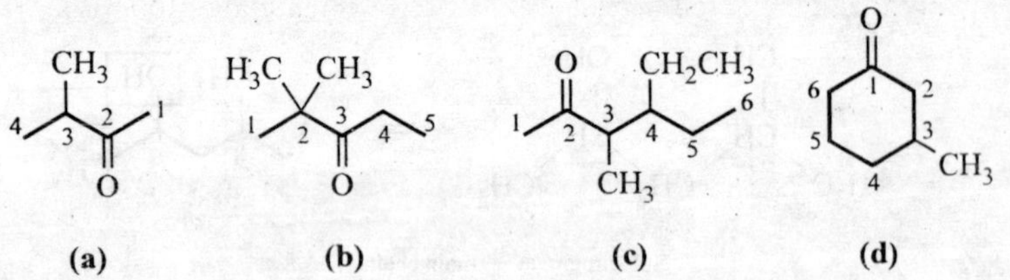

Fig. 3.29. (a) 3-Methyl-2-butanone; (b) 2,2-dimethyl-3-pentanone; (c) 4-ethyl-3-methyl-2-hexanone; (d) 3-methylcyclohexanone.

3-Methyl-2-butanone can in fact be simplified to 3-methylbutanone because there is only one possible place for the ketone functional group in this molecule. In case the carbonyl C = O group is at the end of the chain; it would be an aldehyde and not a ketone. Numbering is also not necessary in locating an aldehyde group since it can only be at the end of a chain (Fig. 3.30).

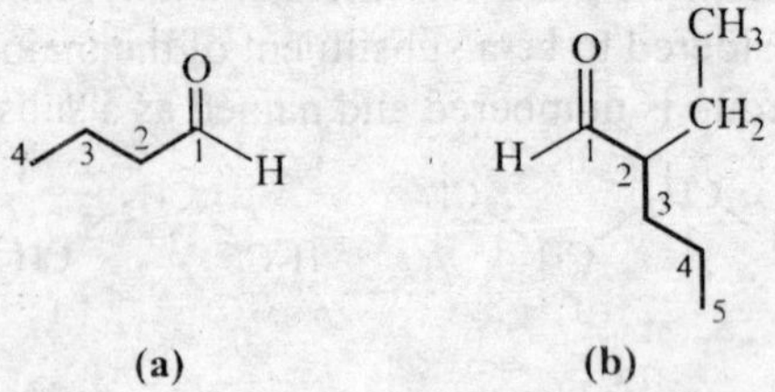

Fig. 3.30. (a) Butanal; (b) 2-ethylpentanal.

Carboxylic Acids and Acid Chlorides

Carboxylic acids and acid chlorides can be identified by adding the suffix -anoic acid and -anoyl chloride respectively. Both these functional groups are always at the end of the main chain and need not be numbered (Fig. 3.31).

Fig. 3.31. (a) 2-Methylbutanoic; (b) 2,3-dimethylpentanoyl chloride.

Esters

For naming an ester, the following procedure is followed:

(i) identify the carboxylic acid (alkanoic acid) from which it was derived;

(ii) change the name to an alkanoate rather than an alkanoic acid;

(iii) identify the alcohol from which the ester was derived and consider this as an alkyl substituent;

(iv) the name becomes an **alkyl alkanoate**.

For example, the ester (Fig. 3.32) is derived from ethanoic acid and methanol. The ester would be an *alkyl ethanoate* since it is derived from ethanoic acid. The alkyl group comes from methanol and is a methyl group. Therefore, the full name is **methyl ethanoate**. (Note that there is a space between both parts of the name).

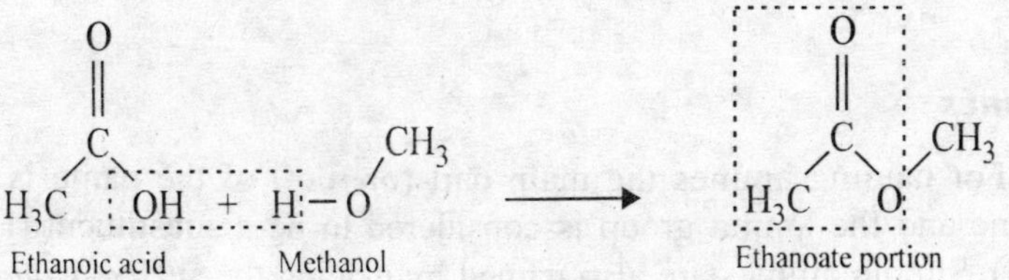

Fig. 3.32. Eater formation.

Amides

Amides are the derivatives of the carboxylic acids. In amides the carboxylic acid is linked with ammonia or an amine. Like esters, the parent carboxylic acid is identified. This is then termed an **alkanamide** and includes the nitrogen atom. For example, linking ethanoic acid with ammonia gives ethanamide (Fig. 3.33).

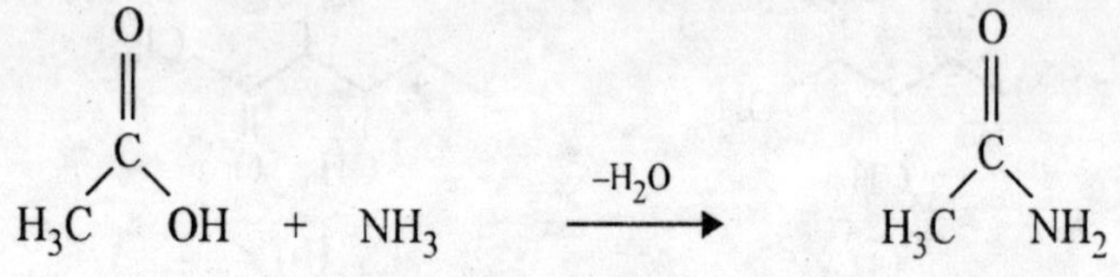

$$H_3C-C(=O)-OH \;+\; NH_3 \xrightarrow{-H_2O} H_3C-C(=O)-NH_2$$

Fig. 3.33. Formation of ethanamide.

If the carboxylic acid is linked with an amine, then the amide will have alkyl groups on the nitrogen. These are considered as *alkyl substituents* and come at the beginning of the name. The symbol *N* is used to show that the substituents are on the nitrogen and not some other part of the alkanamide skeleton. For example, the structure in Fig. 3.34 is named *N*-ethylethanamide.

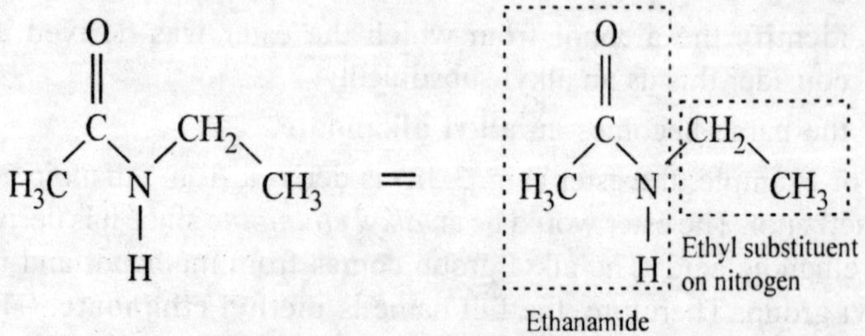

$$H_3C-C(=O)-N(H)-CH_2-CH_3 \;=\; \underbrace{H_3C-C(=O)-N(H)}_{\text{Ethanamide}}-\underbrace{CH_2-CH_3}_{\text{Ethyl substituent on nitrogen}}$$

Fig. 3.34. N-Ethylethanamide.

Amines

For naming amines the main part (or root) of the name is an alkane and the amino group is considered to be a substituent (Fig. 3.35). Simple amines are also named by placing the suffix *-ylamine* after the main part of the name (Fig. 3.36).

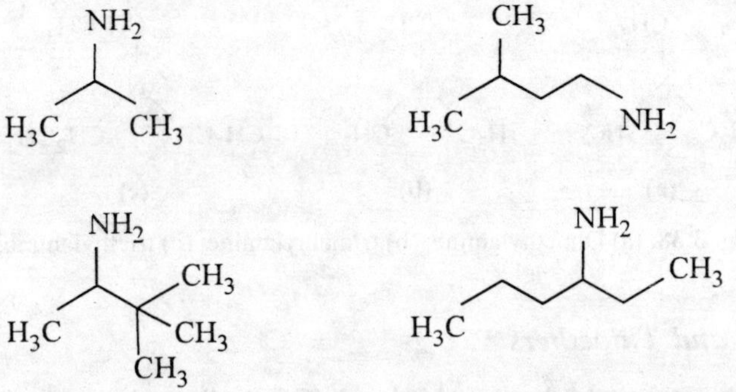

Fig. 3.35. (a) 2-Aminopropane; (b) 1-amino-3-methylbutane; (c) 2-amino-3,3-dimethylbutane; (d) 3-aminohexane.

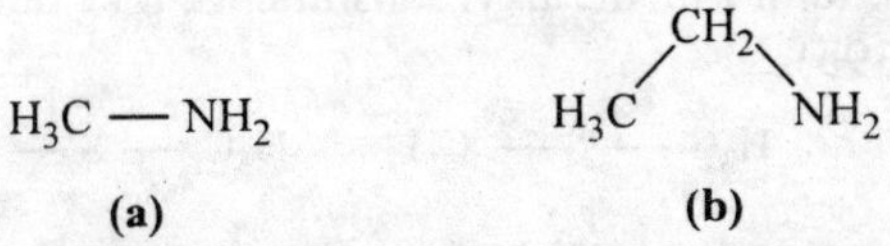

$H_3C — NH_2$ (a)

$H_3C-CH_2-NH_2$ (b)

Fig. 3.36. (a) Methylamine; (b) ethylamine.

Amines containing more than one alkyl group attached are named by selecting the longest carbon chain attached to the nitrogen. In the example (Fig. 3.37), that is an ethane chain and so this molecule is an aminoethane (*N,N*-dimethylaminoethane).

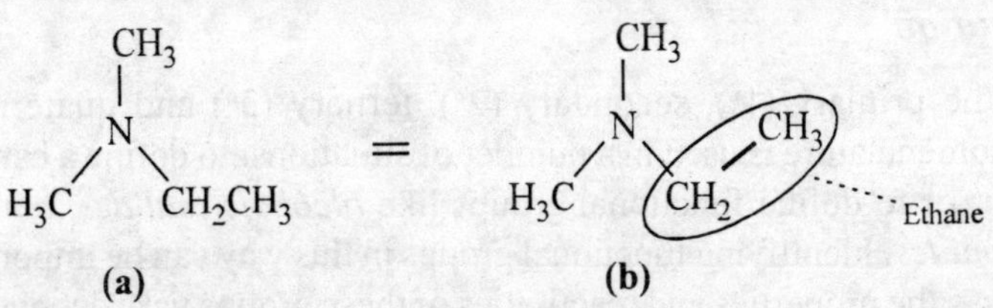

Fig. 3.37. N,N-Dimethylaminoethane.

Some simple secondary and tertiary amines have common names (Fig. 3.38).

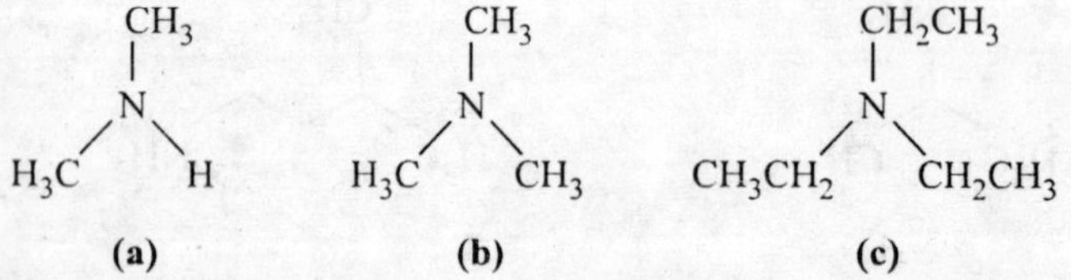

$$\underset{\textbf{(a)}}{H_3C-\overset{CH_3}{\overset{|}{N}}-H}\qquad \underset{\textbf{(b)}}{H_3C-\overset{CH_3}{\overset{|}{N}}-CH_3}\qquad \underset{\textbf{(c)}}{CH_3CH_2-\overset{CH_2CH_3}{\overset{|}{N}}-CH_2CH_3}$$

Fig. 3.38. (a) Dimethylamine; (b) trimethylamine; (c) triethylamine.

Thiols and Thioethers

For naming thiols we add the suffix **thoil** to the name of the parent alkane (Fig. 3.39a). **Thioethers** are named like ethers using the prefix **alkylthio**, for example, 1-(methylthio) propane (Fig. 3.39c). Simple thioethers can be named by identifying the thioether as a **sulfide** and prefixing this term with the alkyl substituents, for example, dimethyl sulfide (Fig. 3.39b).

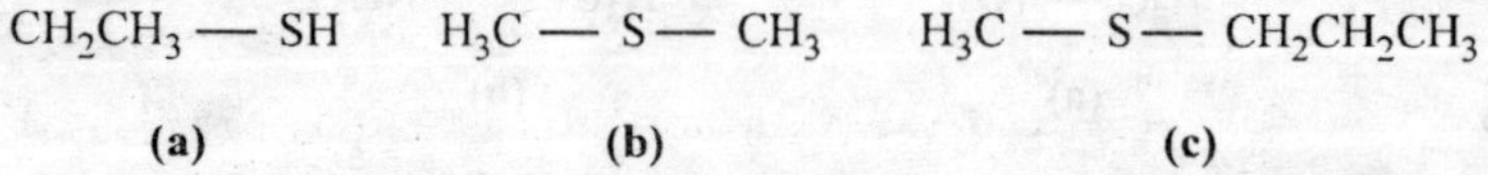

$$\underset{\textbf{(a)}}{CH_2CH_3 - SH}\qquad \underset{\textbf{(b)}}{H_3C - S - CH_3}\qquad \underset{\textbf{(c)}}{H_3C - S - CH_2CH_2CH_3}$$

Fig. 3.39. (a) Ethanethiol; (b) dimethylsulfide; (c) 1-(methylthiopropane).

3.6 Primary, Secondary, Tertiary and Quaternary Nomenclature

Definition

The primary (1°), secondary (2°), tertiary (3°) and quaternary (4°) nomenclature is used in a number of situations: to define a carbon center, or to define functional groups like *alcohols, halides, amines* and *amides*. Identifying functional groups in this way can be important because the properties and reactivities of these groups vary depending on whether they are *primary, secondary, tertiary* or *quaternary*.

Carbon Centers

The easiest ways of determining if a carbon center is 1°, 2°, 3°, or 4° is to count the number of bonds leading from that carbon center to another carbon atom (Fig. 3.40). A **methyl** group (CH_3) is a primary carbon center (attached to one carbon), a **methylene** group (CH_2) is a secondary carbon center (attached to the other carbons), a **methine** group (CH) is a tertiary carbon center (attached to three other carbons) and a carbon center with four alkyl substituents (C) is a quaternary carbon center (attached to four other carbons) (Fig. 3.41).

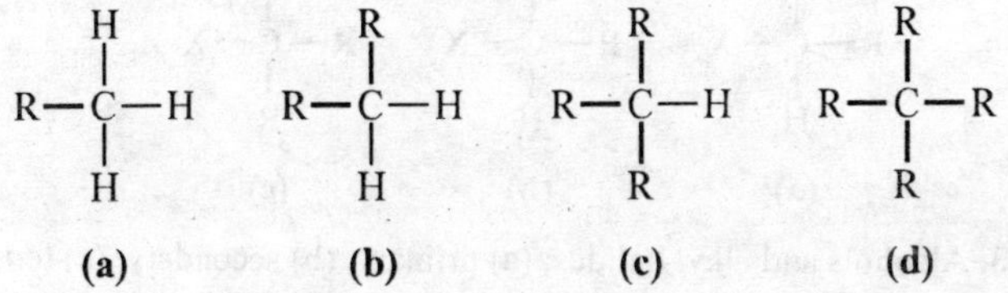

Fig. 3.40. Carbon centers; (a) primary; (b) secondary; (c) tertiary; (d) quaternary.

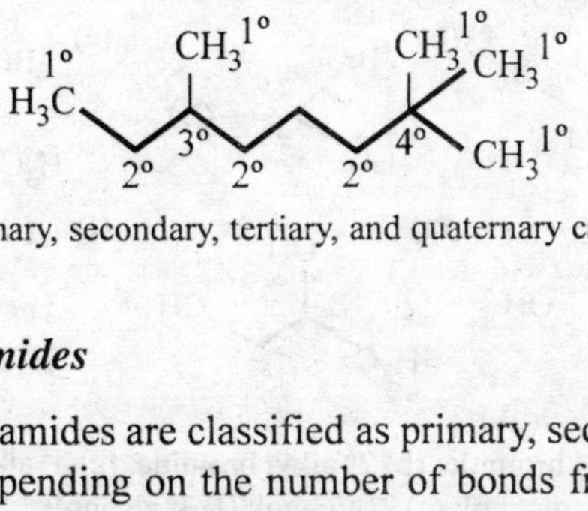

Fig. 3.41. Primary, secondary, tertiary, and quaternary carbon centers.

Amines and Amides

Amines and amides are classified as primary, secondary, tertiary, or quaternary depending on the number of bonds from **nitrogen** to carbon (Fig. 3.42). Note that a quaternary amine is positively charged and is therefore known as a **quaternary ammonium ion. It is not possible to get a quaternary amide.**

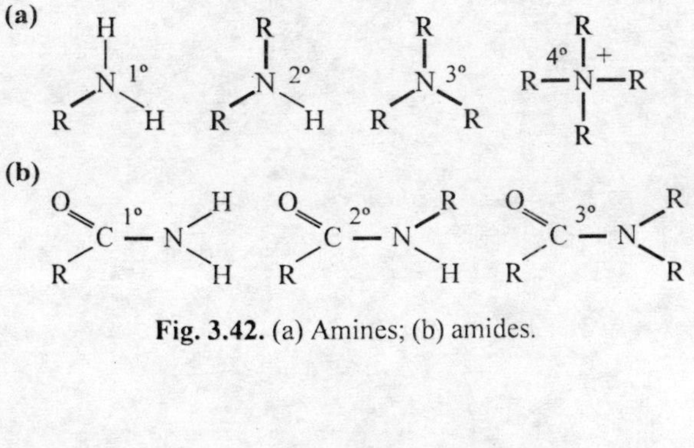

Fig. 3.42. (a) Amines; (b) amides.

Alcohols and Alkyl Halides

Alcohols and alkyl halides can be classified as primary, secondary, or tertiary (Fig. 3.43) depending on the carbon to which the alcohol or halide is attached and it ignores the bond to the functional group. Thus, **quaternary alcohols or alkyl halides are not possible.**

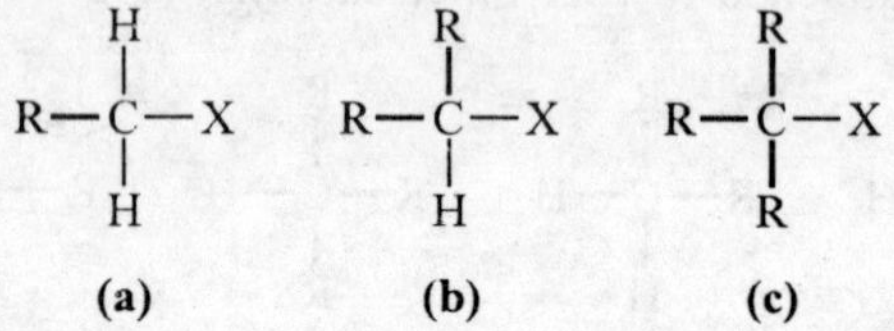

(a) (b) (c)

Fig. 3.43. Alcohols and alkyl halides; (a) primary; (b) secondary; (c) tertiary.

For example (Fig. 3.44) which illustrate different types of alcohols and alkyl halides.

(a) (b) (c)

(d) (e) (f)

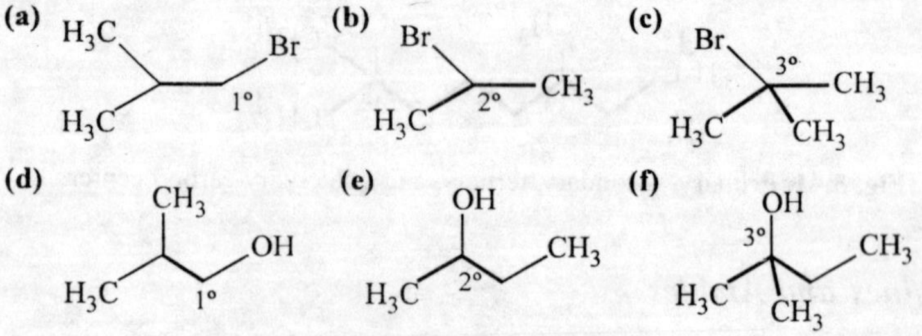

Fig. 3.44. (a) 1° alkyl bromide; (b) 2° alkyl bromide; (c) 3° alkyl bromide; (d) 1° alcohol; (e) 2° alcohol; (f) 3° alcohol.

4

Stereo-Chemistry

4.1 Constitutional Isomers

Introduction

The term **isomers** refers to those compounds that have the same molecular formula (i.e. they have the same atoms), but differ in the arrangement of atoms. The three types of isomers are: *constitutional isomers, configurational isomers,* and *conformational isomers*. **Constitutional isomers** are those isomers in which the atoms are linked together in a different skeletal framework and are different compounds. **Configurational isomers** have structures having the same atoms and bonds, but have different geometrical shapes that cannot be interconverted without breaking covalent bonds. Configurational isomers can be separated and are different compounds with different properties. **Conformational isomers** are different shapes of the same molecule and cannot be separated.

Definition

Constitutional isomers are those compounds that have the same molecular formula but have the atoms joined together in a different way. Thus, they have different carbon skeletons. Constitutional isomers differ in physical and chemical properties.

Alkanes

Alkanes having a particular molecular formula can exist as different constitutional isomers. For example, the alkane having the molecular formula C_4H_{10} can exist as two constitutional isomers–the straight chain alkane (butane) or the branched alkane (2-methylpropane; Fig. 4.1). These are different compounds with different physical and chemical properties.

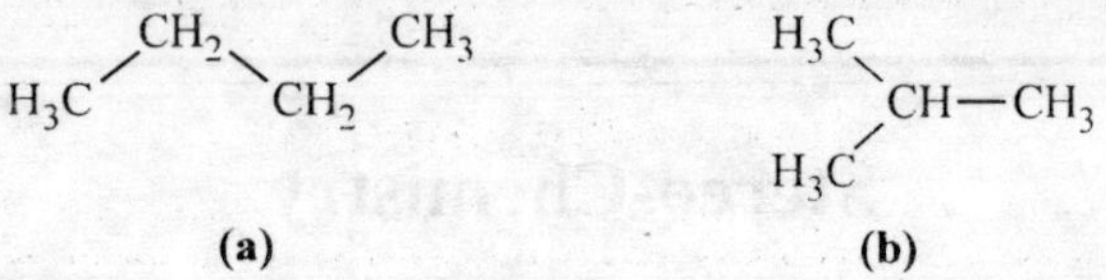

H_3C–CH_2–CH_2–CH_3

(a)

$(H_3C)_2CH$–CH_3

(b)

Fig. 4.1. (a) Butane (C_4H_{10}); (b) 2-methlypropane (C_4H_{10}).

4.2 Configurational Isomers — Alkenes and Cycloalkanes

Definition

Configurational isomers are those isomers in which we find the same molecular formula and the same molecular structure. Thus, they have the same atoms and the same bonds. Such isomers are different because some of the atoms are arranged differently in space, and the isomers cannot be interconverted without breaking and remaking covalent bonds. Thus, configurational isomers are different compounds having different properties. Some examples of configurational isomers are *substituted alkenes* and *substituted cycloalkanes* in which the substituents are arranged differently with respect to each other.

Alkenes –cis *and* Trans *Isomerism*

Alkenes that have identical substituents at either end of the double bond can only exist as one molecule. However, alkenes with different substituents at each end of the double bond can exist as two possible isomers. For example, 1-butene (Fig. 4.2a) has two hydrogens at one end of the double bond and there is only one way of constructing it

On the other hand, 2-butene has different substituents at both ends of the double bond (H and CH_3) and can be constructed in two ways. The methyl groups can be on the same side of the double bond (the *cis* isomer; Fig. 4.2b), or on opposite sides (the *trans* isomer; Fig. 4.2c). The *cis* and *trans* isomers of an alkene are **configurational isomers** (also called **geometric** isomers) as they have different shapes and cannot easily interconvert since the double bond of an alkene cannot rotate. Therefore, the substituents are 'fixed' in space relative to each other. The structures are different compounds with different chemical and physical properties.

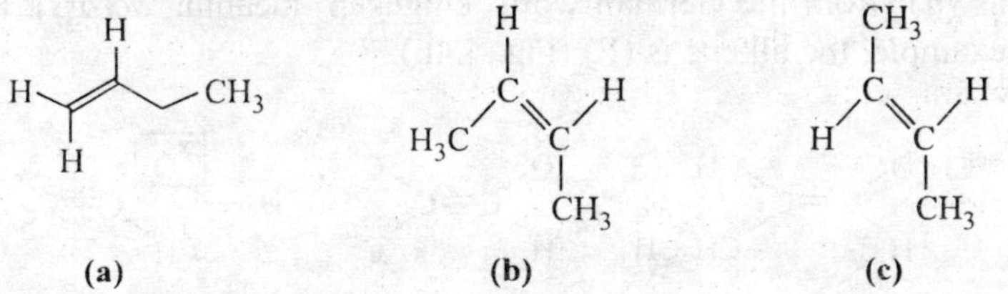

(a) (b) (c)

Fig. 4.2. (a) 1-Butane; (b) cis-2-butene; (c) trans 2-butene.

Alkenes – (Z) and (E) Nomenclature

The *cis* and *trans* nomenclature for alkenes is an old method of classifying the configurational isomers of alkenes and is still in common use. However, it is only suitable for simple 1,2-disubstituted alkenes where we can compare the relative position of the two substituents with respect to each other. When it comes to trisubstituted and tetrasubstituted alkenes, a different nomenclature is needed.

The (Z) / (E) nomenclature permits a clear, unambiguous definition of the configuration of alkenes. The method whereby alkenes are classified as (Z) or (E) is illustrated in Fig. 4.3. First of all, the atoms directly attached to the double bond are identified and given their atomic number (Fig. 4.3b). Next we compare the two atoms at each end of the alkene. The one with the highest atomic number takes priority over the other (Fig. 4.3c). At the left hand side, oxygen has a higher atomic number than hydrogen and takes priority. At the right hand side, both atoms are the same (carbon) and we cannot choose between them.

Therefore, we now have to identify the atom of highest atomic number attached to each of these identical carbons. These correspond to a hydrogen for the methyl substituent and a carbon for the ethyl substituent. These are now different and so a priority can be given (Fig. 4.4a). After identifying which groups have priority, we can now see whether the priority groups are on the same side of the double bond or on opposite sides. If the two priority groups are on the same side of the double bond, the alkene is designated as (Z) (from the German word '*zusammen*' meaning *together*). If the two priority groups are on opposite sides of the double bond, the alkene is called as (E) (from the German word 'entgegen' meaning *across*). In this example, the alkene is (E) (Fig. 4.4b).

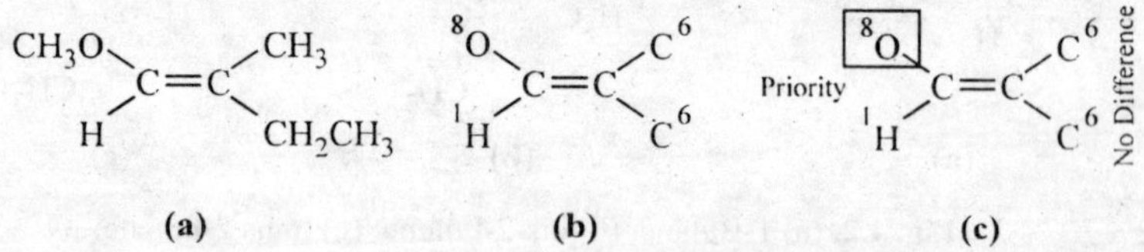

Fig. 4.3. (a) Alkene; (b) atomic numbers; (c) priority groups.

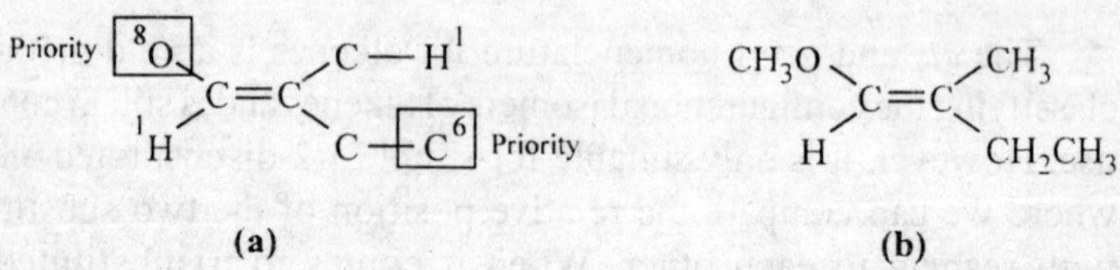

Fig. 4.4. (a) Choosing priority groups; (b) (E)-1-methoxy-2-methyl-1-butene.

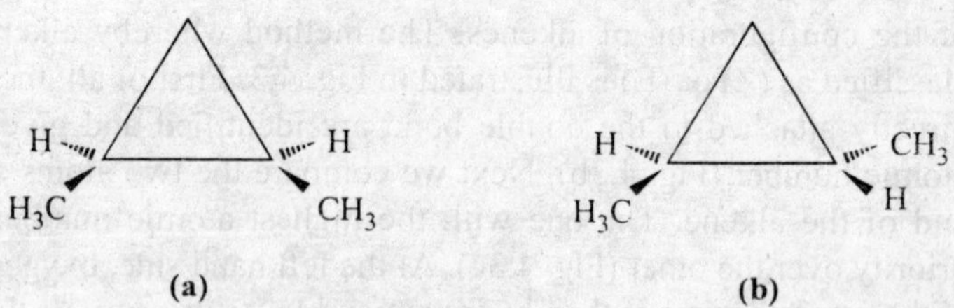

Fig. 4.5. (a) cis-1,2-Dimethylcyclopropane; (b) trans-1,2-dimethylcyclopropane.

Cycloalkanes

Substituted cycloalkanes are also capable of existing as configurational isomers. For example, there are two configurational isomers of 1,2-dimethylcyclopropane depending on whether the methyl groups are on the same side of the ring or on opposite sides (Fig. 4.5). The relative positions of the methyl groups can be defined by bonds. A solid wedge indicates that the methyl group is coming out the page, whereas a hatched wedge indicates that the methyl group is pointing into the page. If the substituents are on the same side of the ring, the structure is classified as *cis*. If they are on opposite sides, the structure is classified as *trans*.

4.3 Configurational Isomers — Optical Isomers

Definition

Optical isomerism is also a type of **configurational isomerism** and is so named because of the ability of optical isomers to rotate plane-polarized light clockwise or counterclockwise. The existence of optical isomers has very important consequences for life, because optical isomers generally have significant *differences in their biological activity*. Apart from their biological activity and their effects on plane-polarized light, optical isomers have identical chemical and physical properties.

Asymmetric Molecules

A molecule like chloroform ($CHCl_3$) is tetrahedral and there is only one way of fitting the atoms together. This is not the case for a molecule like lactic acid. For *lactic acid* there are two ways of constructing a model in such a way that the two structures obtained are nonsuperimposable and cannot be interconverted without breaking covalent bonds. Thus, they represent two different molecules that are configurational isomers. The difference between the two possible molecules is in the way the substituents are attached to the central carbon. This can be represented by the diagrams shown in Fig. 4.6 in which the bond to the hydroxyl group comes out of the page in one isomer but goes into the page in the other isomer.

The two isomers of lactic acid are *mirror images* (Fig. 4.7). A molecule that exists as two nonsuperimposable mirror images has optical activity if only one of the mirror images is present.

Fig. 4.6. Lactic acid.

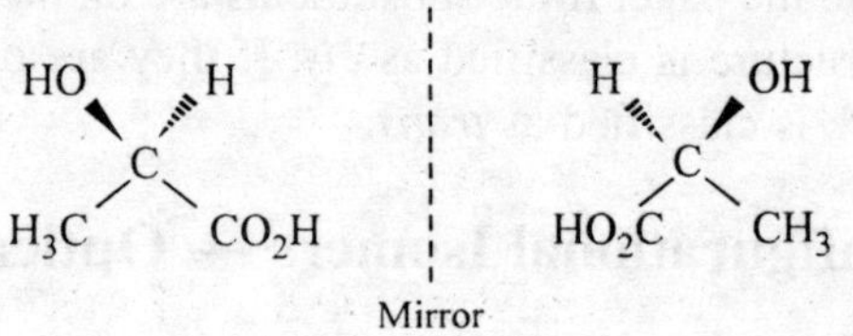

Fig. 4.7. Nonsuperimposable mirror images of lactic acid.

Lactic acid exists as two nonsuperimposable mirror images as it is **asymmetric** i.e. it has no element of symmetry. Asymmetric molecules can also be called a **chiral**, and the ability of molecules to exist as two optical isomers is called **chirality**. In fact, a molecule need not have to be totally asymmetric to be **chiral**. *Molecules containing a single axis of symmetry can also be chiral.*

Asymmetric Carbon Centers

A chiral molecule can be identified by what are known as **asymmetric carbon centers.** This is true for most chiral molecules, but it is not foolproof and that there are several cases where it will not work. For example, some chiral molecules have no asymmetric carbon centers, and some molecules having more than one asymmetric carbon center are not chiral.

Generally, a compound will have optical isomers if it has four different substituents attached to a central carbon (Fig. 4.8). in such cases, the mirror images are nonsuperimposable and the structure will exist as two configurational isomers known as **enantiomers**. The carbon center that contains these four different substituents is called a **stereogenic** or an asymmetric center. A solution of each enantiomer or optical isomer can rotate plane-polarized light. One enantiomer will rotate plane-polarized light clockwise while the other (the mirror image) will rotate it counterclockwise by the same amount. A mixture of the two isomers (a **racemate**) will not rotate plane-polarized light at all. In all other respects, the two isomers are identical in physical and chemical properties and are therefore indistinguishable. The asymmetric centers in the molecules shown (Fig. 4.9) have been identified with as asterisk. The structure lacking the asymmetric center is symmetric or **achiral** and does not have optical isomers. A structure can also have more than one asymmetric center.

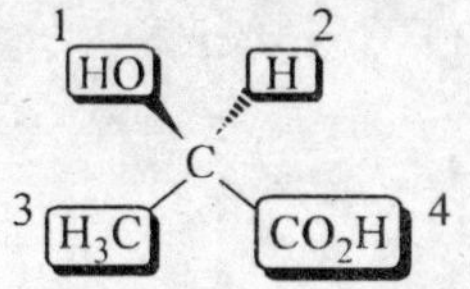

Fig. 4.8. Four different substituents of lactic acid.

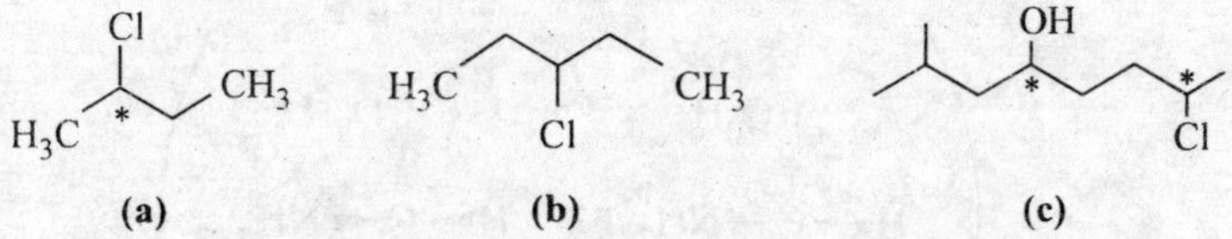

Fig. 4.9. Chiral and achiral structures: (a) chiral; (b) achiral; (c) chiral.

Asymmetric centers are only possible on sp^3 carbons. An sp^2 center is planar and cannot be asymmetric. Similarly, an sp center cannot be asymmetric.

Fischer Diagrams

A chiral molecule can be represented by a *Fischer diagram* (Fig. 4.10). The molecule is drawn in such a way that the carbon chain is vertical with the functional group positioned at the top. The vertical C–C bonds from the asymmetric center point into the page while the horizontal bonds from the asymmetric center come out of the page. This is generally drawn without specifying the wedged and hatched bonds.

The Fischer diagrams of **alanine** permit the structures to be defined as L- or D- from the position of the amino group. If the amino group is to the left, then it is the L- enantiomer. If it is to the right, it is the D-enantiomer. This is an old fashioned nomenclature and is only used for *amino acids* and *sugars*. The L- and D- nomenclature depends on the *absolute configuration* at the asymmetric center and not the direction in which the enantiomer rotates plane-polarized light. It is not possible to predict which way a molecule will rotate plane-polarized light and this can only be found out by experimentation.

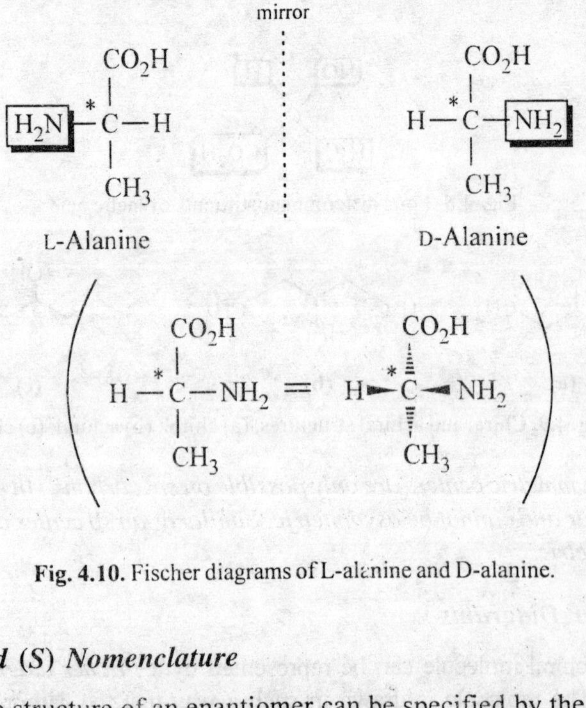

mirror

L-Alanine	D-Alanine
CO_2H / H_2N—C*—H / CH_3	CO_2H / H—C*—NH_2 / CH_3

CO_2H / H—C*—NH_2 / CH_3 = CO_2H / H—C*—NH_2 / CH_3

Fig. 4.10. Fischer diagrams of L-alanine and D-alanine.

(*R*) and (*S*) Nomenclature

The structure of an enantiomer can be specified by the (R) and (S) nomenclature, determined by the **Cahn-Ingold-Prelog** rules. In this (Fig. 4.11) method of nomenclature, first of all, the atoms directly attached to the asymmetric center and their atomic numbers are identified. Then the attached atoms are assigned a priority based on their atomic numbers. In this example (Fig. 4.11) there are two carbon atoms with the same atomic numbers so they cannot be given a priority. In such a case the next atom of highest atomic number is considered (Fig. 4.12a). This means moving to an oxygen for one of the carbons and to a hydrogen for the other. The oxygen has the higher priority and so this substituents takes priority over the other.

Once the priorities have been decided, the structure is redrawn in such a way that the group of lowest priority is positioned 'behind the page'. In this example (Fig. 4.12b), the group of lowest priority (the hydrogen) is already positioned behind the page (note that hatched bond indicating the bond going away from you). An arc is now drawn connecting the remaining groups, starting from the group of highest priority and finishing at the group of third priority. If the arc is drawn clockwise, the assignment is (*R*) (*rectus*). If the arc is drawn counterclockwise, the assignment is (*S*) (*sinister*). In this example the arrow is drawn clockwise. Therefore, the molecule is (*R*)-lactic acid.

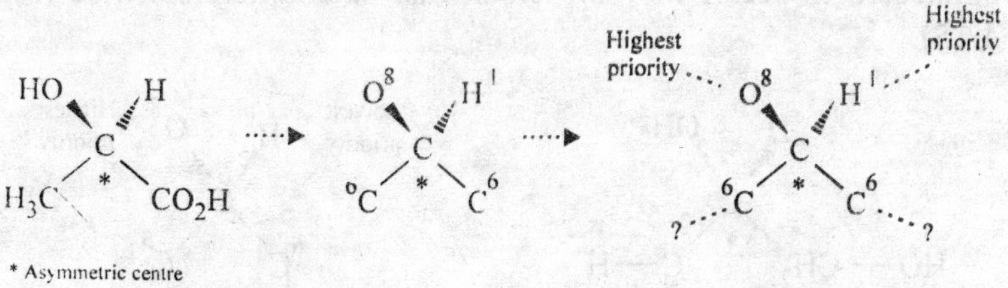

* Asymmetric centre

Fig. 4.11. Assigning priorities to substituents of an asymmetric center.

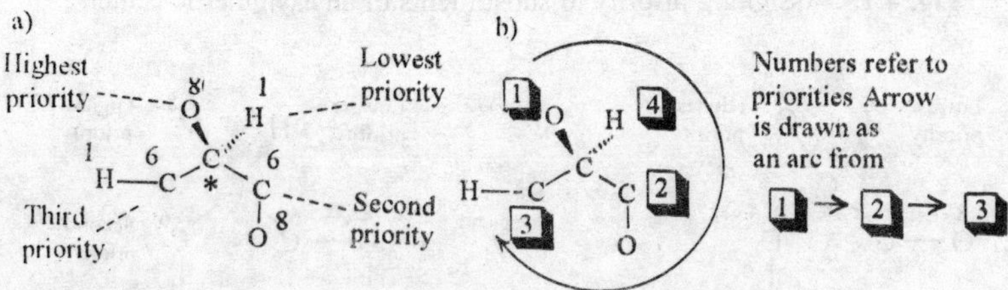

Fig. 4.12. (a) Assigning priorities to substitutes; (b) assigning an asymmetric center as (R) or (S).

Another example (Fig. 4.13) that illustrates another rule involving substituents with double bonds. The asymmetric center is marked with an asterisk. The atoms directly attached to the asymmetric center are shown on the right with their atomic numbers. Now, it is possible to define the group of highest priority (the oxygen) and the group of lowest priority (the hydrogen). There are two identical carbons attached to the asymmetric center so move to the next stage and identify the atom with the highest atomic number joined to each of the identical carbons (Fig. 4.14). This still does not distinguish between the CHO and CH_2OH groups since both carbon atoms have an oxygen atom attached, next look at the second most important atom attached to the two carbon atoms. However, if there is a double bond present, you are allowed to 'visit' the same atom twice. The next most important atom in the CH_2OH group is the hydrogen. In the CHO group, the oxygen can be 'revisited' since there is a double bond. Therefore, this group takes priority over the CH_2OH group. The priorities have been determined, and thus the group of lowest priority is placed behind the page and three most important groups are connected to see if they are clockwise or counterclockwise (Fig. 4.15).

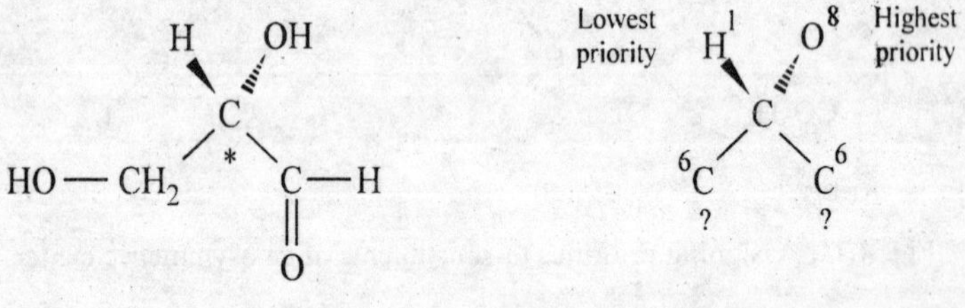

Fig. 4.13. Assigning priority to substituents of an asymmetric center.

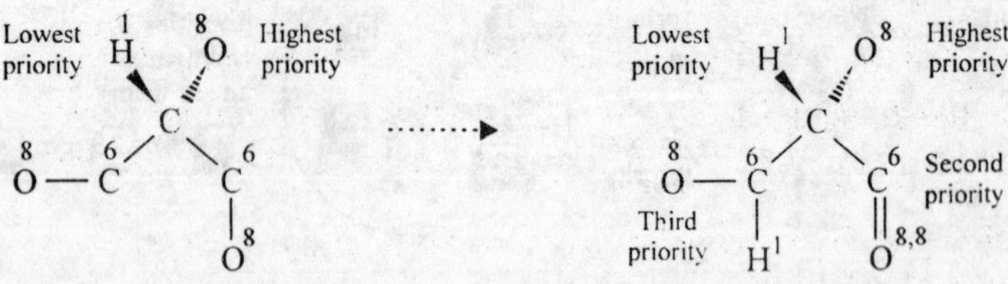

Fig. 4.14. Assigning priority to substituents of an asymmetric center.

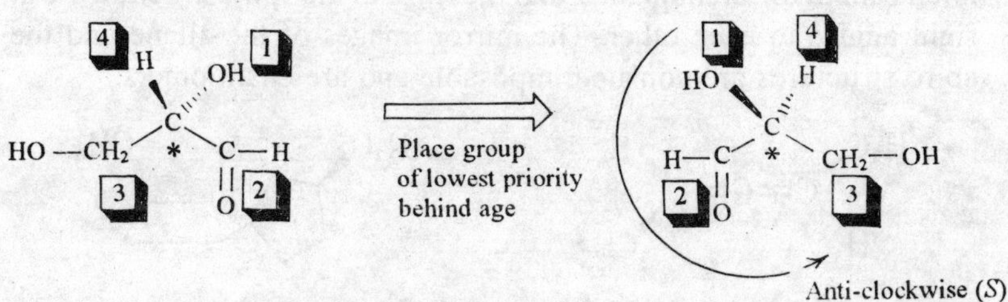

Fig. 4.15. Assigning an asymmetric center as (R) or (S).

(+) *and* (–)

The assignment of an asymmetric center as (*R*) or (*S*) has nothing to do with whichever direction the molecule rotates plane-polarized light. Optical rotation can only be determined experimentally. By convention, molecules which rotate plane-polarized light clockwise are written as (+) or *d*. Molecules which rotate plane-polarized light counterclockwise are written as (–) or *l*. The (*R*) enantiomer of lactic acid is found to rotate plane-polarized light counterclockwise and so this molecule is defined as (*R*)-(–)-lactic acid.

Optical Purity

The optical purity is a measure of enantiomeric purity of a compound and is given in terms of its **enantiomeric excess** (ee). A pure enantiomer would have an optical purity and enantiomeric excess of 100%. A fully racemized compound would have an optical purity of 0%. If the enantiomeric excess is 90%, it signifies that 90% of the sample is pure enantiomer and the remaining 10% is a racemate containing equal amounts of each enantiomer (i.e. 5% + 5%). Therefore the ratio of enantiomers in a sample having 90% optical purity is 95.5.

Allenes and Spiro Compounds

All chiral molecules do not have asymmetric centers. For example, some substituted allenes and spiro structures have no symmetric center but are still chiral (Fig. 4.16). The substituents at either end of the allene are in different planes, and the rings in the spiro structure are at right angles to each other. The mirror images of the allene and the spiro structures are nonsuperimposable and are enantiomers.

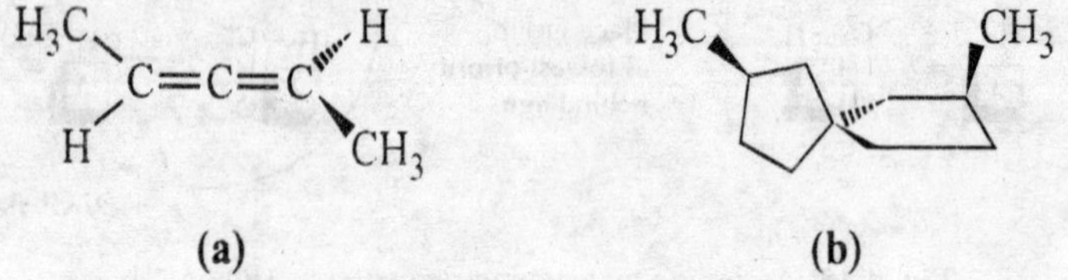

(a) (b)

Fig. 4.16. (a) Allkene; (b) spiro structure.

The rule for determining whether a molecule is chiral or not is to study the symmetry of the molecule. A molecule will be chiral if it is asymmetric (i.e. has no elements of symmetry) or if it has no more than one axis of symmetry.

Meso *structures*

The molecule in (Fig. 4.17a) which there are two identical asymmetric centers but which is not chiral. The mirror images of such a structure are superimposable and so the compound cannot be chiral. This is because the molecule contains a plane of symmetry as shown in Fig. 4.17 in which the molecule has been rotated around the central C–C bond. A structure like this is known as a ***meso*** structure.

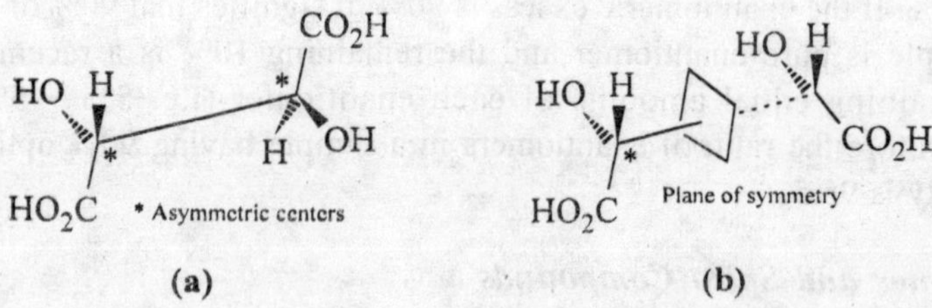

Fig. 4.17. (a) Meso structure showing asymmetric centers; (b) plane of symmetry in meso structure.

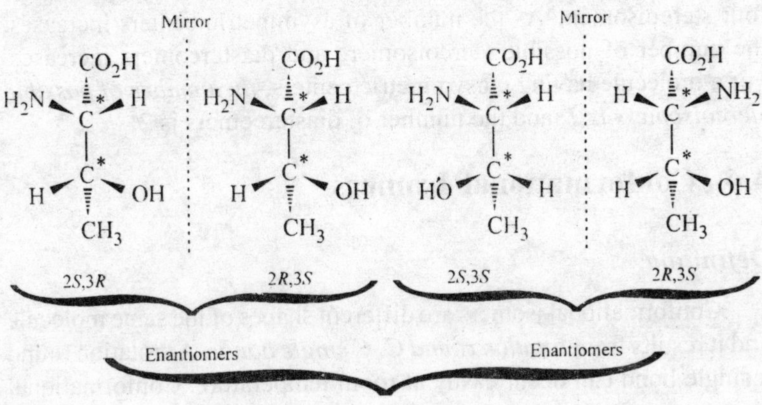

Fig. 4.18. Stereoisomers of threonine.

Diastereomers

If a molecule has two or more asymmetric centers then there are various possible structures that are possible and we use terms like **stereoisomers, diastereomers** and **enantiomers** to discuss them. To illustrate the relative meaning of these terms, let us look at the possible structures of the amino acid *threonine* (Fig. 4.18). This molecule has two asymmetric centers. Thus, four different structures are possible which arises from the two different possible configurations at each center. These are shown with the asymmetric centers at positions 2 and 3 defined as (*R*) or (*S*). The four different structures are called as stereoisomers. The (2S,3R) stereoisomer is a *nonsuperimposable mirror image* of the (2R,3S) stereoisomer and so these structures are **enantiomers** having the same chemical and physical properties. The (2S,3R) stereoisomer is the *nonsuperimposable mirror image* of the (2R,3S) stereoisomer and so these structures are also **enantiomers** having the same chemical and physical properties. *Each set of enantiomers is called a diastereomer.* **Diastereomer** are not mirror images of each other and are completely different compounds with different physical and chemical properties. Thus, threonine has two asymmetric centers i.e. there are two possible diastereomers consisting of two enantiomers each, making a total of four stereoisomers. As the number of asymmetric centers increases, the number of possible stereoisomers and diastereomers increases. For a molecule having n asymmetric centers, *the number of possible stereoisomers is* 2^n and the number of diastereomers is 2^{n-1}

4.4 Conformational Isomers

Definition

Conformational isomers are different shapes of the same molecule and it results from *rotation round C–C single bonds*. As rotation round a single bond can occur easily at room temperature. Conformational isomers are freely interconvertable and they cannot be separated.

Alkanes

Conformational isomers of alkanes arise from the rotation of C–C single bonds. Different shapes of alkanes can be adopted by a molecule like ethane by rotation around the C–C bond. The most distinctive ones (Fig. 4.19) are I and II are known as 'staggered' and 'eclipsed' respectively. In conformation I, the C–H bonds on carbon 1 are staggered with respect to the C–H bonds on carbon 2. In conformation II, they are eclipsed. **Newman projections** (Fig. 4.20) shows the view along the C_1–C_2 bond. In it carbon 1 is represented by the small black circle and carbon 2 is represented by the larger sphere. Viewed in this manner we can see that the C–H bonds on carbon 1 are eclipsed with the C–H bonds on carbon 2 in conformation II.

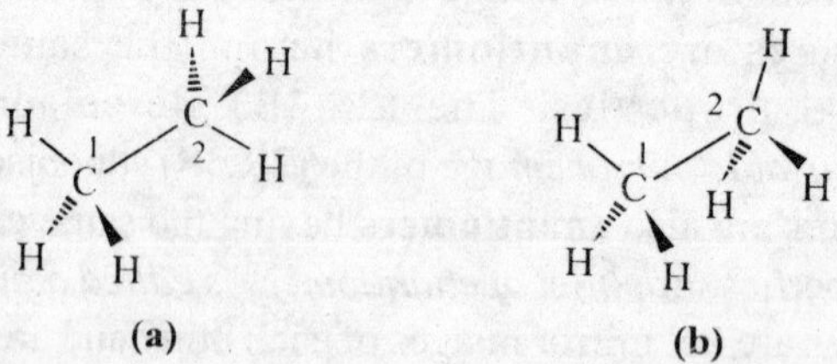

Fig. 4.19. (a) 'Staggered' conformation of ethane; (b) 'eclipsed' conformation of ethane.

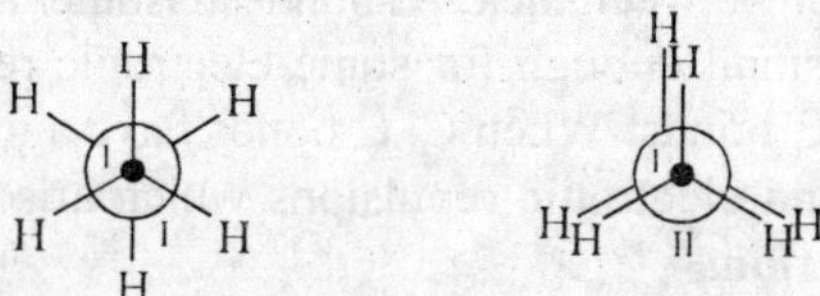

Fig. 4.20. Newman projections of the staggered (I) and eclipsed (II) conformations of ethane.

Out of the two conformations, the *staggered conformation is the more stable* as the C–H bonds and hydrogen atoms are as far apart from each other as possible. In the eclipsed conformation, both the bonds and the atoms are closer together and it can cause strain due to electron repulsion between the eclipsed bonds and between the eclipsed atoms. Therefore, majority of the ethane molecules are in the staggered conformation at any point of time. Since the energy difference between the staggered and eclipsed conformations is small and so it allows each ethane molecule to pass through an eclipsed conformation (Fig. 4.21).

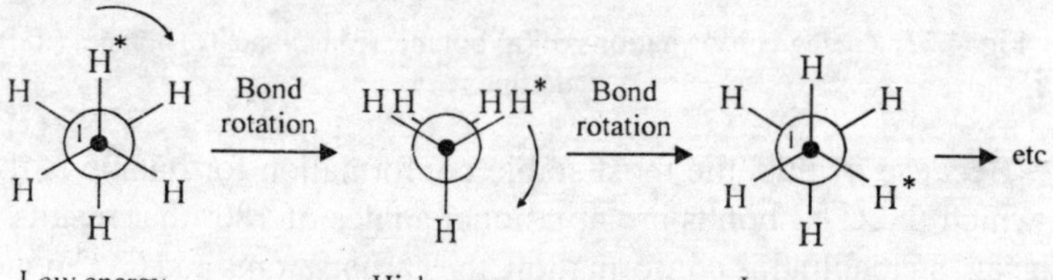

Bond rotation → Bond rotation → etc

Low energy High energy Low energy

Fig. 4.21. Bond rotation of ethane.

Ethane has only one type of staggered conformation, but different staggered conformations are possible with larger molecules like butane (Fig. 4.22). The first and the third C–C bonds in isomer I are at an angle 60° with respect to each other when viewed along the middle C–C bond. In isomer II, these bonds are at an angle of 180°. This angle is called as the **torsional angle** or **dihedral angle**. Isomer II is more stable than isomer I because the methyl groups and the C–C bonds in this **conformer** are as far apart from each other as possible. The methyl groups are bulky and in conformation I they are close enough to interact with each other and lead to some strain. There is also an interaction between the C–C bonds in isomer I since a torsional angle of 60° is small enough for some electronic repulsion to exist between the C–C bonds. When C–C bonds have a torsional angle of 60°, the steric and electronic repulsions which arise are called as a **gauche interaction**.

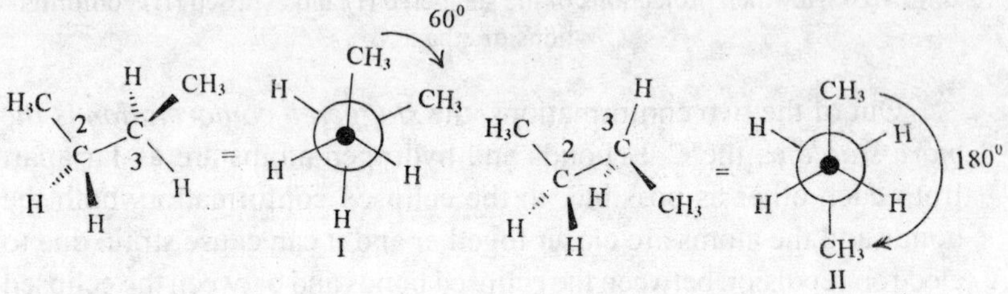

Fig. 4.22. Gauche conformation (I) and anti conformation (II) of butane.

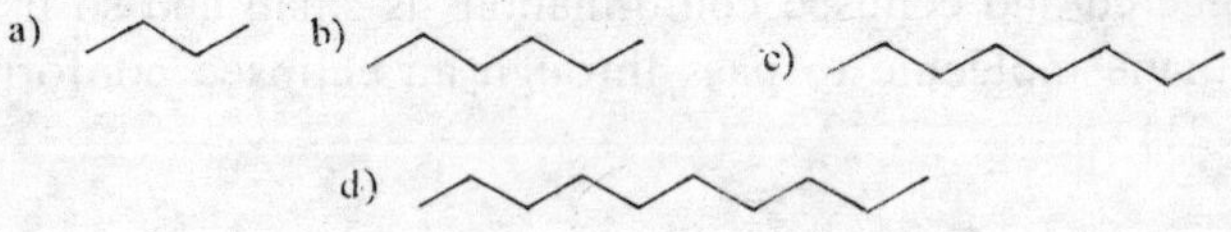

Fig. 4.23. Zigzag conformations of (a) butane; (b) hexane; (c) octane; (d) decane.

Because of this, the most stable conformation for butane is that in which the C–C bonds are at trsional angles of 180° that results in 'zigzag' shape. In this conformation, the carbon atoms and C–C bonds are as far apart from each other as possible. The most stable conformations for longer chain hydrocarbons will also be zigzag (Fig. 4.23). However, since bond rotation is taking place all the time for all the C–C bonds, it is unlikely that many molecules will be in a perfect zigzag shape at any specific time.

Cycloalkanes

Cyclopropane (Fig. 4.24) is a flat molecule in respect of C–atoms, with the hydrogen atoms situated above and below the plane of the ring, so it has no conformational isomers. Cyclobutane (Fig. 4.25) can form three distinct shapes– *a planar shape* and *two 'butterfly'* shapes. Cyclopentane (Fig. 4.26) can also form a number of shapes or conformations. The planar structures for cyclobutane and cyclopentane are too strained to exist in practice because of eclipsed C–H bonds.

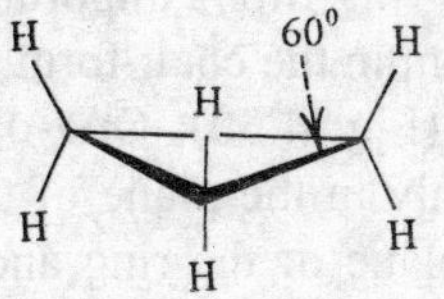

Fig. 4.24. Cyclopropane.

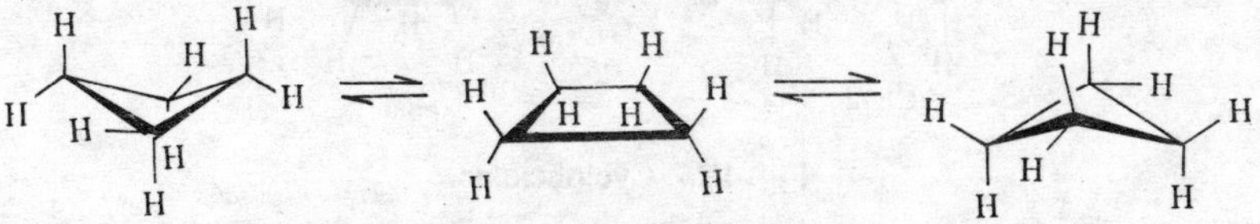

Fig. 4.25. Cyclobutane.

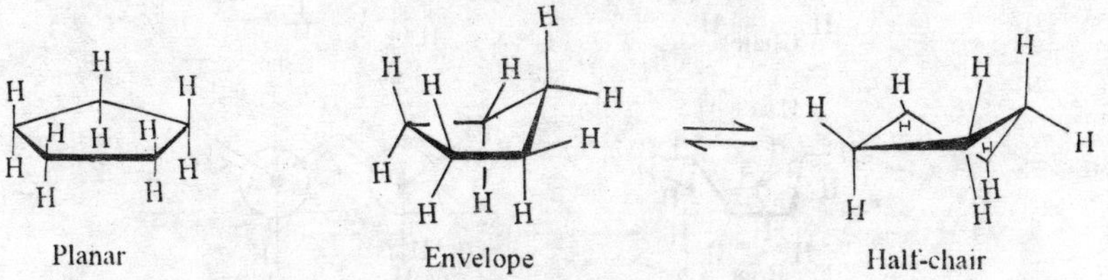

Planar Envelope Half-chair

Fig. 4.26. Cyclopentane.

The two main conformational shapes for cyclohexane are called the chair and the boat (Fig. 4.27). The chair form is more stable than the boat form since the latter has eclipsed C–C and C–H bonds. This can be observed in the Newman projections (Fig. 4.28) that have been drawn in such a way that we are looking along two bonds at the same time–bonds 2-3 and 6-5. In the chair conformation, there are no eclipsed C–C bonds. But, in the boat conformation, bond 1-2 is eclipse with bond 3-4, and bond 1-6 is eclipsed with bond 5-4. This indicates that the boat conformation is less stable than the chair conformation and the majority of cyclohexane molecules exist in the chair conformation. However, the energy barrier is small and the cyclohexane molecules can pass through the boat conformation in a process known as '**ring flipping**' (Fig. 4.29). The ability of a cyclohexane molecule to ring-flip is important when substituents are present. Each carbon atom in the chair form has two C–H bonds, but these are not identical (Fig. 4.30). One of these bonds is called **equatorial** as it is roughly in the plane of the ring. The other C–H bond is vertical to the plane of the ring and is known as the **axial** bond.

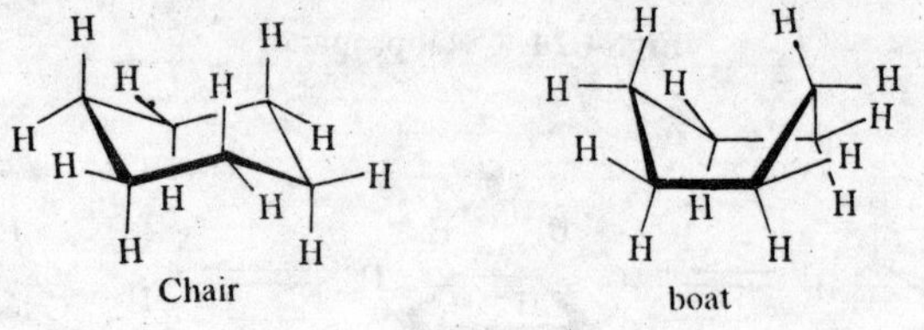

Chair   boat

Fig. 4.27. Cyclohexane.

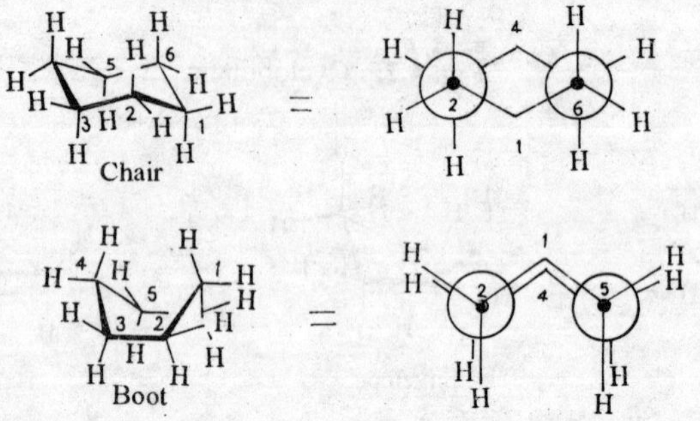

Chair

Boot

Fig. 4.28. Newman projections of the chair and boat conformations of cyclohexane.

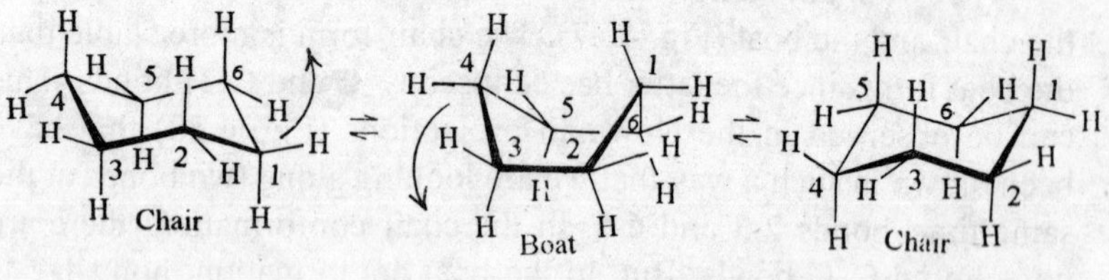

Chair   Boat   Chair

Fig. 4.29. Ring flipping of cyclohexane.

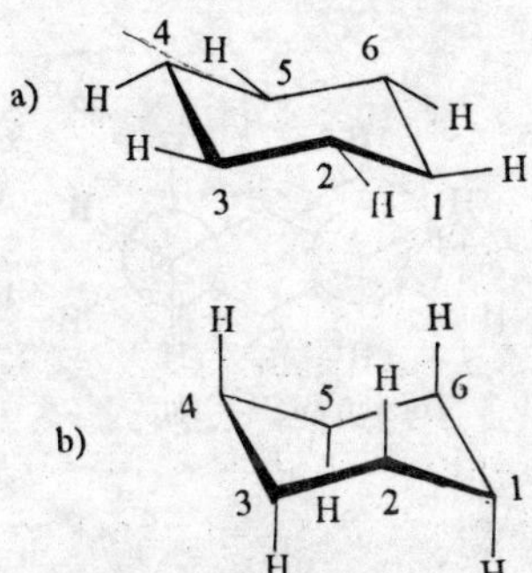

Fig. 4.30. (a) Equatorial C–H bonds; (b) axial C–H bonds.

When ring flipping occurs from one chair to another, all the axial bonds become equatorial bonds and all the equatorial bonds become axial bonds. This does not matter for cyclohexane itself, but it becomes important when there is a substituent present in the ring. For example, methycylclohexane can have two chair structures where the methyl group is either on an equatorial bond or an axial bond (Fig. 4.31).

These are different orientation of the same molecule that are interconvertable because of rotation of C–C single bonds (the ring flipping process). The two chair structures are conformational isomers but they are not of equal stability. *The more stable conformation is the one where the methyl group is in the equatorial position.* In this position, the C–C bond connecting the methyl group to the ring has a torsional angle of 180° with respect to bonds 5-6 and 3-2 in the ring. In the axial position, however, the C–C bond has a torsional angle of 60° with respect to these same two bonds. This can be shown by comparing Newman diagrams of the two methylcyclohexane conformations (Fig. 4.32).

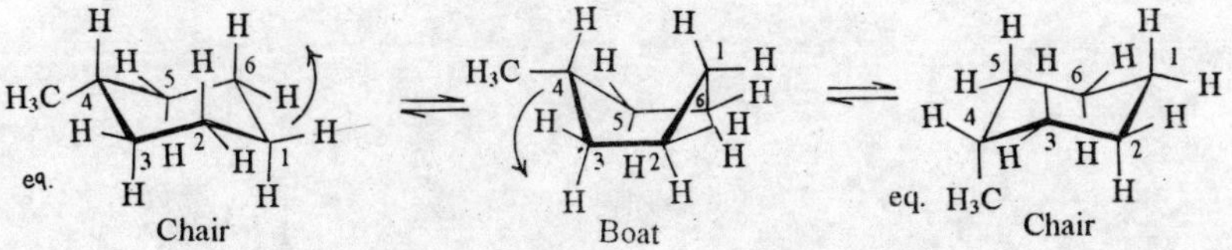

Fig. 4.31. Ring flipping of methylcyclohexane.

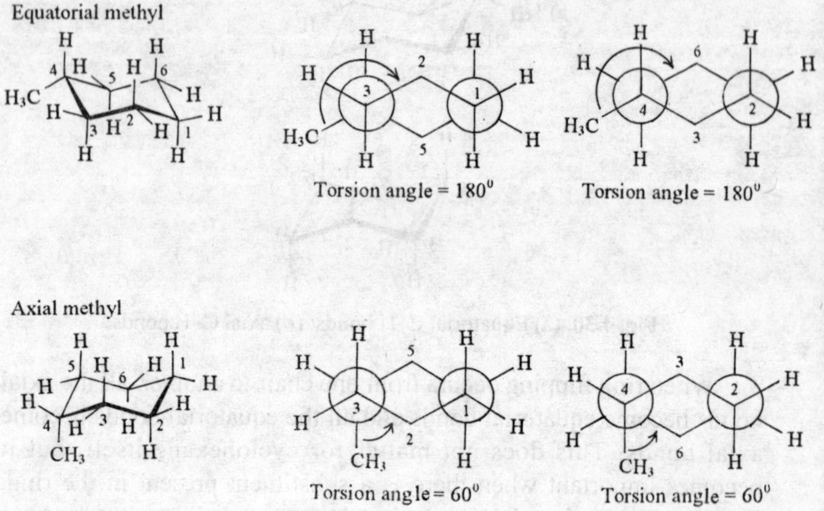

Equatorial methyl

Torsion angle = 180°

Torsion angle = 180°

Axial methyl

Torsion angle = 60°

Torsion angle = 60°

Fig. 4.32. Newman projections of the chair conformations of methylcyclohexane.

A torsion angle of 60° between C–C bonds represents a **gauche interaction** and so an axial methyl substituents experiences two gauche interactions with the cyclohexane ring whereas the equatorial methyl substituents experiences none. Due to this, the latter chair conformation is preferred and about 95% of methylcyclohexane molecules are in this conformation at any point of time, compared to 5% in the other conformation.

5

Nucleophiles and Electrophiles

5.1 Definitions

Nucleophiles

Majority of organic reaction between a molecule that is rich in electrons and a molecule that is deficient in electrons. Such a reaction involves the formation of a new bond in which the electrons are provided by the electron-rich molecule. Electron-rich molecules are known as **nucleophiles** (meaning nucleus-loving). The easiest nucleophiles to identify are negatively charged ions with lone pairs of electrons (e.g., the hydroxide ion), but neutral molecules can also act as nucleophiles if they contain electron-rich functional groups (e.g., an amine).

Thus **neucleophiles (Nu)** are neutral or negatively charged species which are capable of donating a pair of electrons to some other molecule. These species act as **Lewis bases** and attack the electron deficient centers of organic molecule. Some examples are:

$:\bar{Cl},\ :\bar{O}R,\ :\bar{C}N,\ :\bar{N}O_2,\ :RCO\bar{O},\ :\bar{S}H,\ :\bar{O}H,\ :\bar{C} \equiv CR$ **(Charged nucleophiles)**

$H_2\ddot{O}:,\ R\ddot{\underset{..}{O}}H,\ \ddot{N}H_3,\ R\ddot{\underset{..}{O}}R,\ R\ddot{N}H_2,\ R_2\ddot{N}H,\ R_3\ddot{N}$ **(Neutral nucleophiles)**

Nucleophilic Center

In nucleophiles there is a specific atom or region of the molecule that is electron rich. This is known as the **nucleophilic center.** The nucleophilic center of an ion is the atom bearing a lone pair of electrons and the negative charge. The nucleophilic center of a neutral molecule is generally an atom with a lone pair of electrons (e.g., nitrogen or oxygen), or a multiple bond (e.g., alkene, alkyne, aromatic ring).

Electrophiles

Electron-deficient molecules are known as **electrophiles** (electron-loving) and react with nucleophiles. Positively charged ions can easily be identified as electrophiles (e.g., a carbocation), but neutral molecules can also act as electrophiles if they have certain types of functional groups (e.g., carbonyl groups or alkyl halides).

Thus **electrophiles (E)** are neutral or positively charged species which are capable of accepting a pair of electrons from the substrate molecule. These species act as **Lewis acids** and attack the electron rich centers of the organic molecule. Some examples are:

$\overset{+}{H}, \overset{+}{Cl}, \overset{+}{Br}, Ar{-}\overset{+}{N}{\equiv}H, \overset{+}{NO_2}, \overset{+}{NO}$ **(Charged electrophiles)**

$BF_3, AlCl_3, SO_3.$ **(Neutral electrophiles)**

Electrophilic Center

In electrophiles there is a specific atom or region of the molecule that is electron deficient. This region is known as the **electrophilic center**. In a positively charged ion, the electrophilic center is the atom bearing the positive charge (e.g., the carbon atom of a carbocation). In a neutral molecule, the electrophilic center is an electron-deficient atom within a functional group (e.g., a carbon or hydrogen atom linked to an electronegative atom like oxygen or nitrogen).

5.2 Charged Species

Anions

A negatively charged molecule like a hydroxide ion (OH^-) (Fig. 5.1) is electron rich and it acts as a **nucleophile**. The atom that bears the negative charge and a lone pair of electrons is the *nucleophilic center.* In case of the hydroxide ion is the oxygen atom acts as nucleophilic center. Some ions (e.g., the carboxylate ion) are capable of sharing the negative charge between two or more atoms through a process called **delocalization**. In such a case, the negative charge is shared between both oxygen atoms and so both of these atoms can act as nucleophilic centers (Fig. 5.1).

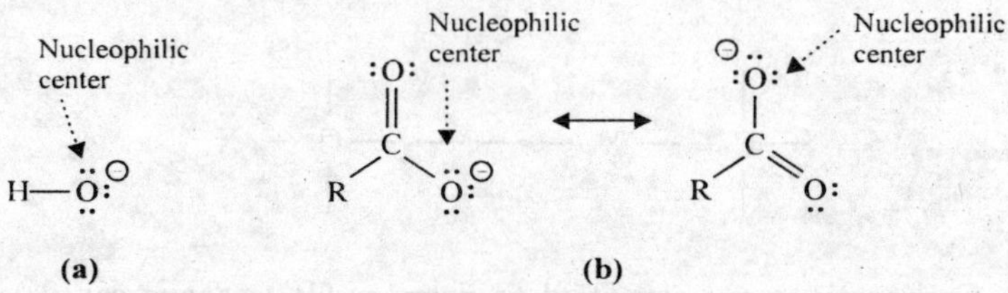

Fig. 5.1. Examples of nucleophiles; (a) hydroxide ion; (b) carboxylate ion.

Cations

A positively charged ion is electron deficient and acts as an electrophile. The atom that bears the positive charge is the electrophilic center. In case of a carbocation (Fig. 5.2), this is the carbon atom. Some molecules (e.g. the allylic cation) are able to delocalize their positive charge between two or more atoms in which case all the atoms capable of sharing the charge are electrophilic centers (Fig. 5.2).

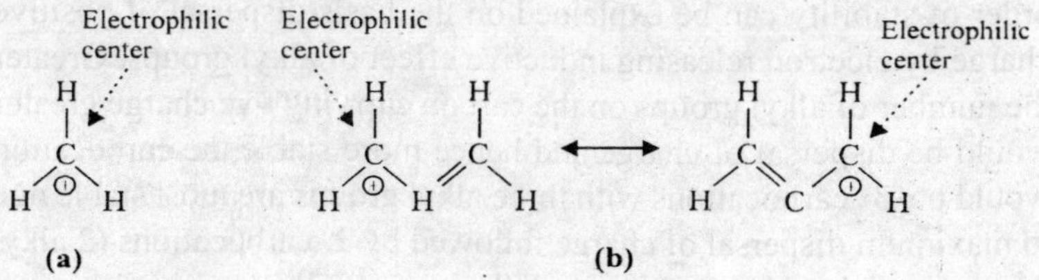

Fig. 5.2. Examples of electrophiles: (a) carbocation; (b) allylic cation.

$$\underset{\text{(a)}}{Me_2N-\overset{\overset{O}{\|}}{C}-H} \qquad \underset{\text{(b)}}{Me-\overset{\overset{O}{\|}}{S}-Me}$$

Fig. 5.3. (a) Dimethylformamide (DMF); (b) dimethylsulfoxide (DMSO).

Carbocations

Carbocation may be defined as a group of atoms that contains a carbon atom bearing positive charge and containing only six electrons in its valence shell. These ions are formed by heterolytic cleavage in which the leaving group takes along with it the shared pair of electrons.

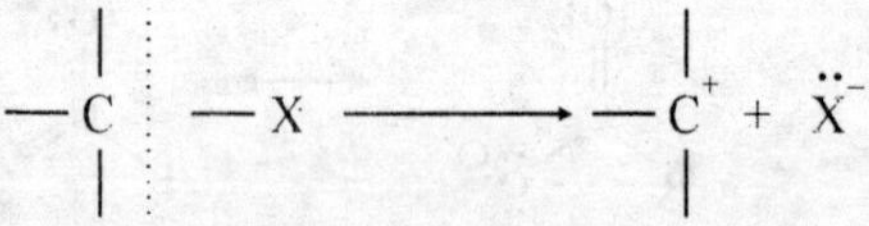

$$-\overset{|}{\underset{|}{C}} \vdots -X \longrightarrow -\overset{|}{\underset{|}{C^+}} + \ddot{X}^-$$

Carbocations are classified as *primary* (1°), *secondary* (2°) or *tertiary* (3°) depending upon the carbon bearing the positive charge. For example, if the positive charge is on primary carbon then the carbocation is primary.

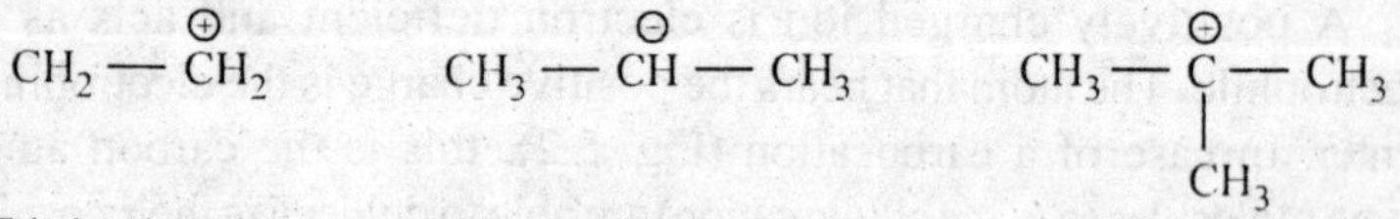

$$\underset{\substack{\text{Ethyl carbocation}\\(1^\circ)}}{CH_2-\overset{\oplus}{C}H_2} \qquad \underset{\substack{\textit{iso}\text{-Propyl carbocation}\\(2^\circ)}}{CH_3-\overset{\ominus}{C}H-CH_3} \qquad \underset{\substack{\textit{tert}\text{-Butyl carbocation}\\(3^\circ)}}{CH_3-\overset{\ominus}{\underset{\underset{CH_3}{|}}{C}}-CH_3}$$

Decreasing order of stability of carbocations is 3° > 2° > 1°. This order of stability can be explained on the basis dispersal of positive charge by electron releasing inductive effect of alkyl groups. Greater the number of alkyl groups on the carbon carrying +ve charge greater would be dispersal of charge and hence more stable the carbocation would be .3° carbocations with three alkyl groups are most stable due to maximum dispersal of charge followed by 2° carbocations (2 alkyl groups) and 1° carbocations (1 alkyl group) as illustrated below:

$$\mathrm{R} \rightarrow \underset{\substack{\uparrow \\ \mathrm{R}}}{\overset{\substack{\mathrm{R} \\ \downarrow}}{\mathrm{C}^{\oplus}}} \; > \; \mathrm{R} \rightarrow \underset{\substack{\uparrow \\ \mathrm{R}}}{\overset{\substack{\mathrm{R} \\ \downarrow}}{\mathrm{C}^{\oplus}}} \; > \; \mathrm{R} \rightarrow \underset{\substack{| \\ \mathrm{H}}}{\overset{\substack{\mathrm{H} \\ |}}{\mathrm{C}^{\oplus}}} \; > \; \mathrm{H}-\underset{\substack{| \\ \mathrm{H}}}{\overset{\substack{\mathrm{H} \\ |}}{\mathrm{C}^{\oplus}}}$$

Carbocations are highly reactive species because they contain a carbon atom which has only six electrons in its valence shell and has tendency to complete its octet. Order of reactivity of carbocations is reverse the order of stability.

$1^\circ > 2^\circ > 3^\circ$ — order of reactivity

$3^\circ > 2^\circ > 1^\circ$ — order of stability

Orbital Structure of Carbocations

Electron deficient carbon in carbocations is bonded to three other atoms directly and for this bonding it uses sp^2 hybridised orbitals. This part of the carbocation is therefore planar. In addition to this there is a vacant *p*-orbital perpendicular to this plane with one lobe above the plane and the other below the plane is illustrated.

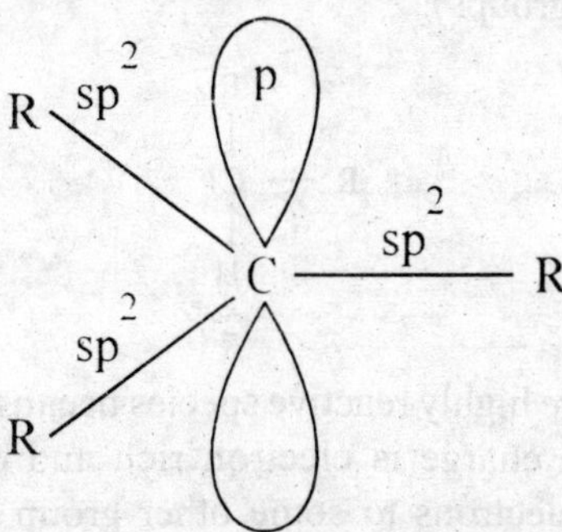

Carbanions

Carbanion may be defined as a group of atoms that contain a carbon atom bearing negative charge. Carbanions are formed by heterolytic cleavage in which shared pair remains with carbon atom.

$$-\overset{|}{\underset{|}{C}}-X \longrightarrow -\overset{|}{\underset{|}{C}}:^- + X^+$$

$$H-\overset{O}{\overset{||}{C}}-\overset{H}{\underset{H}{\overset{|}{\underset{|}{C}}}}-H \longrightarrow H-\overset{O}{\overset{||}{C}}-\overset{H}{\underset{H}{\overset{|}{\underset{|}{C}}}}:^- + H_2O$$

Carbanions are classified as primary (1°), secondary (2°) or tertiary (3°) after the carbon bearing the negative charge. For example, if the negative charge is on primary carbon then the carbanion is primary.

Decreasing Order of Stability of Carbanions is 1° > 2° > 3°.

This order of stability can be explained on the basis of electron releasing inductive effect of alkyl groups. Greater the number of alkyl groups on the carbon bearing negative charge, greater would be the concentration of negative charge on carbon resulting in instability. Primary (1°) carbanion with only one alkyl group is maximum stable followed by secondary (2°) (with two alkyl groups) and tertiary (3°) (with three alkyl groups).

$$R \rightarrow \overset{R\downarrow}{\underset{H\uparrow}{C}}:^- \;(1°) \quad > \quad R \rightarrow \overset{R\downarrow}{\underset{H\uparrow}{C}}:^- \;(2°) \quad > \quad R \rightarrow \overset{R\downarrow}{\underset{R\uparrow}{C}}:^- \;(3°)$$

Carbanions are highly reactive species because in them the carbon carrying negative charge is electron rich and can donate its non-bonding pair of electrons to some other group for sharing. Hence, carbanions behave as nucleophiles. Order of reactivity of carbanions is reverse the order of stability.

3° > 2° > 1°—order of reactivity

Orbital Structure of Carbanions

In carbanions the carbon carrying negative charge is sp^3 hybridised. Three sp^3 hybrid orbitals are used to form three σ bonds with three different atoms whereas the fourth sp^3 hybrid orbital accommodates lone pair.

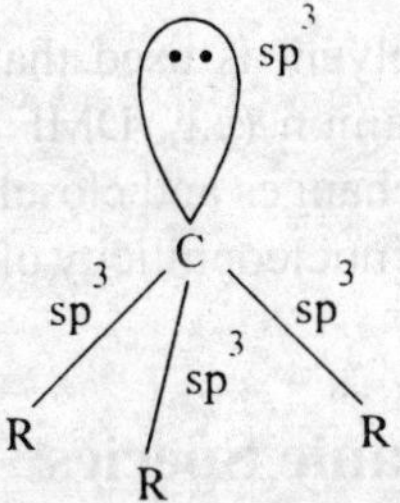

Relative Nucleophilicity

In any series of anions, nucleophilic strength parallels basicity if the nucleophilic center is the same atom e.g. the nucleophilic strengths of the following oxygen compounds ($RO^- > HO^- >> RCO_2^-$) matches their order of basicity.

The same is true for anions where the nucleophilic center is an element in the same row of the periodic table (e.g., C,N,O,F). Thus, the order of nucleophilicity of the following anions ($R_3C^- > R_2N^- > RO^- > F^-$) is the same as their order of basicity. This trend can be related to the electronegativities of these atoms. *The more electronegative the atom (e.g., F), the more tightly it holds on to its electrons and the less available these electrons are for forming new bonds (less nucleophilic).*

It becomes complex if we try to compare anions having nucleophilic centers from different parts of the periodic table. In such a case, relative nucleophilicity does not match relative basicity. This is because the solvent used in a reaction has an important effect. In protic solvents like water or alcohol, the stronger nucleophiles are those which have a large nucleophilic center, i.e., an atom lower down the periodic table (e.g. S^- is more nucleophilic than O^- but is less basic). This is because protic solvents can form hydrogen bonds to the anion. *The smaller the anion, the stronger the solvation and the more difficult it is for the anion to react as a nucleophilic.*

The order of nucleophilicity of same common anions in protic solvents is as follows: $SH^- > CN^- > I^- > OH^- > N_3^- > Br^- > CH_3CO_2^- > Cl^- > F^-$.

When an organic solvent is used that is incapable of forming hydrogen bonds to the anion (e.g. DMF or DMSO; Fig. 5.3), the order of nucleophilicity changes and closely matches that of basicity. For example, the order of nucleophilicity of the halides in DMSO is $F^- > Cl^- > Br^- > I^-$.

5.3 Neutral Inorganic Species

Polar Bonds

When two atoms across a bond are of quite different electronegativities then the bond connecting them will be *polar covalent* and the bonding electrons are biased towards the more electronegative atom. This gives the latter a slightly negative charge and thus makes it a *nucleophilic center*. Conversely, the less electronegative atom gains a slightly positive charge and because an **electrophilic center** (Fig. 5.4). The further right an element is in periodic table, the more electronegative it is. Thus, fluorine is more electronegative than oxygen, which in turn is more electronegative than nitrogen. All the nucleophilic atoms have lone pair of electrons.

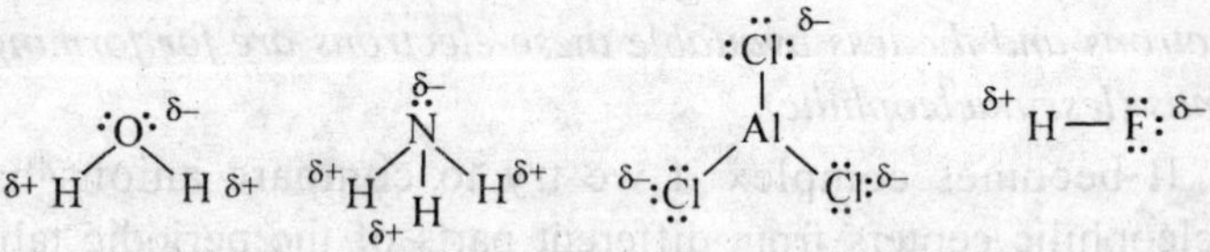

Fig. 5.4. Nucleophilic (δ–) and electrophilic (δ+) centers in neurtal inorganic molecules.

Nucleophilic Strength

We have learnt that the molecules can have both *nucleophilic* and *electrophilic centers* and so can act as *nucleophiles* or as *electrophiles*. However, it is generally found that there is a preference to react as one rather than the other. This can be explained by considering the relative strengths of nucleophilic and electrophilic centers. Let us consider the relative strengths of nucleophilic centers by comparing N, O and F. If we look at the relative positions of these atoms in the periodic table, we find that fluorine is more electronegative than oxygen, which in turn is more electronegative than nitrogen. However, when we compare the nucleophilic strengths of these atoms, we observe that the nitrogen is more nucleophilic than oxygen, which in turn is more nucleophilic than fluorine.

The relative nucleophilic strengths of these atoms can be explained by looking at the products which would be formed if these atoms **were** to act as nucleophiles. Let us compare the three molecules HF, H_2O, and NH_3 and observe what happens if they form a bond to a proton (Fig. 5.5). Since the proton has no electrons, both electrons for the new bond must come from the nucleophilic centers (i.e. the F, O and N). Due to this, these atoms will gain a positive charge. If hydrogen fluoride acts as a nucleophilic, then the fluorine atom gains a positive charge. Because the fluorine atom is strongly electronegative, it does not tolerate a positive charge. Hence, this reaction does not occur. Oxygen is less electronegative and is able to tolerate the positive charge slightly better, such that an equilibrium is possible between the charged and uncharged species. Nitrogen is the least electronegative of the three atoms and tolerates the positive charge so well that the reaction is irreversible and a slat is formed.

Thus, nitrogen is strongly nucleophilic and will generally react as such, whereas halogens are weakly nucleophilic and will rarely react as such.®

Lastly, if all these molecules are weaker nucleophiles than their corresponding anions, i.e. HF, H_2O, and NH_3 are weaker nucleophiles than F^-, OH^- and NH_2^- respectively.

Electrophilic Strength

Using the above argument in reverse when looking at the relative electrophilic strengths of atoms in different molecules, we can compare the electrophilic strengths of the hydrogens in HF, H_2O, and NH_3. In this case, reaction with a strong nucleophile or base would produce anions (Fig. 5.6). Fluorine being the most electronegative atom is best able to stabilize a negative charge and so the fluoride ion is the most stable ion of the three. Oxygen is also able to stabilize a negative charge, though not as well as fluorine. Nitrogen is the least electronegative of the three atoms and has the least stabilizing influence on a negative charge and so the NH_2^- ion is unstable. *The more stable the anion, the more easily it is formed and hence the hydrogen, which is lost, will be strongly electrophilic.* This is the case for HF. In contrast, the hydrogen in ammonia is a very weak electrophilic center since the anion formed is unstable. Due to this, nitrogen anions are only formed with very strong bases.

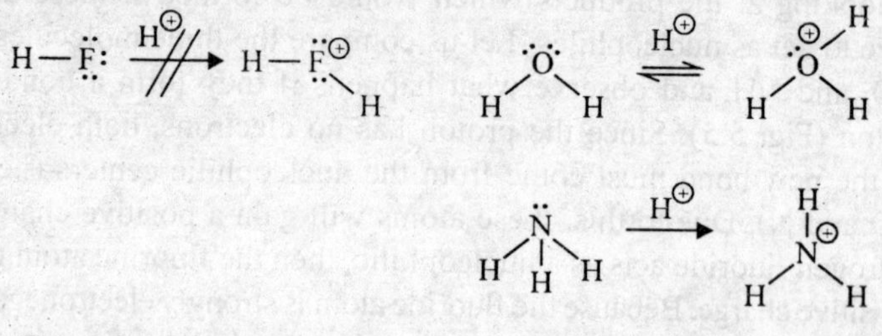

Fig. 5.5. Cations formed if HF, H_2O, and NH_3 act as nucleophiles.

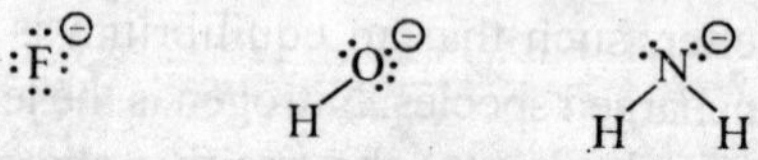

Fig. 5.6. Anions generated when HF, H_2O and NH_3 act as electrophiles.

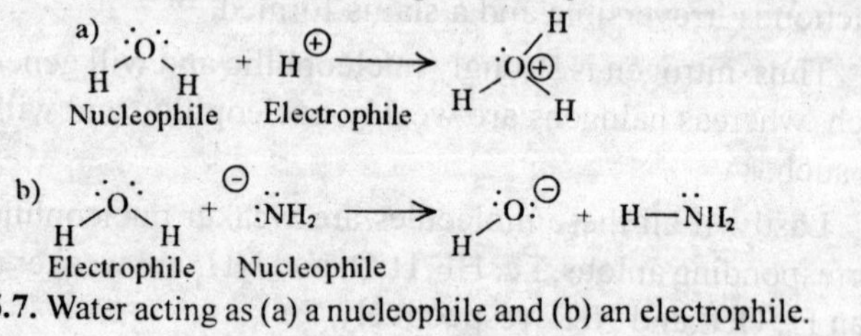

a) Nucleophile Electrophile

b) Electrophile Nucleophile

Fig. 5.7. Water acting as (a) a nucleophile and (b) an electrophile.

Properties

We can predict whether molecules are more likely to react as nucleophiles or electrophiles depending on the strength of the nucleophilic and electrophilic centers present. For example, ammonia has both electrophilic and nucleophilic centers. However, it generally reacts as a nucleophile because the nitrogen atom is a strong nucleophilic center and the hydrogen atom is a weak electrophilic center. By contrast, molecules like hydrogen fluoride or aluminum chloride prefer to react as electrophiles. This is because the nucleophilic centers in both these molecules (halogen atoms) are weak, whereas the electrophilic centers (H or Al) are strong. Water is a molecule that can react equally well as a nucleophile or as an electrophile. For example, water reacts as a nucleophile with a proton and as an electrophile with an anion (Fig. 5.7).

5.4 Organic Structures

Alkanes

Alkanes consist of carbon-carbon and carbon-hydrogen single bonds and are unreactive compounds. This is due to the fact that C–C and C–H bonds are covalent in nature and so there are no electrophilic or nucleophilic centers present. Since most reagents react with nucleophilic or electrophilic centers. so the alkanes are unreactive molecules.

Polar Functional Groups

We can identify the nucleophilic and electrophilic centers in common functional groups, based on the relative electronegatives of the atoms present. The following guidelines are helpful in this:-

(i) C–H and C–C bonds are covalent. So, neither carbon nor hydrogen is a nucleophilic or electrophilic center;

(ii) Nitrogen is immediately to the right of carbon in the periodic table. The nitrogen is more electronegative but the difference in electronegativity between these two atoms is small and so the N–C bond is not quite polar. Therefore, the carbon atom can generally be ignored as an electrophilic center;

(iii) N–H and O–H bonds are *polar covalent*. Nitrogen and oxygen are strong nucleophilic centers. Hydrogen is an electrophilic center;

(iv) C = O, C = N and C ≡ N bonds are *polar covalent*. The O and N are nucleophilic centers and the carbon is an electrophilic center;

(v) C–O and C–X bonds (X = halogen) are *polar* covalent. The oxygen atom is moderately nucleophilic whereas the halogen atom is weakly nucleophilic. The carbon atom is an electrophilic center.

Making use of the above guidelines, the nucleophilic and electrophilic centers of the common functional groups can be identified, where atoms having a slightly negative charge are nucleophilic centers and atoms having a slightly positive charge are electrophilic centers (Fig. 5.8).

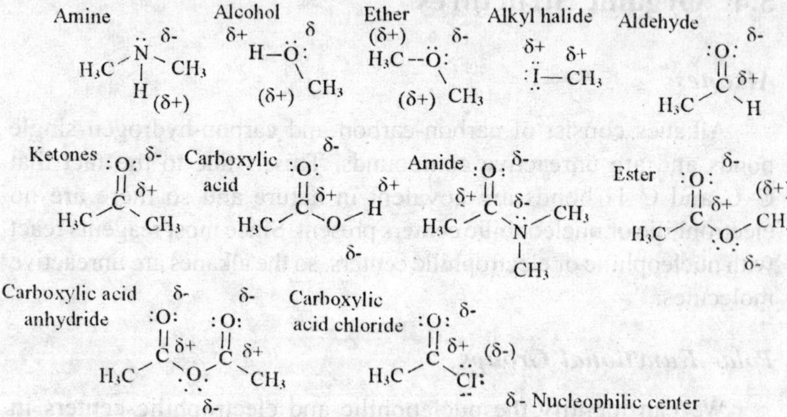

Amine Alcohol Ether Alkyl halide Aldehyde

Ketones Carboxylic acid Amide Ester

Carboxylic acid anhydride Carboxylic acid chloride

δ - Nucleophilic center
δ - Nucleophilic center
(δ+) Weak electrophilic center
(δ+) Weak electrophilic center

Fig. 5.8. Nucleophilic and electrophilic centers of common functional groups.

Please note that all the nucleophilic and electrophilic centers are not of equal importance. E.g. a nitrogen atom is more nucleophilic than an oxygen atom. Also halogen atoms are very weakly nucleophilic and will not generally react with electrophiles if there is a stronger nucleophilic center present. Hydrogen atoms attached to halogens are more electrophilic than hydrogen atoms attached to oxygen. Hydrogen atoms attached to nitrogen are very weakly electrophilic.

Thus we find some functional groups are more likely to react as nucleophiles while some functional groups are more likely to react as electrophiles, e.g., amines, alcohols and ethers are more likely to reat as nucleophiles, because they have strong nucleophilic centers and weak electrophilic centers. Alkyl halides are more likely to react as electrophiles because they have strong electrophilic centers and weak nucleophilic centers. Aldehydes and ketones can react as nucleophiles or electrophiles because in both electrophilic and nucleophilic centers are strong.

Some functional groups contain several nucleophilic and electrophilic centers. For example, carboxylic acids and their derivatives and so there are several possible centers where a nucleophile or an electrophile could react.

Unsaturated Hydrocarbons

All functional groups do not have polar bonds, e.g. alkenes, alkynes, and aromatic compounds have covalent multiple bonds and since space between the multiple bonded carbons is rich in electrons and is therefore nucleophilic. Thus, the nucleophilic center in these molecules is not a specific atom, but the multiple bond (Fig. 5.9).

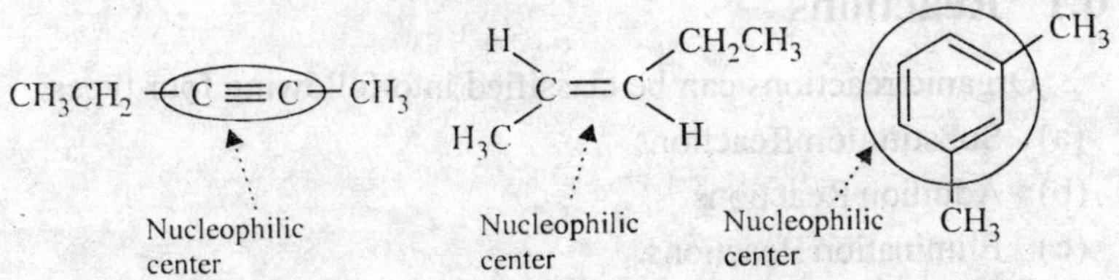

$CH_3CH_2-C\equiv C-CH_3$ Nucleophilic center

H, CH_2CH_3, H_3C, H: C=C — Nucleophilic center

CH_3, CH_3 — Nucleophilic center

Fig. 5.9. Nucleophilic centers in (a) an amine; (b) an alkene; (c) an aromatic compound.

6

Reactions and Mechanisms

6.1 Reactions

Organic reactions can be classified into following four types:

(a) Substitution Reactions

(b) Addition Reactions

(c) Elimination Reactions.

(d) Rearrangement Reactions.

All reactions involve the bond cleavage and the bond formation. Let us discuss bond formation.

Bond Formation

Basically, most reactions involve electron-rich molecules forming bonds to electron deficient molecules (i.e. nucleophiles forming bonds to electrophiles). The bond will be formed particularly between the *nucleophilic center of the nucleophile* and the *electrophilic center of the electrophile.*

Classification of Reactions

We can also classify reactions as:-

(a) acid/base reactions;

(b) functional group transformations;

(c) carbon-carbon bond formations.

The reaction of type (a) are relatively simple and involves the reaction of an acid with a base to give a salt. The reaction of type (b) are one functional group can be converted into another. Normally these reactions are relatively straightforward and proceed in high yield. The reactions of type (c) are extremely important to organic chemistry as these are the reactions that allow the chemist to construct complex molecules from simple starting materials. In general, these reactions are the most difficult and temperamental to carry out. Some of these reactions are so important that they are named after the scientists who developed them (e.g. Grignard and Aldol reactions).

These reactions can also be classified by grouping together, depending on the process or mechanism involved. This is particularly useful since specific functional groups will undergo certain types of reaction category. Table 6.1 serves as a summary of the types of reactions which functional groups normally undergo.

Table 6.1. Different categories of reaction undergone by functional groups

Reaction Category	Functional Group
Electrophillic addition	Alkenes and alkynes
Electrophilic Substitution	Aromatic
Nucleophilic addition	Aldehydes and ketones
Nucleophilic Substitution	Carboxylic acid derivatives Alkyl halides
Elimination	Alcohols and alkyl halides
Reduction	Alkenes, alkynes, aromatic, aldehydes, ketones, nitriles, carboxylic acids, and carboxylic acid derivatives,
Oxidation	Alkenes, alcohols, aldehydes
Acid/base reactions	Carboxylic acids, phenols, amines

(a) ***Substitution Reactions:*** These reactions involve the replacement of an atom or group from the organic molecule by some other atom or group without changing the remaining part of the molecule. The product formed as a result of replacement is called **substitution product** e.g.,

(i) $CH_3CH_2OH \xrightarrow{PCl_5} CH_3CH_2Cl + POCl_3 + HCl$
Ethanol → Ethyl chloride

(ii) $R—I + KOH\ (aq) \longrightarrow R—OH + KI$
Iodoalkane → (alcohol)

(iii) $C_6H_5H + HNO_3 \xrightarrow{Conc.\ H_2SO_4} C_6H_5NO_2 + H_2O$
Benzene (Conc.) → Nitorbenzene

(b) ***Addition Reactions:*** These reactions are generally given by the organic molecule containing multiple bonds. They involve combination of two molecules to form a single molecule. In general in these reactions one p-bond is cleaved and two sigma bonds are formed. The product formed is known as **addition product** or **adduct.** Some examples are:

(i) $CH_2 = CH_2 + HBr \longrightarrow CH_3CH_2Br$
Ethene → Ethyl bromide

(ii) $CH \equiv CH + 2H_2 \xrightarrow[575\ K]{Ni} CH_3—CH_3$
Ethyne → Ethane

(iii) $CH_3C(H) = O + HCN \longrightarrow CH_3CH(CN)(OH)$
Acetaldehyde → Cynohydrin

(c) ***Elimination Reactions:*** These reactions involve the removal of two or more atoms/groups from the organic molecule under suitable conditions to form a product with multiple bond. Elimination can be considered as reverse of addition. Some examples are:

(i) $H_2C(H)—CH_2(OH) \xrightarrow{H_2SO_4} H_2C = CH_2 + \mathbf{H_2O}$
Ethanol → Ethene

$$\text{(ii)}\quad \underset{\text{Ethyl bromide}}{\overset{\;\;\text{H}\qquad\text{Br}}{\text{H}_2\text{C}-\text{CH}_2}} \xrightarrow{\text{KOH}\,(alc.)} \underset{\text{Ethene}}{\text{H}_2\text{C}=\text{CH}_2} + \textbf{HBr}$$

(d) ***Rearrangement Reactions:*** These reactions involve the rearrangement of atoms within the molecule under suitable conditions to form the product with different properties. Some examples are:

$$\text{(i)}\quad \underset{\text{Ammonium cyanate}}{NH_4CNO} \xrightarrow{\text{Heat}} \underset{\text{Urea}}{(H_2N)_2C=O}$$

$$\text{(ii)}\quad \underset{n\text{-Butane}}{CH_3CH_2CH_2CH_3} \xrightarrow[\text{Heat}]{AlCl_3} \underset{\text{Isobutane}}{CH_3-\overset{\overset{\displaystyle CH_3}{|}}{CH}-CH_3}$$

6.2 Mechanisms

Definition

A clear understanding of electrophilic and nucleophilic centers permits us to predict where reactions might occur but not what sort of reaction will occur. To understand and predict the outcome of reactions, it is essential to understand what goes on at the electronic level. This process is a **mechanism.**

A mechanism tells us as to how a reaction occurs. It explains how molecules react together to give the final product. The mechanism tells us how bonds are formed and how bonds are broken and in what order. It explains what is happening to the valence electrons in the molecule as it is the movement of these electrons that result in a reaction. Consider the reaction between a hydroxide ion and a proton to form water (Fig. 6.1). The hydroxide ion is a nucleophile and the proton is an elecrophile. A reaction occurs between the nucleophilic center (the oxygen) and the electrophilic center (the hydrogen) and water is formed. A new bond is formed between the oxygen of the hydroxide ion and the proton. The mechanism of this reaction suggests that a lone pair of electrons from oxygen is used to form a bond to the proton. In this way, the oxygen effectively 'loses' one electron and the proton effectively gains one electron. Because of this, the oxygen loses its negative charge and the proton loses its positive charge.

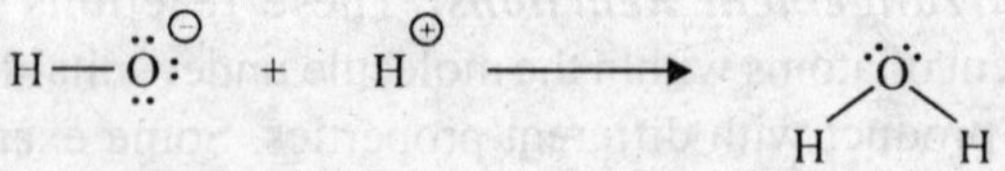

$$H-\ddot{\underset{..}{O}}:^{\ominus} + H^{\oplus} \longrightarrow H-\ddot{\underset{..}{O}}-H$$

Fig. 6.1. Reaction of hydroxide ion and a proton form water.

Curly Arrows

To understand what happens to the valence electrons during a reaction mechanism there is a diagrammatic way making use of curly arrows. For example, the above mechanism can be explained by using a curly arrow to show what happens to the lone pair of electrons (Fig. 6.2). In this case, the arrow starts from a lone pair of electrons on the oxygen (the source of the two electrons) and points to where the **center** of the new bond will be formed.

Sometimes the arrow is written directly to the proton (6.3). Formally, this is incorrect. Arrows should only be drawn directly to an atom if the electrons are going to end up to that atom as a lone pair of electrons.

The following rules are useful when drawing arrows:

(i) curly arrows show the movement of electrons, **not** atoms;

(ii) curly arrows start from the source of two electrons (i.e. a lone pair of electrons on an atom or the middle of a bond which is about to be broken);

(iii) curly arrows point to an **atom** if the electrons are going to end up as a lone pair on that atom;

(iv) curly arrows point to where a new bond will be formed if the electrons are being used to form a new bond.

Fig.6.4 is a demonstration of how arrows should be drawn. One of the lone pairs of electrons on the hydroxide ion is used to form a bond to the acidic proton of the carboxylic acid. The curly arrow representing this starts from a lone pair of electrons and points to the space between the two atoms to show that a bond is being formed.

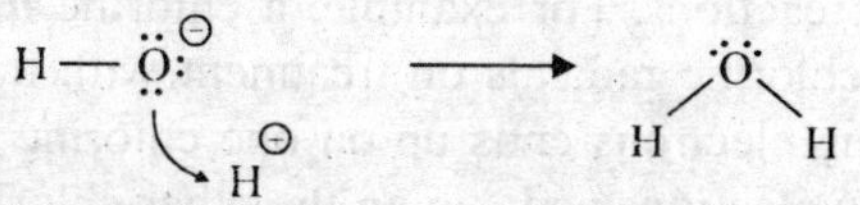

Fig. 6.2. Mechanism for the reaction of a hydroxide ion with a proton.

Fig. 6.3. Incorrect way of drawing a curly arrow.

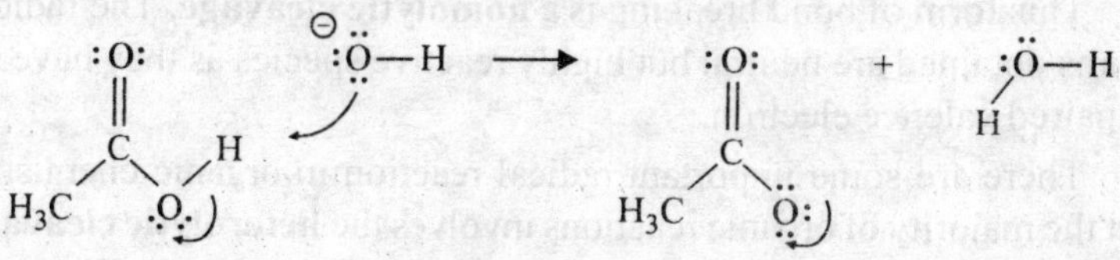

Fig. 6.4. Mechanism for the reaction of a hydroxide ion with ethanoic acid.

At the same time as this new bond is being formed, the O–H bond of the carboxylic acid must break. This is because the hydrogen atom can form only one bond. The electrons in this bond end up on the carboxylate oxygen as a third lone pair of electrons. The arrow representing this starts from the **center** of the bond being broken and points directly to the atom where the electrons will end up as a lone pair.

In the process, the negatively charged oxygen of the hydroxide ion ends up as a neutral oxygen in water, because one of the oxygen's lone pairs is used to form the new bond. Both electrons are now shared between two atoms and so the oxygen effectively loses one electron and its negative charge. The oxygen in the carboxylate ion (which was originally neutral in the carboxylic acid) becomes negatively charged since it now has three lone pairs of electrons and has effectively gained an extra electron.

Half Curly Arrows

Sometimes reactions take place that involve the movement of single electrons rather than pairs of electrons. Such reactions are called **radical** reactions. For example, a chlorine molecule can be split into two chlorine radicals on treatment with light. One of the original bonding electrons ends up on one chlorine radical and the second bonding electrons ends up on the other chlorine radical. The movement of these single electrons can be illustrated by using half curry arrows rather than full curly arrows (Fig. 6.5).

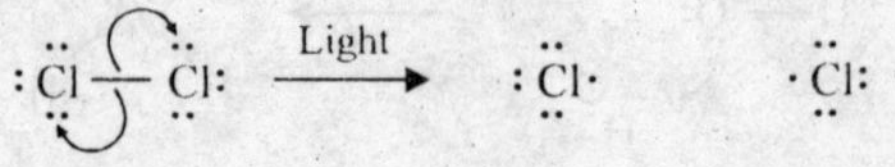

$$:\ddot{\underset{..}{Cl}}-\ddot{\underset{..}{Cl}}: \xrightarrow{\text{Light}} :\ddot{\underset{..}{Cl}}\cdot \quad \cdot\ddot{\underset{..}{Cl}}:$$

Fig. 6.5. Use of half curly arrows in a mechanism (homolytic cleavage).

This form of bond breaking is a **homolytic cleavage**. The radical atoms obtained are neutral but highly reactive species as they have an unpaired valence electron.

There are some important radical reaction in organic chemistry, but the majority of organic reactions involves the **heterolytic cleavage** of covalent bonds where electrons move together as a pair (Fig. 6.6).

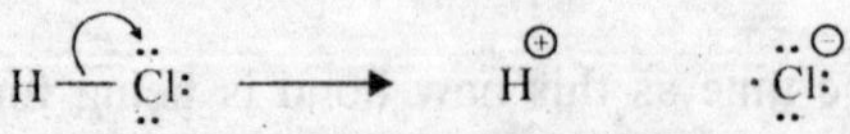

$$H-\ddot{\underset{..}{Cl}}: \longrightarrow H^{\oplus} \quad :\ddot{\underset{..}{Cl}}:^{\ominus}$$

Fig. 6.6. Heterolytic cleavage of a bond.

Free radicals: These are the neutral species having an unpaired electron, e.g. Cl, Br, OR, R CH_3.

7

Acid-Base Reaction

7.1 Br∅nsted-Lowry Acids and Bases

Definition

According to Br∅nsted-Lowry concept an acid is a molecule that can donate a proton and a base is a molecule that can accept that proton.

An example of a simple acid/base reaction is the reaction of ammonia with water (Fig. 7.1). In it, water loses a proton and acts as an acid. Ammonia accepts the proton and acts as the base.

$$H_2\ddot{O}: \; + \; \ddot{N}H_3 \rightleftharpoons \overset{\ominus}{:\ddot{O}H} \; + \; \overset{\oplus}{NH_4}$$

Fig. 7.1. Reaction of ammonia with water.

In it, the ammonia uses its lone pair of electrons to form a new bond to the proton and so it is acting as a nucleophile, thus, the water is acting as an electrophile.

As the nitrogen uses its lone pair of electrons to form the new bond, the bond between hydrogen and oxygen must break because hydrogen can form only one bond. The electrons making up the O–H bond will move onto oxygen to produce a third lone pair of electrons, thus giving the oxygen a negative charge (Fig. 7.2). Since the nitrogen atom or ammonia has used its lone pair of electrons to form a new bond, it now has to share the electrons with hydrogen and so nitrogen gains a positive charge.

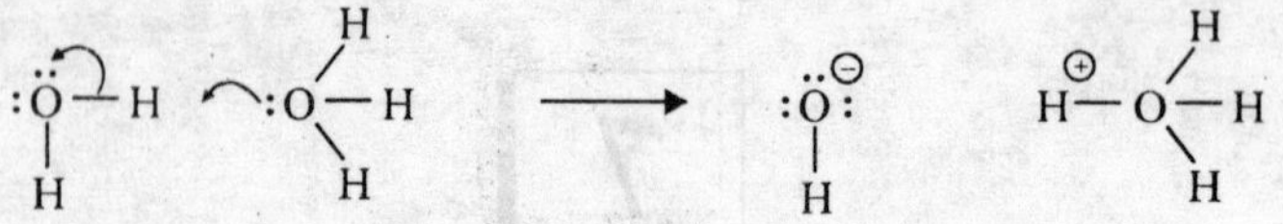

Fig. 7.2. Mechanism for the reaction of ammonia with water.

BrØnsted-Lowry Acids

According to this concept an acid is a molecule that contains an acidic hydrogen. In order to be acidic, the hydrogen must be slightly positive or electrophilic. This is possible if hydrogen is attached to an electronegative atom like a halogen, oxygen, or nitrogen. The following mineral acids and functional groups contain hydrogen's that are potentially acidic (Fig. 7.3).

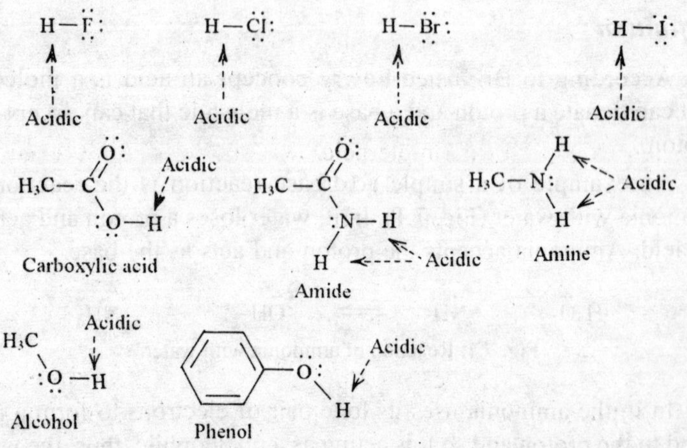

Fig. 7.3. Acidic protons in mineral acids and common functional groups.

Hydrogens attached to carbon are not normally acidic. However, in some special cases hydrogens attached to carbon **are** acidic.

BrØnsted-Lowry Bases

According to BrØnsted-Lowry concept a base is a molecule that can form a bond to a proton. They may be negatively charged ions with a lone pair of electrons (Fig. 7.4).

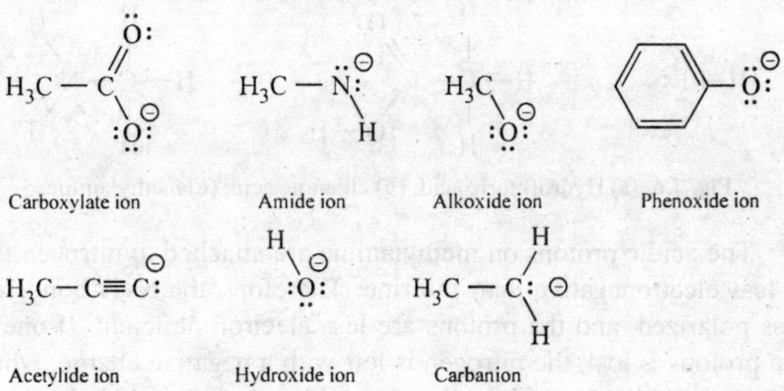

Carboxylate ion Amide ion Alkoxide ion Phenoxide ion

Acetylide ion Hydroxide ion Carbanion

Fig. 7.4. Examples of BrÆnsted-Lowry bases.

Neutral molecules can also act as bases if they contain an oxygen or nitrogen atom. The most common examples are amines. However, water, ethers and alcohols are also capable of acting as bases (Fig. 7.5).

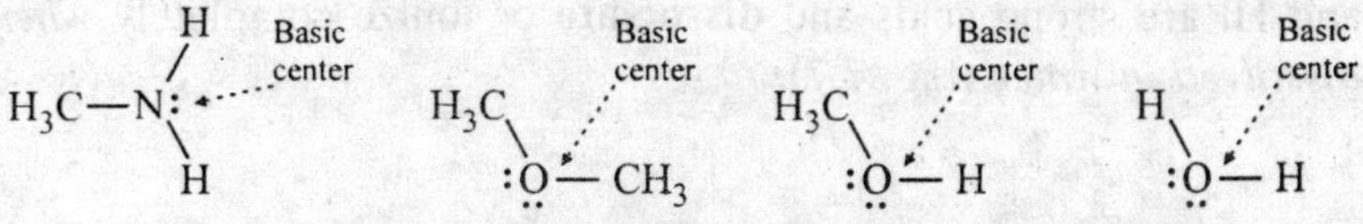

Basic center Basic center Basic center Basic center

Fig. 7.5. Examples of neutral BrØnsted-Lowry bases.

7.2 Acidic Strength

Electronegativity

The acidic protons of various molecules depends on various factors, such as the electronegativity of the atom to which they are attached. For example, if we consider hydrofluoric acid, ethanoic acid, and methylamine (Fig. 7.6). Hydrofluoric acid has the most acidic proton because the hydrogen is attached to a strongly electronegative fluorine. The fluorine strongly polarizes the H–F bond such that the hydrogen becomes highly electron deficient and can be easily lost. Once the proton is lost, the fluoride ion can stabilize the resulting negative charge.

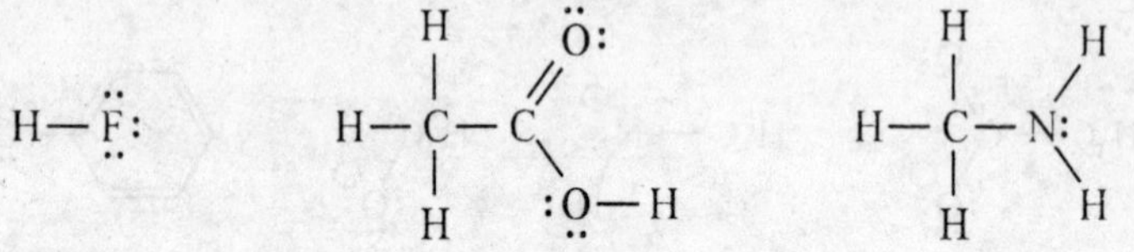

Fig. 7.6. (a) Hydrofluoric acid; (b) ethanoic acid; (c) methylamine.

The acidic protons on methylamine are attached to nitrogen that is less electronegative than fluorine. Therefore, the N–H bonds are less polarized, and the protons are less electron deficient. If one of the protons is lost, the nitrogen is left with a negative charge, which it cannot stabilize as efficiently as a halide ion. Thus, methylamine is a much weaker acid than hydrogen fluoride. Ethanoic acid is more acidic than methylamine but less acidic than hydrofluoric acid. This is because the electronegativity of oxygen lies between that of a halogen and that of a nitrogen atom.

These differences in acid strength can be shown if the three molecules above are placed in water. Mineral acids like HF, HCl, HBr, and HI are strong acids and **dissociate** or ionize completely *when dissolved in water* (Fig. 7.7).

Fig. 7.7. Ionization of hydrochloric acid.

Ethanoic acid (acetic acid) partially dissociates in water and an equilibrium is set up between the carboxylic acid (called the **free acid**) and the carboxylate ion (Fig. 7.8). An acid that only partially ionizes in this way is called a **weak acid**.

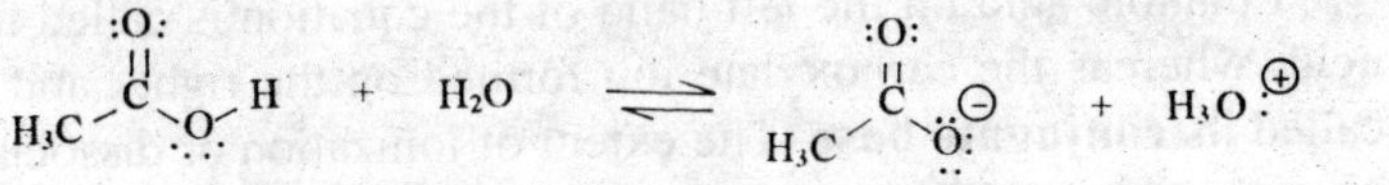

$$H_3C-C(=O)-O-H + H_2O \rightleftharpoons H_3C-C(=O)-O^{\ominus} + H_3O^{\oplus}$$

Fig. 7.8. Partial ionization of ethanoic acid.

When methylamine is dissolved in water, none of the acidic protons are lost at all and so the amine behaves as a weak base instead of an acid, and is in equilibrium with its protonated form (Fig. 7.9).

$$H_3C-NH_2 + H_2\ddot{O} \rightleftharpoons H_3C-\overset{\oplus}{N}H_3 + HO^{\ominus}$$

Fig. 7.9. Equilibrium acid-base reaction of methylamine with water.

Methylamine can act as an acid when it is treated as a strong base like butyl lithium (Fig. 7.10).

The hydrogen atoms attached to carbon are not generally acidic because carbon atoms are not electronegative. There are some exceptions to this rule.

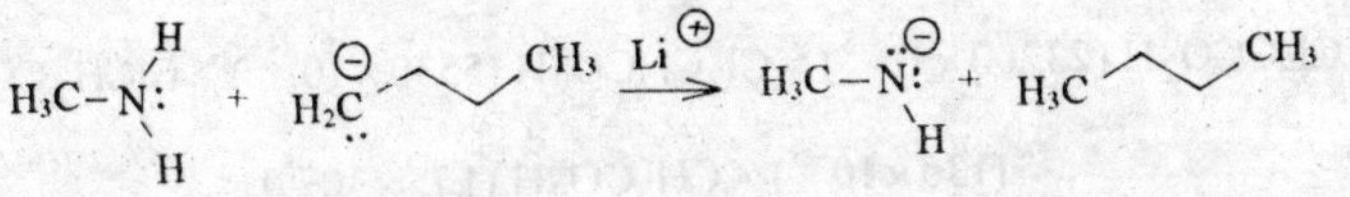

$$H_3C-NH_2 + {}^{\ominus}H_2C-CH_2-CH_2-CH_3\ Li^{\oplus} \longrightarrow H_3C-\overset{\ominus}{N}H + H_3C-CH_2-CH_2-CH_3$$

Fig. 7.10. Methylamine acting as an acid with a strong base (butyl lithium).

pK_a

Acids can be called as being weak or strong comparing by their pK_a values. Dissolving acetic acid in water, results in an equilibrium between the carboxylic acid and the carboxylate ion (Fig. 7.11).

$$H_3C-C(=O)-O-H + H_2O \rightleftharpoons H_3C-C(=O)-O^{\ominus} + H_3O^{\oplus}$$

Fig. 7.11. Equilibrium acid-base reaction of ethanoic acid with water.

Ethanoic acid on the left hand of the equation is called the **free acid**, whereas the carboxylate ion formed on the right hand side is called its **conjugate base**. The extent of ionization or dissociation is given by the **equilibrium constant** (K_{eq});

$$K_{eq} = \frac{[\text{PRODUCTS}]}{[\text{REACTANTS}]} = \frac{[CH_3CO_2^-][H_3O^+]}{[CH_3CO_2][H_2O]}$$

K_{eq} is generally measured in a dilute aqueous solution of the acid and therefore the concentration of water is high and assumed to be constant. Hence, we can rewrite the equilibrium equation in a simpler form where K_a is the acidity constant and includes the concentration of pure water (55.5M).

$$K_a = K_{eq}[H_2O] = \frac{[CH_3CO_2^-][H_3O^+]}{[CH_3CO_2H]}$$

The acidity constant can also be a measure of dissociation and of how much acidic a particular acid is. The stronger be acid, is more ionized and therefore the greater the concentration of products in the above equation. Hence, a strong acid has a high K_a value. The K_a values for the following ethanoic acid are in brackets and these show that the strongest acid in the series in trichloroacetic acid.

$$Cl_3CCO_2H\ (23200 \times 10^{-5}) > Cl_2CHCO_2H\ (5530 \times 10^{-5}) > ClCH_2CO_2H$$

$$(136 \times 10^{-5}) > CH_3CO_2H\ (1.75 \times 10^{-5}).$$

K_a values are less commonly used and it is more usual to measure the acidic strength as a pK_a value rather than K_a. The pK_a is *the negative logarithm of K_a* ($pK_a = -\log_{10} K_a$) and results in more manageable numbers. The pK_a values for each of the above ethanoic acids is shown in brackets below. *The strongest acid (trichloroacetic acid) has the lowest pK_a value.*

$$Cl_3CCO_2H\ (0.63) < Cl_2CHCO_2H\ (1.26) < ClCH_2CO_2H\ (2.87)$$

$$< CH_3CO_2H\ (4.76)$$

Thus the stronger the acid, the **higher** the value of K_a, and the **lower** the value of pK_a. An amine like ethylamine ($CH_3CH_2NH_2$) is a very weak acid ($pK_a = 40$) compared to ethanol ($pK_a = 16$). This is because of the relative electronegativities of oxygen and nitrogen. However, the electronegativity of neighboring atoms is not the only influence on acidic strength. For example, the pK_a values of ethanoic acid (4.76), ethanol (16), and phenol (10) show that ethanoic acid is more acidic than phenol, and that phenol is more acidic than ethanol. The difference in acidity is quite marked, yet hydrogen is attached to oxygen in all three structures.

Similarly, the ethanoic acids Cl_3CCO_2H (0.63), Cl_2CHCO_2H (1.26), $ClCH_2CO_2H$ (2.87), and CH_3CO_2H (4.76) have significantly different pK_a values and yet the acidic hydrogen is attached to an oxygen in each of these structures. Therefore, factors other than electronegativity also play a role in determining acidic strength.

Inductive Effect

Permanent displacement of electrons along a certain chain when some atom or group of atoms with different electronegativity than carbon is attached to carbon chain is called inductive effect. For illustration consider a carbon chain which has a chlorine atom attached to one end.

$$-C_4 \longrightarrow \overset{\delta''+}{C_3} \longrightarrow \overset{\delta'+}{C_2} \longrightarrow \overset{\delta+}{C_1} \longrightarrow \overset{\delta-}{Cl}$$

δ- *(-I) – Effect*

Since chlorine atom has higher electronegativity than carbon so the shared pair of electrons between C_1 and chlorine atom lies closer to chlorine atom as a result of which chlorine atom acquires a small negative charge and carbon atom C_1 acquires small +ve charge. As C_1 is now somewhat positively charged it attracts towards itself the shared pair of electrons between C_1 and C_2. This results in a small positive charge on C_2 also, but this charge is less than on C_1. In this way electrons are displaced towards chlorine in the carbon chain. This type of displacement of electrons along a carbon chain is known as **Electron Withdrawing Inductive Effect** or **(–I) effect.**

If the electronegativity of the atom or group of atoms attached to the carbon chain is less than the electronegativity of the carbon then the displacement of electrons takes place away from group along the

C-chain and this effect is called the **Electron Releasing Inductive Effect** (or +I)–effect. This inductive effect can be represented as shown below:

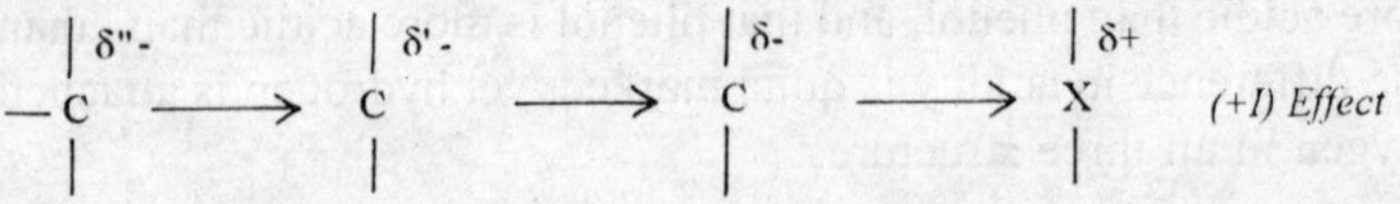

$$-\overset{|}{\underset{|}{C}}^{\delta''-} \longrightarrow \overset{|}{\underset{|}{C}}^{\delta'-} \longrightarrow \overset{|}{\underset{|}{C}}^{\delta-} \longrightarrow \overset{|}{\underset{|}{X}}^{\delta+} \quad (+I)\ Effect$$

Inductive effect is a permanent effect and decreases rapidly as the distance from the source increases. Some of the groups which show Inductive effect are shown below:

– I effect: $-NO_2 > -F > -Cl > -Br > -I > -OCH_3 > -C_6H_5$

+ I effect: $-C(CH_3)_3 > -C(CH_3)_2 > -C_2H_5 > -CH_3$

Any effect that stabilizes the negative charge of the conjugate base will increase the acid strength. Substituents can stabilize a negative charge and they do so by an **inductive effect**. This can be illustrated by comparing the pK_a values of the alcohols CF_3CH_2OH and CH_3CH_2OH (12.4 and 16, respectively) where CF_3CH_2OH is more acidic than CH_3CH_2OH. This means that the anion $CF_3CH_2O^-$ is more stable than $CH_3CH_2O^-$ (Fig. 7.12).

a) $F_3C-CH_2-\ddot{\underset{..}{O}}:^{\ominus}$ b) $H_3C-CH_2-\ddot{\underset{..}{O}}:^{\ominus}$

Fig. 7.12. (a) 2,2,2-Trifluoroethoxy; (b) ethoxy ion.

Fluorine atoms are strongly electronegative and therefore each C–F bond is strongly polarized. In such a way that the carbon bearing the fluorine atoms becomes strongly electropositive. As this carbon atom is now electron deficient, so it will 'demand' a greater share of the electrons in the neighboring C–C bond. Due to this the electrons are being withdrawn from the neighboring carbon, making it electron deficient too. This inductive effect will continue to be felt through the various bonds of the structure. It will decrease through the bonds but it is still significant enough to be felt at the negatively charged oxygen. As the inductive effect is electron withdrawing it will decrease the negative charge on the oxygen and help to stabilize it. Therefore the original fluorinated alcohol will lose its proton more readily and will be a stronger acid.

The inductive effect can also explain the relative acidities of the chlorinated ethanoic acids Cl_3CCO_2H (0.63), Cl_2CHCO_2 (1.26), $ClCH_2CO_2H$ (2.87), and CH_3CO_2H (4.76). Trichloroethanoic acid is the strongest acid as its conjugate base (the carboxylate ion) is stabilized by the inductive effect created by three electronegative chlorine atoms. As the number of chlorine atoms decrease, the inductive effect also decreases.

Inductive effect can also explain the difference between the acid strengths of ethylamine ($pK_a \sim 40$) and ammonia ($pK_a \sim 33$). The pK_a values show that ammonia is a stronger acid than ethylamine. In this case, the inductive effect is electron donating. The alkyl group of ethylamine increases the negative charge of the conjugate base and so destabilizes it thus, making ethylamine a weaker acid than ammonia (Fig. 7.13).

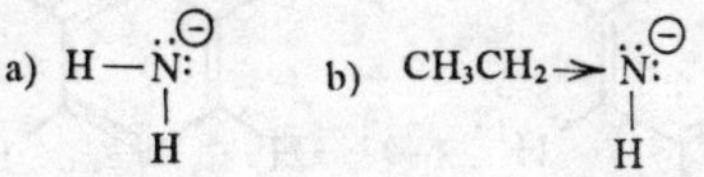

a) $H-\ddot{N}:^{\ominus}$ (with H on N) b) $CH_3CH_2 \rightarrow \ddot{N}:^{\ominus}$ (with H on N)

Fig. 7.13. Conjugate bases (a) ammonia and (b) ethylamine.

Resonance

Sometimes it is not possible to assign a single electronic structure to a molecule that may account for all its properties. In such a case the molecule is represented by two or more electronic structures but none of these represents the actual structure of the molecule. The actual structure of the molecule lies somewhere in between these structures but can not be expressed on paper. Such a molecule is said to exhibit resonance. The various structures assigned to the molecule are called **contributing structures** or **canonical structures** where as the intermediate structures is called the **resonance hybrid**. For example benzene (C_6H_6) may be assigned the following two structures:

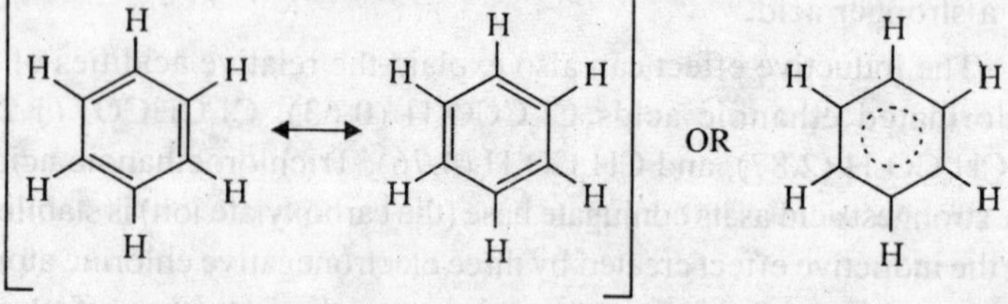

Any of these structures alone cannot account for all the properties of benzene. According to these structures there should be three single bonds (1.54Å) and three double bonds (1.34Å) between carbon atoms. But actually it has been found that all the six carbon-carbon bonds in benzene have same bond length (1.39Å). The actual structure of benzene is resonance hybrid of the above two structures and may be represented as under:

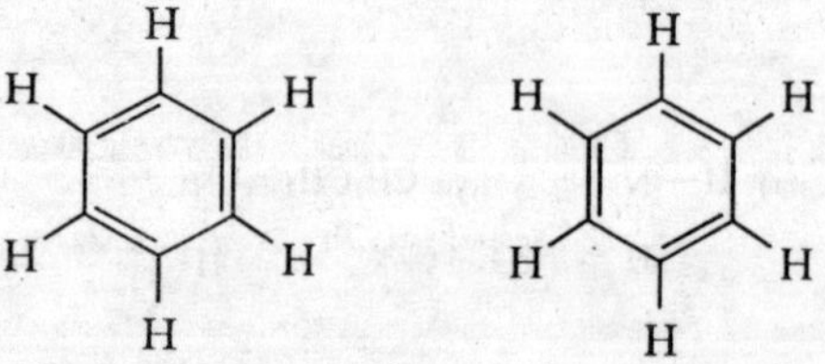

I II

The resonance hybrid is more stable than any of the contributing structures and the difference between the energy of the most stable contributing structure and resonance hybrid is called ***resonance energy***.

Carboxylic acids and carboxylate ions also exhibit resonance.

Carboxylic acid

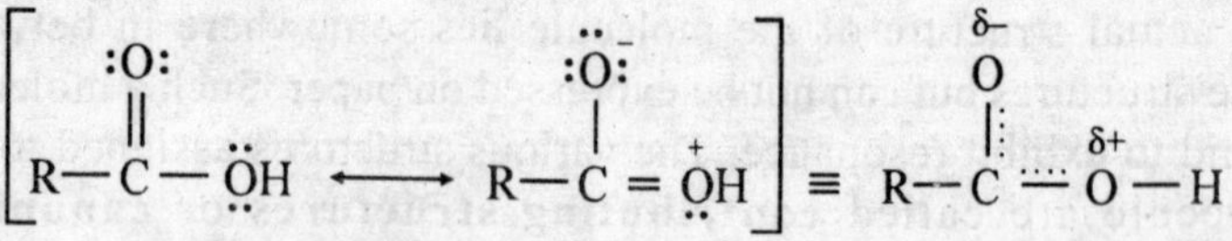

Contributing structures Hybrid structure

Carboxylate ion

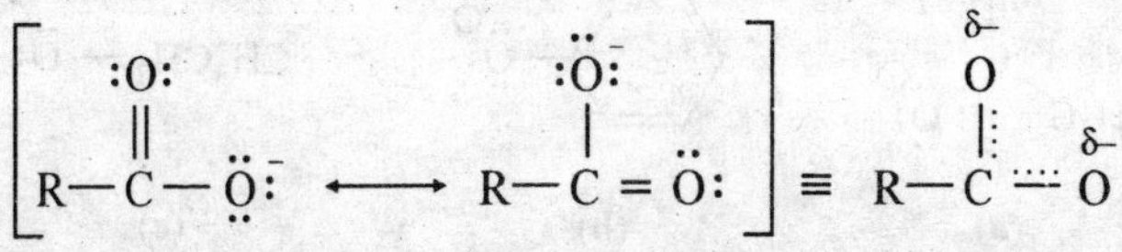

$$\left[R-\overset{\overset{:\ddot{O}:}{\|}}{C}-\ddot{\underset{\cdot\cdot}{O}}:^{-} \longleftrightarrow R-\overset{\overset{:\ddot{O}:^{-}}{|}}{C}=\ddot{O}: \right] \equiv R-\overset{\overset{\delta-}{O}}{C}\cdots\overset{\delta-}{O}$$

Contributing structures Hybrid structure

Some important features of resonance are:

1. The various contributing structures may differently in electronic arrangement but should have same arrangement of atoms.
2. The number of unpaired electrons should be same in all the contributing structures.
3. All contributing structures should have almost same energy.
4. The more stable contributing structure makes more contribution.
5. The bond distances of hybrid structure are intermediate of those of resonating forms.

Resonance Effect

It may be noted that in **conjugated systems** (*having alternate single and double bonds*) resonance causes displacement of electrons from one part of the system to another part creating centers of high and low electron density. This is called resonance effect.

The negative charge on some conjugate bases can be stabilized by **resonance**. Resonance involves the movement of valence electrons around a structure, which results in the sharing of charge between different atoms. This process is called **delocalization**. The effects of resonance can be shown by comparing the acidities of ethanoic acid (pK_a 4.76), phenol (pK_a 10.0) and ethanol (pK_a 12.4). The pK_a values show that ethanoic acid is a stronger acid than phenol, and that phenol is a stronger acid than ethanol.

The varying acidic strengths of ethanoic acid, phenol and ethanol can be explained by considering the relative stabilities of their conjugate bases (Fig. 7.14).

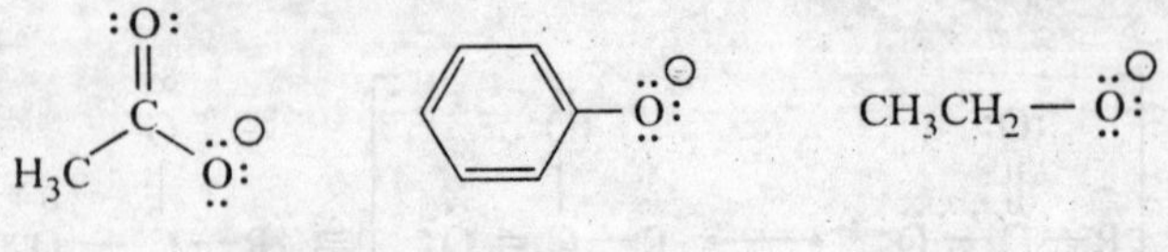

Fig. 7.14. Conjugate bases of (a) ethanoic acid; (b) phenol; (c) ethanol.

The charge of the carboxylate ion is on an oxygen atom, and because oxygen is electronegative so the charge is stabilized. However, the charge can be shared with the other oxygen leading to delocalization of the charge. This arises by a resonance interaction between a lone pair of electrons on the negatively charged oxygen and the π electrons of the carbonyl group (Fig. 7.15). a lone pair of electrons on the 'bottom' oxygen forms a new π bond to the neighboring carbon. Simultaneously, the weak π bond of the carbonyl group breaks. This is essential or else the carbonyl carbon would end up with five bonds and that is not allowed. Both electrons in the original π bond now end up on the 'top' oxygen that means that this oxygen ends up with three lone pairs and gains a negative charge. Please note that the π bond and the charge have effectively 'swapped places'. Both the structures involved are referred to as **resonance structures** and are easily interconvertible. The negative charge is now shared or delocalized equally between both oxygens and is stabilized. Hence, ethanoic acid is a stronger acid than one would expect based on the electronegativity of oxygen alone.

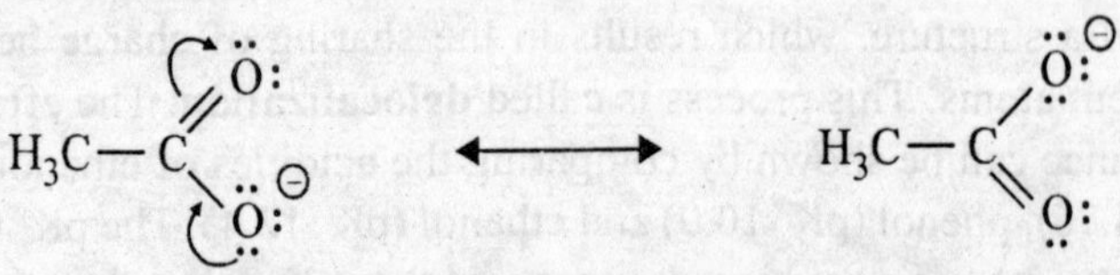

Fig. 7.15. Resonance interaction for the carboxylate ion.

Phenol is less acidic than ethanoic acid but is more acidic than ethanol. Once again the resonance concept can explain the differences. The conjugate base of phenol is known as the **phenolate ion**. In this case, the resonance process can be carried out several times to place the negative charge on four separate atoms i.e. the oxygen atom and three of the aromatic carbon atoms (Fig. 7.16). Since the negative charge can be spread over four atoms might suggest that all the phenolate anion should be more stable than the carboxylate anion, since the charge is spread over more atoms. However, with the phenolate ion, three of the resonance structures place the charge on a carbon atom that is much less electronegative than an oxygen atom. These resonance structures will therefore be far less important than the resonance structure having the charge on oxygen. Because of this, delocalization is weaker for the phenolate ion than for the ethanoate ion. However, a certain amount of delocalization still occurs that is why a phenolate ion is more stable than an ethoxide ion.

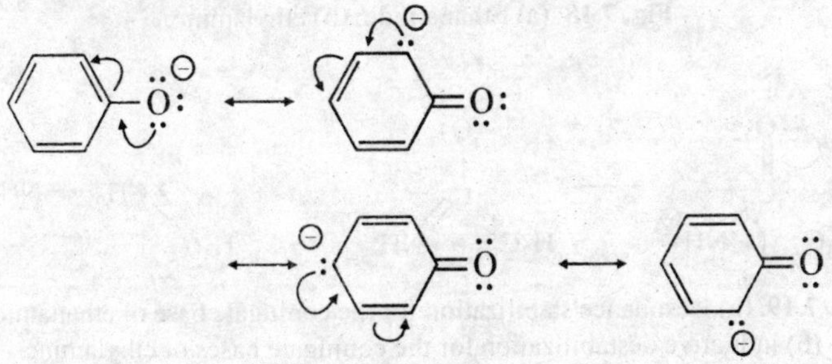

Fig. 7.16. Resonance interactions for the phenolate ion.

In case of ethanol, the conjugate base is the ethoxide ion that cannot be stabilized by delocalizing the charge, because resonance is not possible. There is no π bond available to participate in resonance. Thus, the negative charge is localized on the oxygen. Moreover, the inductive donating effect of the neighboring alkyl group (ethyl) increases the charge and destabilizes it (Fig. 7.17). This makes the ethoxide ion the least stable (or most reactive) of the three anions that we have studied. Due to this, ethanol is the weakest acid.

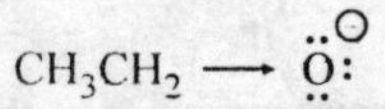

$$CH_3CH_2 \longrightarrow \ddot{\underset{..}{O}}:^{\ominus}$$

Fig. 7.17. Destabilizing inductive effect of the ethoxide ion.

Amines and Amides

Amines and amides are very weak acids and they only react with very strong bases. The pK_a values for ethanamide and ethylamine are 15 and 40, respectively, which means that ethanamide has the more acidic proton (Fig. 7.18). This can be explained by making use of resonance and inductive effects (Fig. 7.19).

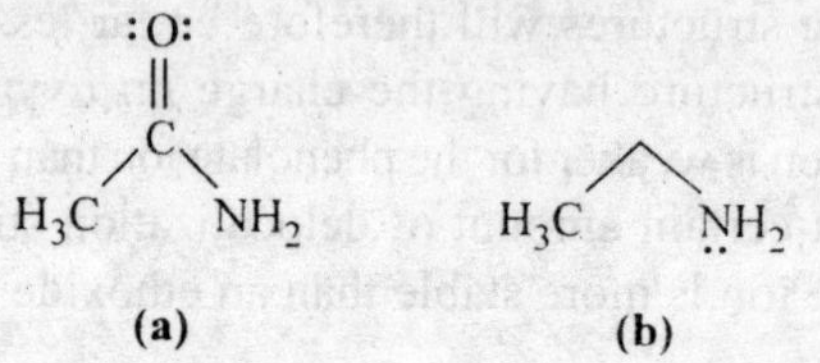

(a) (b)

Fig. 7.18. (a) Ethanamide; (b) ethylamine.

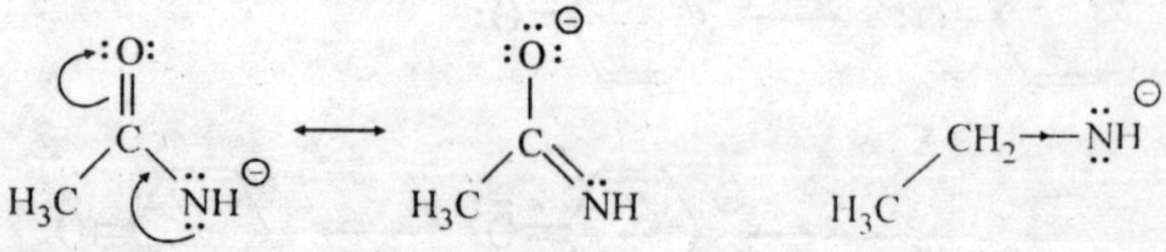

Fig. 7.19. (a) Resonance stabilization for the conjugate base of ethanamide; (b) inductive destabilization for the conjugate bases of ethylamine.

7.3 Base Strength

Electronegativity

Electronegativity influences the basic strength of the compound. If we compare the fluoride ion, hydroxide ion, amide ion and the methyl carbanion, then the order of basicity is as shown (Fig. 7.20).

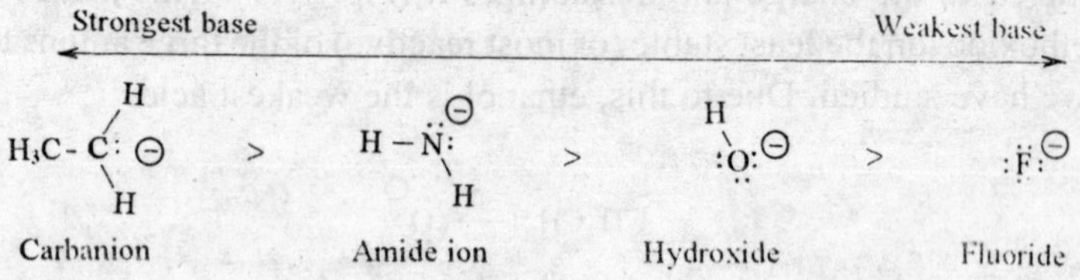

Fig. 7.20. Comparison of basic strength.

The strongest base is the carbanion as this has the negative charge situated on the least electronegative atom i.e. the carbon atom. The weakest base is the fluoride ion which has the negative charge situated on the most electronegative atom i.e. the fluorine atom. Strongly electronegative atoms like fluorine are able to stabilize a negative charge making the ion less reactive and less basic. The order of basicity of the anions formed from alkanes, amines, and alcohols follows a similar order because of the same reason (Fig. 7.21).

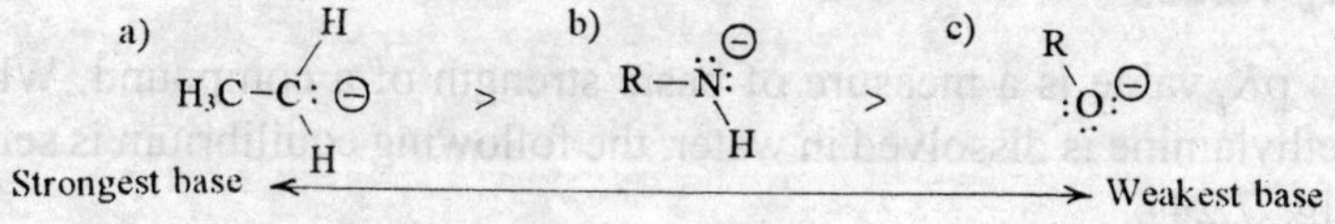

a) $H_3C-\ddot{C}H_2^{\ominus}$ > b) $R-\ddot{N}H^{\ominus}$ > c) $R-\ddot{O}:^{\ominus}$

Strongest base ⟵⟶ Weakest base

Fig. 7.21. Comparison of basic strengths: (a) a Carbanion; (b) an amide ion; (c) an alkoxide ion.

Electronegativity can also explain the order of basicity for neutral molecules like amines, alcohols, and alkyl halides (Fig. 7.22).

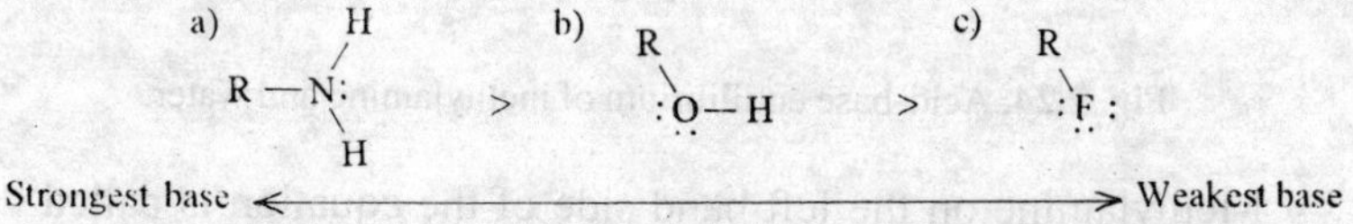

a) $R-\ddot{N}H_2$ > b) $R-\ddot{O}-H$ > c) $R-\ddot{F}:$

Strongest base ⟵⟶ Weakest base

Fig. 7.22. Comparison of basic strengths: (a) an amine; (b) an alcohol; (c) an alkyl fluoride.

These neutral molecules are much weaker bases than their corresponding anions, but the order of basicity is still the same and can be explained by considering the relative stability of the cations that are formed when these molecules bind a proton (Fig. 7.23).

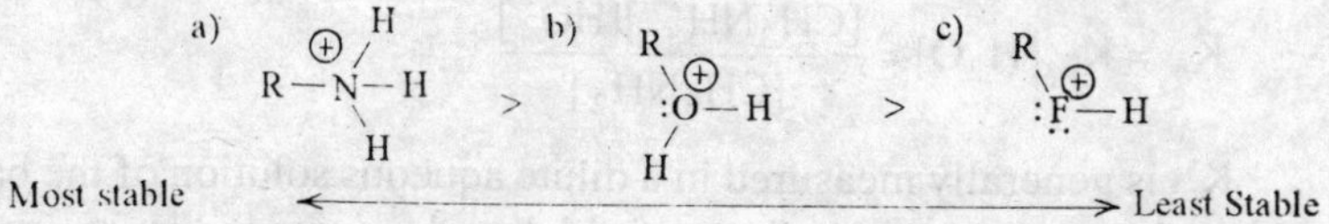

a) $R-\overset{\oplus}{N}H_3$ > b) $R-\ddot{O}^{\oplus}H_2$ > c) $R-\ddot{F}^{\oplus}-H$

Most stable ⟵⟶ Least Stable

Fig. 7.23. Relative stability of the carbons fromed form (a) an amine; (b) an alcohol; (c) an alkyl fluoride.

A nitrogen can stabilize a positive charge better than a fluorine atom because the former is less electronegative. Electronegative atoms prefer to have a negative charge rather than a positive charge. Fluorine is so electronegative that its basicity is negligible. Therefore, amines act as weak bases in aqueous solution and are partially ionized. Alcohols only act as weak bases in acidic solution. Alkyl halides are essentially nonbasic even in acidic solutions.

pK_b Values

pK_b value is a measure of basic strength of a compound. When methylamine is dissolved in water, the following equilibrium is set up (Fig. 7.24).

$$H_3C{-}\ddot{N}H_2 + H_2\ddot{O}: \rightleftharpoons H_3C{-}N^{\oplus}H_3 + HO:^{\ominus}$$

Fig. 7.24. Acid-base equilibrium of methylamine and water.

Methylamine on the left hand side of the equation is called the **free base**, whereas the methyl ammonium ion formed on the right hand side is called the **conjugate acid**. The extent of ionization or dissociation in the equilibrium reaction is defined by the **equilibrium constant** (K_{eq});

$$K_{eq} = \frac{[\text{Products}]}{[\text{Reactants}]} = \frac{[CH_3NH_3^+][HO^-]}{[CH_3NH_2][H_2O]}$$

$$K_b = K_{eq}[H_2O] = \frac{[CH_3NH_3^+][HO^-]}{[CH_3NH_2]}$$

K_{eq} is generally measured in a dilute aqueous solution of the base and so the concentration of water is high and assumed to be constant. Therefore, we can rewrite the equilibrium equation in a simpler form where K_b is the basicity constant and includes the concentration of pure water (55.5M). pK_b is the negative logarithm of K_b and is used as a measure of basic strength ($pK_b = -Log_{10}K_b$).

A large pK_b indicates a weak base. For example, the pK_b values of ammonia and methylamine are 4.74 and 3.36, respectively, which indicates that ammonia is a weaker base than methylamine.

pK_b and pK_a are related by the equation $pK_a + pK_b = 14$. Therefore, if we know the pK_a of an acid, the pK_b of its conjugate base can be calculated and *vice versa*.

Inductive Effects

Inductive effects affect the strength of a charged base by influencing the negative charge. For example, *an electron-withdrawing group helps to stabilize a negative charge*, which results in a weaker base. *An electron-donating group will destabilize a negative charge,* which results in a stronger base. Amongst Cl_3CCO_2H, Cl_2CHCO_2H, $ClCH_2CO_2H$, and CH_3CO_2H, trichloroacetic acid is a strong acid as its conjugate base (the carboxylate ion) is stabilized by the three electronegative chlorine groups (Fig. 7.25).

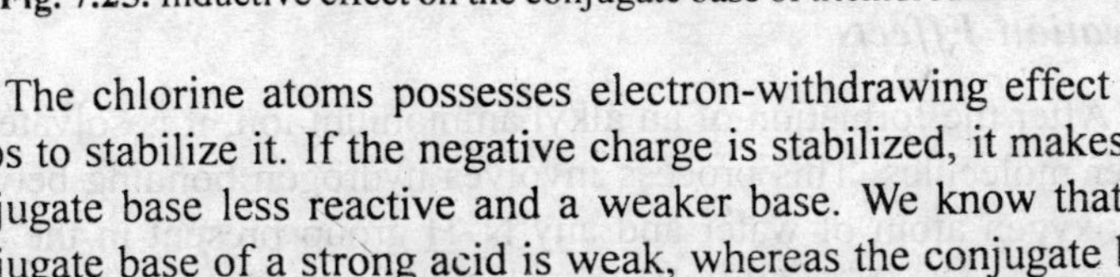

$Cl_3C-C(=O)OH \rightleftharpoons Cl_3C \leftarrow C(=O)O^{\ominus}$

Strong acid $pK_a = 0.63$ Weak Conjugate base (stabilized)

Fig. 7.25. Inductive effect on the conjugate base of trichloroacetic acid.

The chlorine atoms possesses electron-withdrawing effect that helps to stabilize it. If the negative charge is stabilized, it makes the conjugate base less reactive and a weaker base. We know that the conjugate base of a strong acid is weak, whereas the conjugate base of a weak acid is strong. Therefore, the order of basicity for the ethanoate ions $Cl_3CCO_2^-$, $Cl_2CHCO_2^-$, $ClCH_2CO_2^-$, and $CH_3CO_2^-$ is the opposite to the order of acidity for the corresponding carboxylic acids, i.e. the ethanoate ion is the strongest base, while the trichlorinated ethanoate ion is the weakest base.

Inductive effects can also influence the basic strength of neutral molecules (e.g. amines). The pK_b for ammonia is 4.74, which compares with pK_b values for methylamine, ethylamine, and propylamine of 3.36, 3.25 and 3.33 respectively. The alkylamines are stronger bases than ammonia due to the inductive effect of an alkyl group on the **alkyl ammonium ion** (RNH_3^-; Fig. 7.26). Alkyl groups donate electrons towards a neighboring positive center gets partially dispersed over the alkyl group. If the ion is stabilized, the equilibrium of the acid-base reaction will shift to the ion, that means that the amine is more basic. The larger the alkyl group, the more significant this effect.

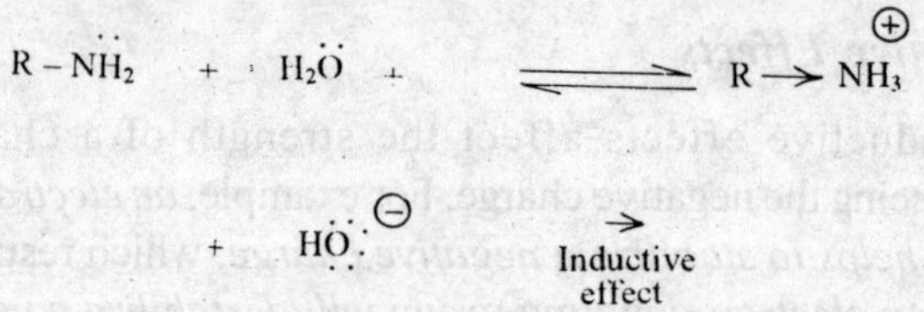

$$R-\ddot{N}H_2 + H_2\ddot{O} \rightleftharpoons R \rightarrow \overset{\oplus}{N}H_3 + H\ddot{O}:^{\ominus}$$

Inductive effect

Fig. 7.26. Inductive effects of an alkyl group on the alkyl ammonium ion.

If one alkyl group can influence the basicity of an amine, then further alkyl groups should have an even greater inductive effect. Therefore, one might expect secondary and tertiary amine is to be stronger bases than primary amines. In fact, this is not necessarily the case. There is no easy relationship between basicity and the number of alkyl groups attached to nitrogen. Although the inductive effect of more alkyl groups is certainly greater, this effect is counterbalanced by a solvation effect.

Solvation Effects

After the formation of an alkyl ammonium ion, it is solvated by water molecules. This process involves hydrogen bonding between the oxygen atom of water and any N–H group present in the alkyl ammonium ion (Fig. 7.27). Water solvation is a stabilizing factor that is as important as the inductive effect of the alkyl substituents and the more hydrogen bonds that are possible, the greater the stabilization. Solvation is stronger for the alkyl ammonium ion formed from a primary amine than for the alkyl ammonium ion formed from a tertiary amine. This is due to the fact that the former ion has three N–H hydrogens available for H-bonding, compared with only one such

N–H hydrogen the latter. Because of this there is more solvent stabilization experienced for the alkyl ammonium ion of a primary amine compared to that experienced by the alkyl ammonium ion of a tertiary amine. This means that tertiary amines are generally weaker bases than primary or secondary amines.

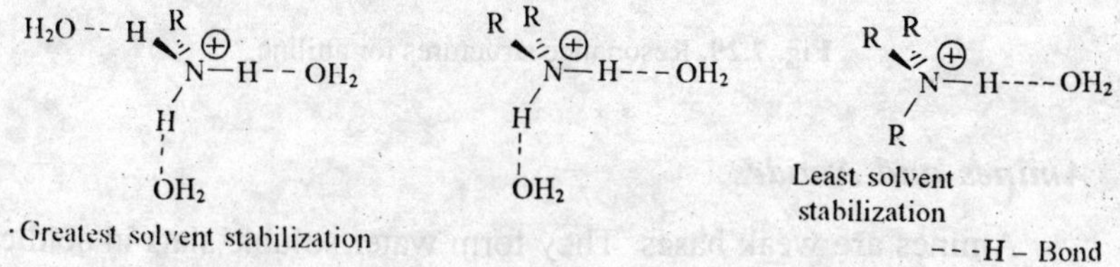

Fig. 7.27. Solvent effect on alkyl ammonium ions from primary, secondary, and tertiary amines.

Resonance

We have learnt that resonance can stabilize a negative charge by delocalizing it over two or more atoms. Resonance explains why a carboxylate ion is more stable than an alkoxide ion. The negative charge in the former can be delocalized between two oxygens whereas the negative charge on the former is localized on the oxygen. We used this to explain why a carboxylic acid is a stronger acid than an alcohol. We can use the same argument in reverse to explain the difference in basicities between a carboxylate ion and an alkoxide ion (Fig. 7.28). Because the latter is less stable, it is more reactive and is therefore a stronger base.

Resonance effects can also explain why aromatic amines (arylamines) are weaker bases than alkylamines. The lone pair of electrons on nitrogen can interact with the π system of the aromatic ring resulting in the possibility of three **zwitterionic** resonance structures (Fig. 7.29). (A zwitterion is a neutral molecule containing a positive and a negative charge). Since nitrogen's lone pair of electrons is involved in this interaction, it is less available to form a bond to a proton and so the amine is less basic.

Fig. **7.28.** (a) Carboxylate ion; (b) alkoxide ion.

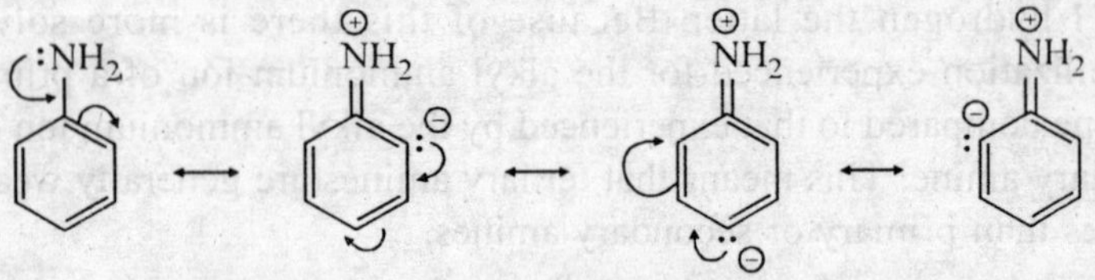

Fig. 7.29. Resonance structures for aniline.

Amines and Amides

Amines are weak bases. They form water soluble slats in acidic solutions (Fig. 7.30a) and in aqueous solution they are in equilibrium with their conjugate acid (Fig. 7.30b).

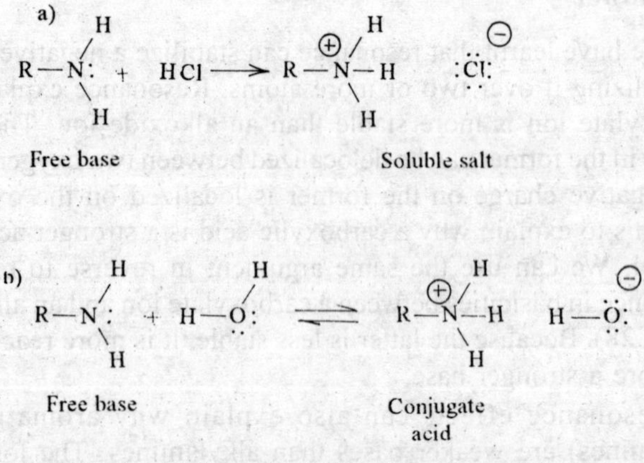

Fig. 7.30. (a) Salt formation; (b) acid-base equilibrium.

Amines are the basic because they have a lone pair of electrons that can form a bond to a proton. Amides also have a nitrogen with a lone pair of electrons, but unlike amines they are not basic. This is because a resonance occurs within the amide structure that involves the nitrogen lone pair (Fig. 7.31). The driving force behind this resonance is the electronegative oxygen of the neighboring carbonyl group that is 'hungry' for electrons. The lone pair of electrons on nitrogen forms a π bond to the neighboring carbon atom. As this occurs, the π bond of the carbonyl group breaks and both electrons move onto the oxygen to give it a total of three lone pairs and a negative charge. Because the nitrogen's lone pair is involved in this resonance, it is unavailable to bind to a proton and therefore amides are not basic.

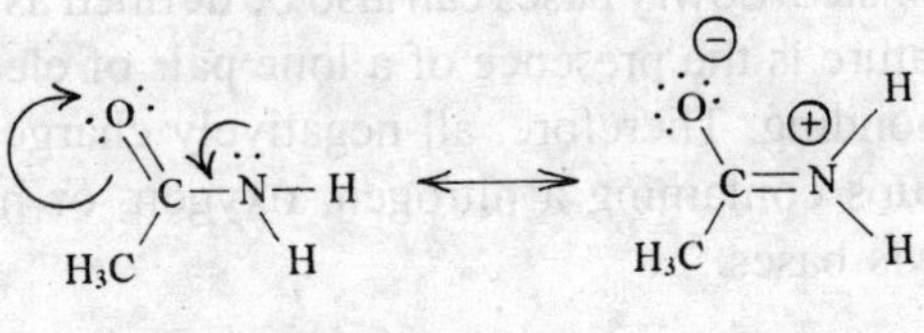

Fig. 7.31. Resonance interaction of an amide.

7.4 Lewis Acids and Bases

Lewis Acids

Lewis acids are ions or electron deficient molecules having an unfilled valence shell. They are known as acids because they can accept a lone pair of electrons from another molecule to fill their valence shell. Lewis acids include all the Bronsted-Lowry acids we have already discussed, as well as ions (e.g. H^+, Mg^{2+}), and neutral species such as BF_3 and $AlCl_3$.

Both Al and B are in Group 3 of the periodic table and have three valence electrons in their outer shell. These elements can form three bonds. However, there is still room for a fourth bond. For example in BF_3, boron is surrounded by six electrons (three bonds containing two electrons each). However, boron's valence shell can accommodate eight electrons and so a fourth bond is possible if the fourth group can provide both electrons for the new bond. Since both boron and aluminum are in Group 3 of the periodic table, they are electropositive and will react with electron-rich molecules so as to obtain this fourth bond. Many transition metal compounds can also act like Lewis acids (e.g. $TiCl_4$ and $SnCl_4$).

Lewis Bases

A Lewis base is a molecule that can donate a lone pair of electrons to fill the valence shell of a Lewis acid (Fig. 7.32). The base can be a negatively charged group such as a halide, or a neutral molecule like water, an amine, or an ether, as long as there is an atom present with a lone pair of electrons (i.e. O, N or a halogen).

All the Bronsted-Lowry bases can also be defined as Lewis bases. The crucial feature is the presence of a lone pair of electrons that is available for bonding. Therefore, all negatively charge ions and all functional groups containing a nitrogen, oxygen, or halogen atom can act as Lewis bases.

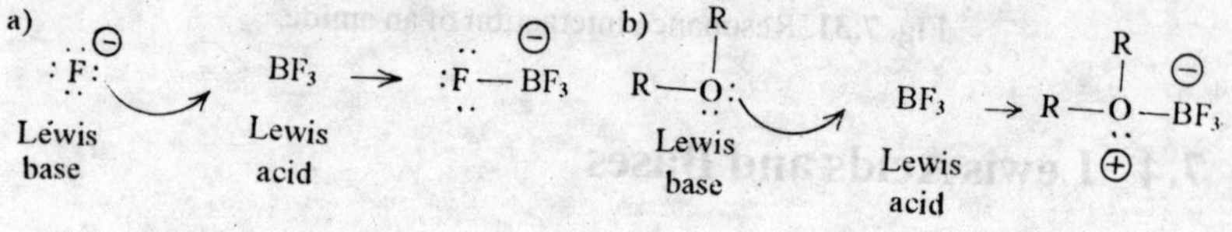

a) $:\ddot{F}:^{\ominus}$ + BF_3 → $:\ddot{F}—BF_3^{\ominus}$

Lewis base, Lewis acid

b) $R—\ddot{O}(R):$ + BF_3 → $R—O(R)^{\oplus}—BF_3^{\ominus}$

Lewis base, Lewis acid

Fig. 7.32. Reactions between Lewis acids and Lewis bases.

7.5 Enolates

Acidic C–H Protons

Most acidic protons are attached to heteroatoms like halogen, oxygen, and nitrogen. Protons attached to carbon are not normally acidic but there are exceptions. One such exception occurs with aldehydes or ketones when there is a C$\underline{H}$R$_2$, C$\underline{H}_2$R or C$\underline{H}_3$ group next to the carbonyl group (Fig. 7.33). The protons indicated are acidic and are attached to the α (alpha) carbon. They are therefore called as α protons.

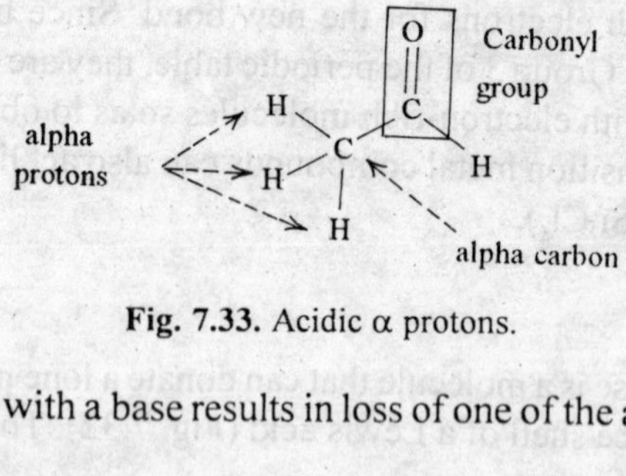

Fig. 7.33. Acidic α protons.

Treatment with a base results in loss of one of the acidic α protons (Fig. 7.34).

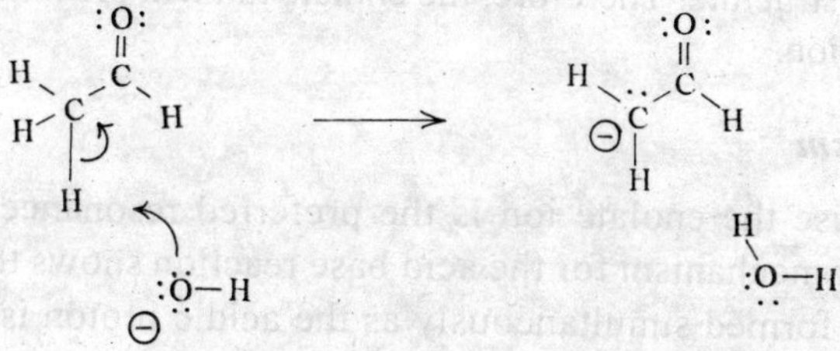

Fig. 7.34. Loss of an a proton and formation of a carbanion.

A lone pair on the hydroxide oxygen forms a new bond to an α proton. Simultaneously, the C–H bond breaks. Both electrons of that bond end up on the carbon atom and give it a lone pair of electrons and a negative charge (a **carbanion**). However, carbanions are generally very reactive, unstable species that are not easily formed. Therefore, some form of stabilization is involved here.

Fig. 7.35. Resonance interaction between carbanion and enolate ion.

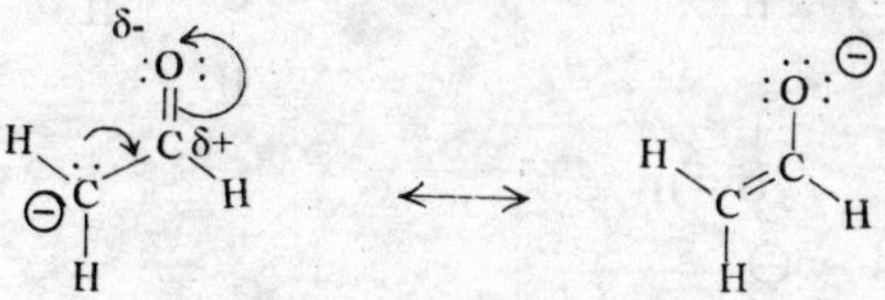

Stabilization

As carbon is not electronegative so it cannot stabilize the charge. However, stabilization is possible through resonance (Fig. 7.35). The lone pair of electrons on the carbanion form a new π bond to the carbonyl carbon. As this bond is formed, the weak π bond of the carbonyl group breaks and both these electrons move onto the oxygen. This results in the negative charge ending up on the electronegative oxygen where it is more stable. This mechanism is exactly the same as the one for the carboxylate ion. However, whereas both resonance structures are equally stable in the carboxylate ion but this is not the case here. The resonance structure having the charge on the oxygen atom (an **enolate ion**) is more stable than the original carbanion resonance structure. Therefore, the enolate ion will predominate over the carbanion.

Mechanism

Because the enolate ion is the preferred resonance structure so a better mechanism for the acid base reaction shows the enolate ion being formed simultaneously as the acidic proton is lost (Fig. 7.36). As the hydroxide ion forms its bond to the acidic proton, the C–H bond breaks, and the electrons in that bond form a π bond to the carbonyl carbon atom. Simultaneously, the carbonyl π bond breaks in such a way that both electrons move onto the oxygen. *The electronegative oxygen is responsible for making the α proton acidic.*

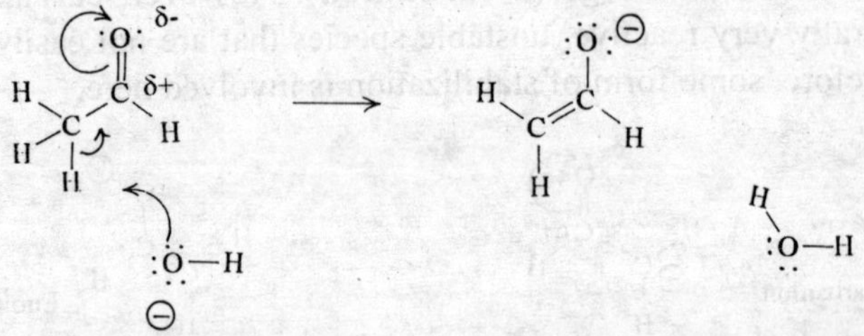

Fig. 7.36. Mechanism for the formation of the enolate ion.

Enolate Ion

Resonance structures represent the extreme possibilities for a particular molecule and the true structure is really a hybrid of both (Fig. 7.37). The 'hybrid' structure shows that the negative charge is 'smeared' or delocalized between three sp^2 hybridized atoms. Since these atoms are sp^2 hybridized, they are planar and have a $2p$ orbital that can interact with its neighbors to form one molecular orbital, thus spreading the charge between the three atoms (Fig. 7.38). Keeping this in mind, we can state which of the methyl hydrogens is most likely to be lost in the formation of an enolate ion. The hydrogen circled (Fig. 7.39a) is the one which will be lost since the σ C–H bond is correctly orientated to interact with the π orbital of the carbonyl bond. The orbital diagram (Fig. 7.39b) illustrates this interaction. A Newman diagram can also be drawn by looking along the C–C bond to indicate the relative orientation of the α hydrogen which will be lost (Fig. 7.39c). In this example, there is no difficulty in the proton being in the corrrect orientation since there is free rotation around the C–C single bond. However, in cyclic systems, the hydrogen atoms are locked in space and the relative stereochemistry becomes important if the α proton is to be acidic.

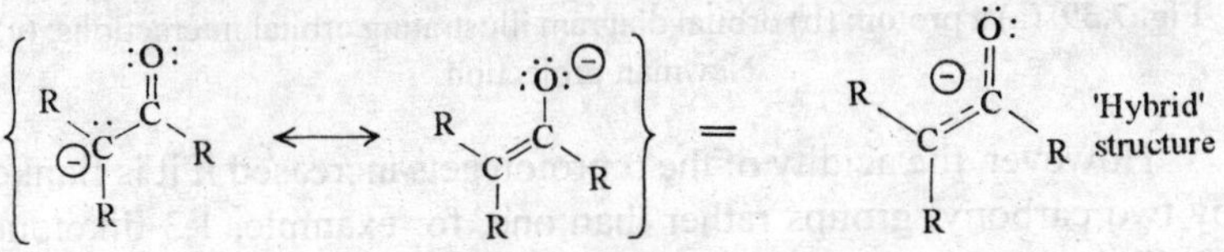

'Hybrid' structure

Fig. 7.37. Resonance structures and 'hybrid' structure for the enolate ion.

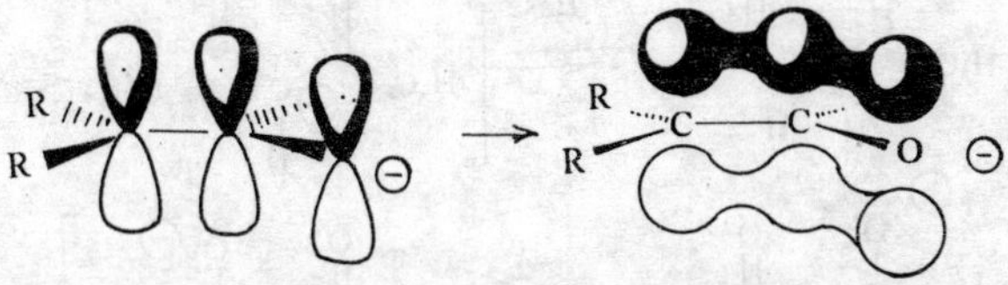

Fig. 7.38. Interaction of 2*p* orbitals to form a molecular orbital.

Enolate ions formed from, ketones or aldehydes are extremely important in the synthesis of more complex organic molecules. The ease with which an enolate ion is formed is related to the acidity of the α proton. The pK_a of propane (acetone) is ≈ 19.3 that means that it is a stronger acid compared to ethane (pK_a ≈ 60) and a much weaker acid than acetic acid (pK_a 4.7) i.e. strong bases like sodium hydride, sodium amide, and lithium diisopropylamide $LiN(i\text{-}C_3H_7)_2$ are needed to form an enolate ion.

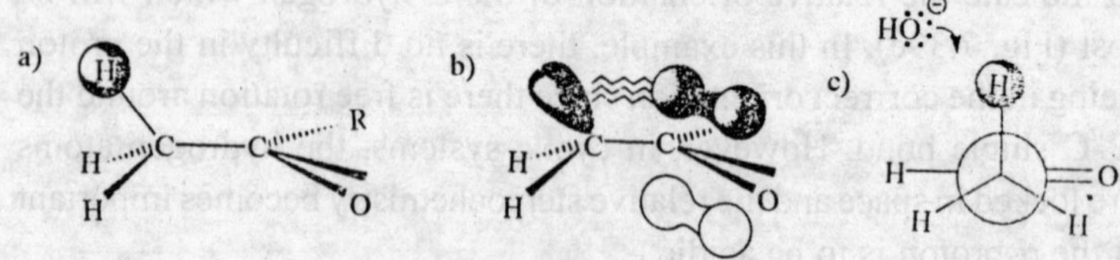

Fig. 7.39. (a) a proton; (b) orbital diagram illustrating orbital interactions; (c) Newman projection.

However, the acidity of the α proton gets increased if it is flanked by two carbonyl groups rather than one, for example, 1,3-diketones (β-diketones) or 1,3-diesters (β-keto esters). This is due to the fact that the negative charge of the enolate ion can be stabilized by both carbonyl groups which results in three resonance structures (Fig. 7.40). For example, the pK_a of 2,4-pentanedione is 9.

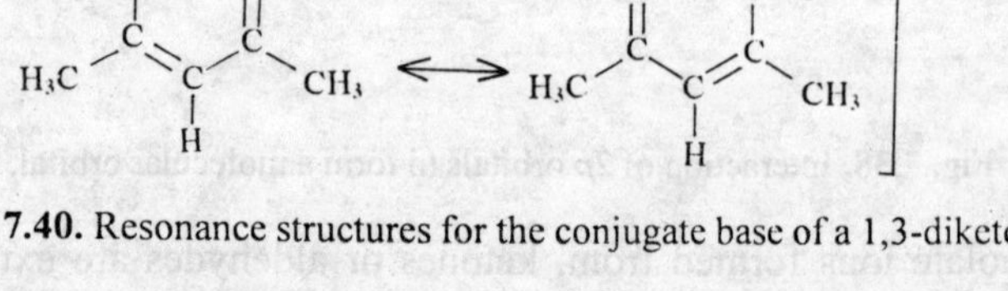

Fig. 7.40. Resonance structures for the conjugate base of a 1,3-diketone.

8

Alkenes and Alkynes

8.1 Alkenes

Preparation of Alkenes

In the laboratory alkenes can be obtained by

(i) *passing alcohol vapours over heated alumina (Al_2O_3)* at about 700k.

$$\underset{\text{Ethyl alcohol}}{CH_3CH_2OH} \xrightarrow{Al_2O_3,\ 650K} CH_2{=}CH_2 + H_2O$$

Dehydration may also be affected by heating in the presence of strong acids like H_2SO_4, H_3PO_4 or p-toluene sulphuric acid. The process is called "*acid-catalyzed dehydration*".

(ii) *Dehydrohalogenation of alkyl halides*

$$C_2H_5 + \text{Alc. KOH} \longrightarrow CH_2{=}CH_2 + KCl + H_2O$$

(iii) *Dehalogenation of vicinal dihalides*

$$\underset{\text{Vincinal dihalide}}{R-\underset{X}{CH}-\underset{X}{CH}-R} + \underset{\text{(zinc dust)}}{Zn} \xrightarrow{\text{Alcohol}} \underset{\text{alkene}}{RCH=CHR} + ZnX_2$$

$$CH_2-CH_2 + Zn \xrightarrow{\text{Alcohol}} CH_2=CH_2 + ZnBr_2$$

The dibromide itself is usually prepared from the same alkene and so the reaction is not particularly useful for the synthesis of alkenes. It is useful, however, in protection strategy. During a lengthy synthesis, it may be necessary to protect a double bond so that it does not undergo any undesired reactions. Bromine can be added to form the dibromide and removed later by debromination in order to restore the functional group.

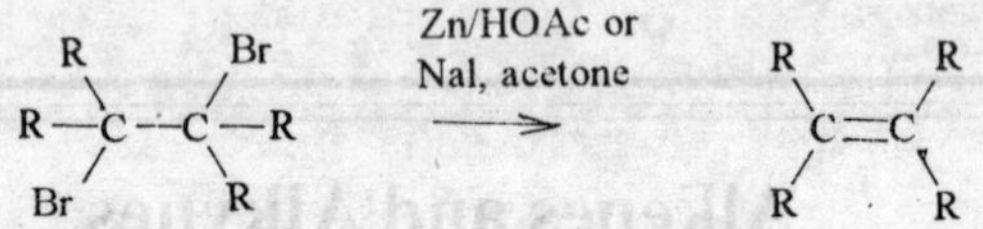

$$R_2C(Br)-C(Br)R_2 \xrightarrow[\text{NaI, acetone}]{\text{Zn/HOAc or}} R_2C{=}CR_2$$

Fig. 8.1. Synthesis of an alkene from a vicinal dibromide.

(iv) *Partial hydrogenation of alkynes*

Partial reduction of alkynes with sodium in liquid ammonia or by catalytic hydrogenation.

$$\underset{\text{Ethyne}}{CH \equiv CH} + H_2 \xrightarrow{\text{Pd, 575K}} \underset{\text{Ethene}}{CH_2 = CH_2}$$

Preparation of Alkynes

Ethyne, $CH \equiv CH$ commonly known as *acetylenes* in the first number of this series and is prepared in the laboratory by the action of water on calcium carbide (CaC_2)

$$CaC_2 + 2H_2O \rightarrow \underset{\text{Ethyne}}{CH \equiv CH} + Ca(OH)_2$$

Higher alkynes can be synthesized from alkenes through a two-step process which involves the **electrophilic addition** of bromine to form a vicinal dibromide then **dehydrohalogenation** with strong base (Fig. 8.2). The second stage involves the loss of two molecules of hydrogen bromide and so two equivalents of base are required.

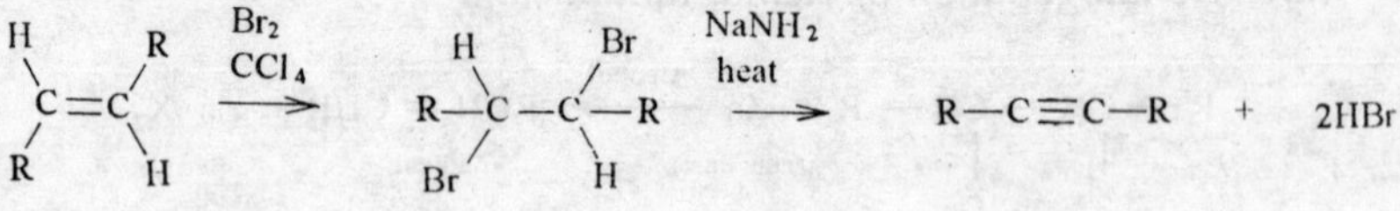

$$HRC{=}CHR \xrightarrow[\text{CCl}_4]{\text{Br}_2} R-CH(Br)-CH(Br)-R \xrightarrow[\text{heat}]{\text{NaNH}_2} R-C{\equiv}C-R + 2HBr$$

Fig. 8.2. Synthesis of an alkyne from an alkene.

Higher homologues of acetylene can also be obtained from acetylene itself. Acetylene is converted into acetylide and the acetylide is reacted with alkyl halide to get a higher homologue.

$$\underset{\text{Ethyne}}{CH \equiv CH} + \underset{\text{Sodamide}}{NaNH_2} \xrightarrow{\text{Liquid Ammonia}} \underset{\text{Sod. acetylide}}{HC \equiv \overset{-}{C}\overset{+}{Na}}$$

$$CH \equiv \overset{-}{C}\overset{+}{Na} + Rx \longrightarrow \underset{\text{alkyne}}{RC \equiv CH} + Nax$$

8.2 Properties of Alkenes and Alkynes

Properties of Alkenes

The functional group in alkenes is (> C = C<). The alkene functional group ($R_2C = CR_2$) is planar in shape with bond angles of 120°. The two carbon atoms involved in the double bond are both sp^2 hybridized. Each carbon has three sp^2 hybridized orbitals which are used for σ bonds while the *p* orbital is used for a π bond. Thus, the double bond is made up of one σ bond and one π bond (Fig. 8.3).

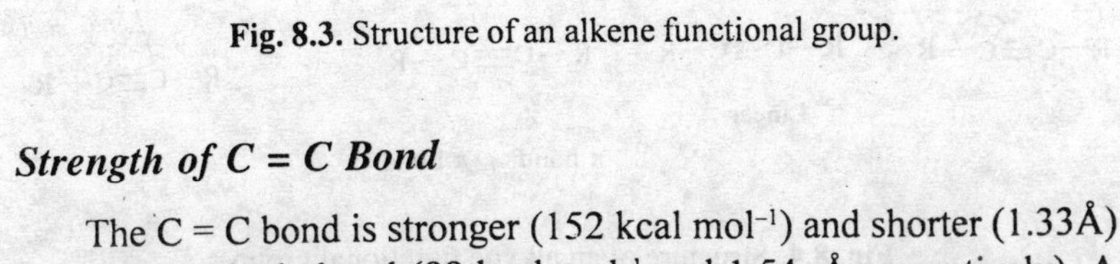

120° 120° 120° Planar σ-bond π-bond sp^2 Center

Fig. 8.3. Structure of an alkene functional group.

Strength of C = C Bond

The C = C bond is stronger (152 kcal mol^{-1}) and shorter (1.33Å) than a C–C single bond (88 kcal mol^{-1} and 1.54. Å respectively). A C = C bond contains one σ bond and one π bond, with the π bond being weaker than the σ bond. This is important with respect to the reactivity of alkenes.

Bond rotation is not possible for a C = C double bond since this would require the π bond to be broken. Therefore, isomers of alkenes are possible depending on the relative position of the substituents. These can be defined as the *cis* or *trans*, but are more properly defined as (Z) or (E).

Alkenes are defined as mono-, di-, tri-, or tetrasubstituted depending on the number of substituents which are present. The more substituents which are present, the more stable the alkene.

Physical Properties

The physical properties of alkenes are similar to those of alkanes. They are relatively *non-polar*, dissolve in non-polar solvents and are not soluble in water.

Since only weak van der Waals interactions are possible between unsaturated molecules such as alkenes so they have comparatively *low boiling points.*

Properties of Alkynes

In them the functional group is (–C ≡ C–). The alkyne functional group consists of a carbon carbon triple bond and is linear in shape with bond angles of 180° (Fig. 8.4).

The two carbon atoms involved in the triple bond are *sp* hybridized, such that each carbon atom has two *sp* hybridized orbitals and two *p* orbitals. The *sp* hybridized orbitals are used for two σ bonds while the *p* orbitals are used for two π bonds. Thus, the triple bond is made up of one σ bond and two π bonds.

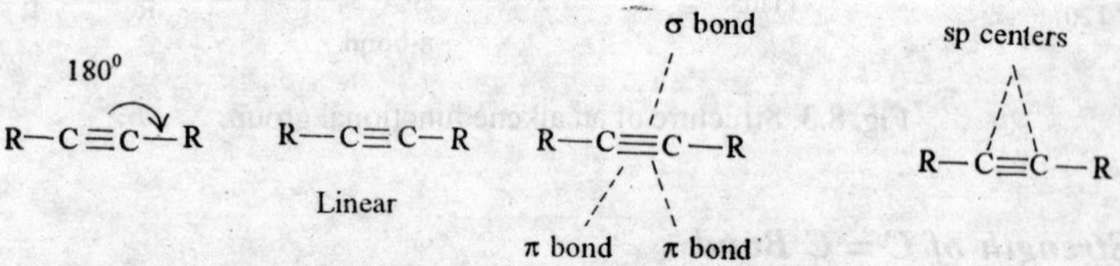

180°
R—C≡C—R
R—C≡C—R
Linear
σ bond
R—C≡C—R
π bond
π bond
sp centers
R—C≡C—R

Fig. 8.4. Structure of an alkyne functional group.

The bond length of a carbon triple bond is 1.20 Å and the bond strength is 200 kcal mol^{-1}. The π bonds are weaker than the σ bond. The presence of the π bonds explains why alkynes are more reactive than alkanes.

Physical Properties of Alkynes

Alkynes have physical properties similar to alkanes. They are relatively nonpolar, dissolve in nonpolar solvents and are not very soluble in water. Only weak van der Waals interactions are possible between unsaturated molecules such as alkynes, and so these structures have low boiling points compared to other functional groups.

Nucleophilicity

Alkenes and alkynes are *nucleophilic* and they react with electrophiles in a reaction called electrophilic addition. The nucleophilic center of the alkene or alkyne is the double bond or triple bond (Fig. 8.5). These are areas of high electron density due to the bonding electrons. The specific electrons which are used to form bonds to attacking electrophiles are those involving in π bonding.

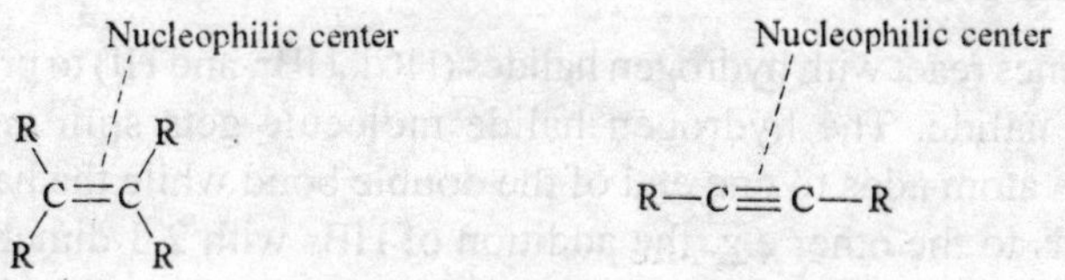

Fig. 8.5. Nucleophilic centers of an alkene and an alkyne.

8.3 Electrophilic Addition to Symmetrical Alkenes

Reactions

Many of the reactions which alkenes undergo take place by a mechanism known as **electrophilic addition** (Fig. 8.6). In these reactions, the π bond of the double bond has been used to form a bond to and incoming electrophile and is no longer present in the product. Furthermore, a new substituents has been added to each of the carbon atoms.

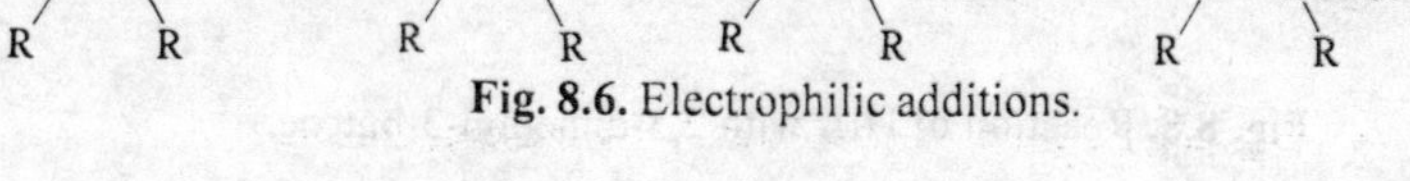

$$R_2C{=}CR_2 \xrightarrow{X_2} X{-}CR_2{-}CR_2{-}X \qquad R_2C{=}CR_2 \xrightarrow{HX} H{-}CR_2{-}CR_2{-}X$$

Fig. 8.6. Electrophilic additions.

Symmetrical and Unsymmetrical Alkenes

Now we shall look at the electrophilic addition of symmetrical alkenes. A symmetrical alkene is an alkene that has the same substituents at each end of the double bond (Fig. 8.7a). Unsymmetrical alkenes have different substituents at each end of the double bond (Fig. 8.7b). Electrophilic addition to unsymmetrical alkenes will be discussed in next section.

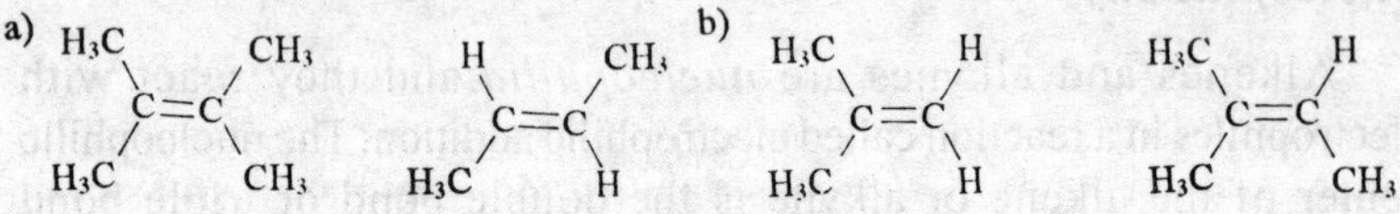

a) $(H_3C)_2C{=}C(CH_3)_2$; $(H)(H_3C)C{=}C(CH_3)(H)$ b) $(H_3C)_2C{=}CH_2$; $(H_3C)_2C{=}C(H)(CH_3)$

Fig. 8.7. (a) Symmetrical alkenes; (b) unsymmetrical alkenes.

Hydrogen Halide

Alkenes react with hydrogen halides (HCl, HBr, and HI) to produce an alkyl halide. The hydrogen halide molecule gets split and the hydrogen atom adds to one end of the double bond while the halogen atom adds to the other e.g. the addition of HBr with 2,3-dimethyl-2-butene (Fig. 8.8). In this reaction, the alkene acts as a nucleophile. It has an electron-rich double bond containing four electrons, two of which make up a strong σ bond and two of which make up a weaker π bond. The *double bond* can be considered as a *nucleophilic center*. Hydrogen bromide has a polar H–Br bond and so the *hydrogen is an electrophilic center* and the *bromine is a nucleophilic center*. Since, halogen atoms are extremely weak nucleophilic center. So this molecule is a more likely to react as an electrophile through its electrophilic hydrogen.

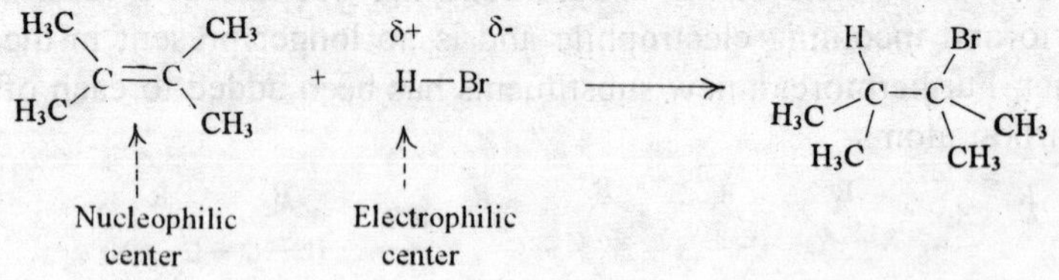

$$(H_3C)_2C{=}C(CH_3)_2 + \overset{\delta+}{H}{-}\overset{\delta-}{Br} \longrightarrow (H_3C)_2CH{-}CBr(CH_3)_2$$

Nucleophilic center (C=C) ; Electrophilic center (H)

Fig. 8.8. Reaction of HBr with 2,3-dimethyl-2-butene.

Various Steps Involved in Electrophilic Addition Reactions

The first step of electrophilic addition (Fig. 8.9) is the one in which alkene acts as a nucleophile and uses its two π electrons to form a new bond to the hydrogen of HBr. As a new bond is formed, the H–Br bond breaks since hydrogen can form one bond. Both electrons in that bond end up on the bromine atom to produce a bromide ion. Since the electrons form the π bond have been used for the formation of a new σ bond, the π bond is no longer present. Because of this, the 'left hand' carbon is left with only three bonds and becomes positively charged. This is called a **carbocation** since the positive charge is on a carbon atom.

This structure is called a **reaction intermediate.** It is a reactive species and will not survive very long with the bromide ion in the vicinity. The *carbocation* is an **electrophile** since it is positively charged. The *bromide* ion is a **nucleophile** since it is negatively charged. Therefore, the bromide ion uses one of its lone pairs of electrons to form a new σ bond to the carbocation and the final product is formed.

The addition of HBr to the alkene is an electrophilic addition since the first step of the mechanism involves the addition of the electrophilic hydrogen to the alkene. The second step involves a nucleophilic addition of the bromide ion to the carbocation intermediate, but *it is the first step which defines this reaction.*

The mechanism is shown (Fig. 8.9), in it the π electrons of the alkene provided the electrons for a new bond between the right hand carbon and hydrogen. They could equally well have been used to form a bond between the left hand carbon and hydrogen (Fig. 8.10).

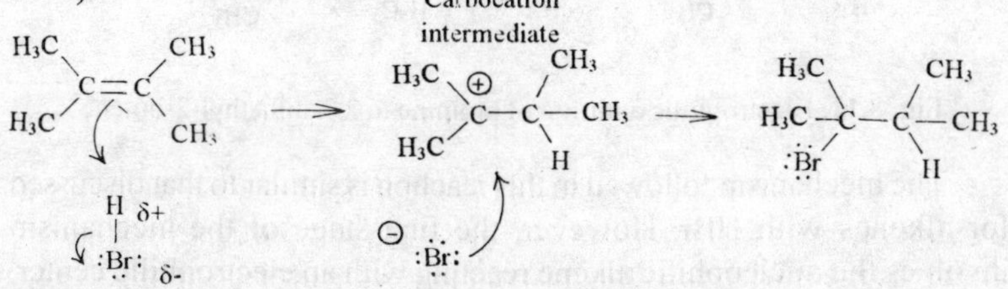

$$(H_3C)_2C{=}C(CH_3)_2 + \overset{\delta+}{H}{-}\overset{\delta-}{Br} \longrightarrow \underset{\text{Carbocation intermediate}}{(H_3C)_2\overset{\oplus}{C}{-}CH(CH_3)_2} + :\!Br^{\ominus} \longrightarrow (H_3C)_2C(Br){-}CH(CH_3)_2$$

Fig. 8.9. Mechanism of electrophilic addition of HBr to 2,3-dimethyl-2-butene.

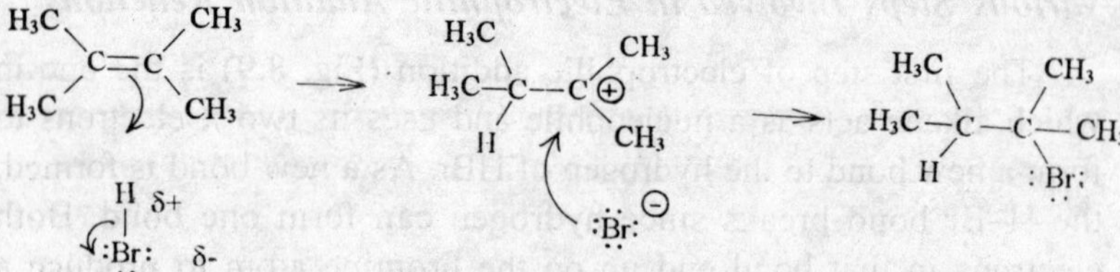

Fig. 8.10. 'Alternative' mechanism for electrophilic addition of HBr to 2,3-dimethyl-2-butene.

In case of a symmetrical alkene, the product is the same and so it does not matter which end of the double bond is used for the new bond to hydrogen. The chances are equal of the hydrogenating to one side or the other.

The electrophilic additions of H–Cl and H–I follow the same mechanism to produce alkyl chlorides and alkyl iodides.

Addition of Halogens

The addition reaction of an alkene with a halogen like bromine or chlorine gives a vicinal dihalide. The halogen molecule is split and the halogens are added to each end of the double bond (Fig. 8.11). Vicinal dibromides are quite useful in the purification or protection of alkenes since the bromine atoms can be removed under different reaction conditions to restore the alkene. Vicinal dibromides can also be converted to alkynes as already discussed in preparation of alkynes.

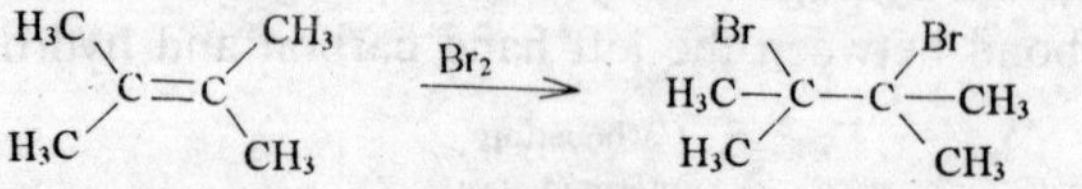

Fig. 8.11. Electrophilic addition of bromine to 2,3-dimethyl-2-butene.

The mechanism followed in this reaction is similar to that discussed for alkenes with HBr. However, the first stage of the mechanism involves the nucleophilic alkene reacting with an electrophilic center, and yet there is no obvious electrophilic center in bromine. The bond between the two bromine atoms is a covalent σ bond with both electrons equally shared between the bromine atoms.

When the bromine molecule approaches end-on to the alkene double bond and an electrophilic center is included (Fig. 8.12). Since the alkene double bond is electron rich it repels the electrons in the bromine molecule and this results in a polarization of the Br–Br bond in such a way that the nearer bromine becomes electron deficient (electrophilic). In this way when an electrophilic center has been generated, the mechanism is the same as before.

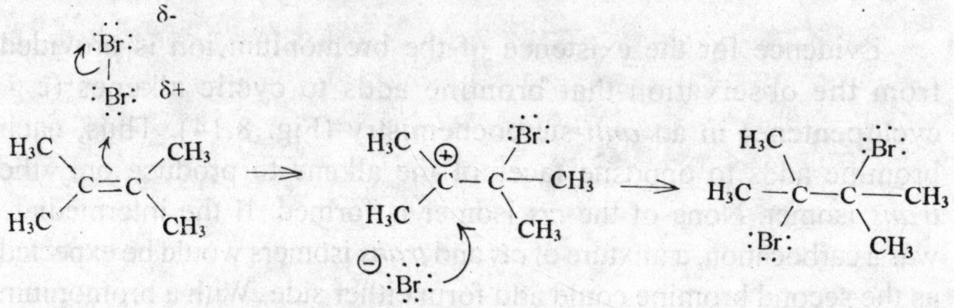

Fig. 8.12. Mechanism for the electrophilic addition of Br_2 with 2.3-dimethyl-2-butene.

The carbocation intermediate can be stabilized by neighboring alkyl groups through *inductive* and *hyper-conjugation effects*. However, it can also be stabilized by sharing the positive charge with the bromine atom and a second carbon atom (Fig. 8.13).

The positively charged carbon is an electrophilic center. The bromine is a weak nucleophilic center. A neutral halogen does not normally act as a nucleophile, but in this case the halogen is held close to the carbocation making reaction more likely. Once the lone pair of electrons on bromine is used to form a bond to the carbocation, a **bromonium** ion is formed in which the bromine gains a positive charge. The mechanism can go in reverse to regenerate the original carbocation. Alternatively, the other carbon-bromine bond can break with both electrons moving onto the bromine. This gives a second carbocation where the other carbon bears the positive charge. Thus, the positive charge is shared between three different atoms and is further stabilized.

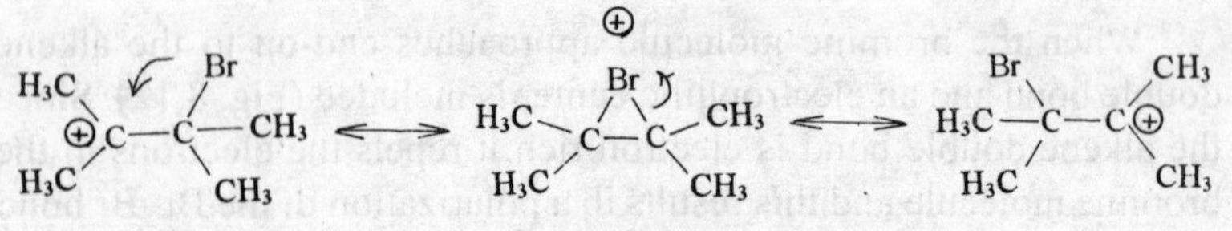

Bromonium ion

Fig. 8.13. Formation of the bromonium ion.

Evidence for the existence of the bromonium ion is provided from the observation that bromine adds to cyclic alkenes (e.g. cyclopentene) in an *anti*-stereochemistry (Fig. 8.14). Thus, each bromine adds to opposite faces of the alkene to produce only the *trans* isomer. None of the *cis* isomer is formed. If the intermediate was a carbocation, a mixture of *cis* and *trans* isomers would be expected as the second bromine could add form either side. With a bromonium ion, the second bromine must approach from the opposite side.

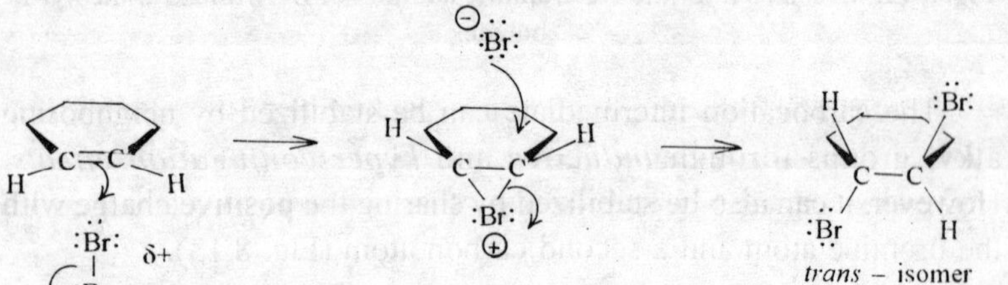

trans – isomer

Fig. 8.14. Anti-stereochemistry of bromine addition to a cyclic alkene.

The reaction of an alkene with a halogen like bromine and chlorine generally gives a vicinal dihalide. However, if the reaction is carried out in water as solvent, the product obtained is a **halohydrin** where the halogen adds to one end of the double bond and a hydroxyl group from water adds to the other (Fig. 8.15).

In this reaction, the first stage of the mechanism proceeds as normal, but then water acts as a nucleophile and 'intercepts' the carbocation intermediate (Fig. 8.16). Because water is the solvent, there are far more molecules of it present compared to the number of bromide ions generated from the first stage of the mechanism.

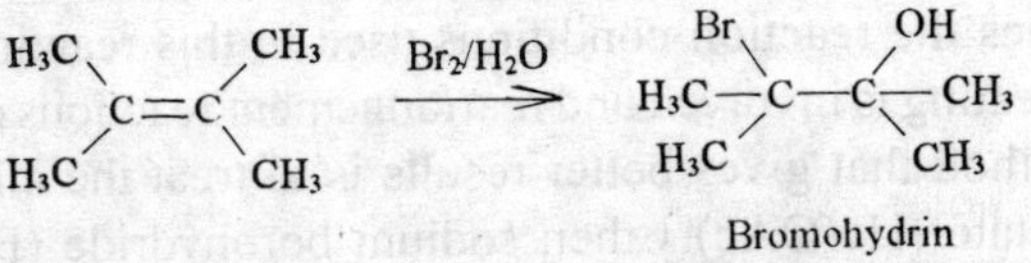

Fig. 8.15. Formation of a bromohydrin from 2,3-dimethyl-2-butene.

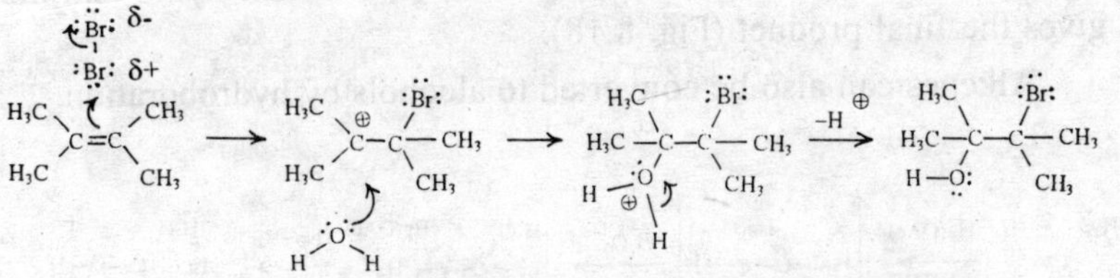

Fig. 8.16. Mechanism of bromohydrin formation.

Water makes use of its a lone pair of electrons on oxygen to form a bond to the carbocation. Because of this, the oxygen effectively 'loses' an electron and gains a positive charge. This charge is lost and the oxygen regains its second lone pair when one of the O–H bonds breaks and both electrons move onto the oxygen.

Addition of H_2O

Alkenes get converted to alcohols by treatment with aqueous acid (sulfuric or phosphoric acid; Fig. 8.17). This electrophilic addition reaction *involves the addition of water across the double bond.* The hydrogen adds to one carbon while a hydroxyl group adds to the other carbon.

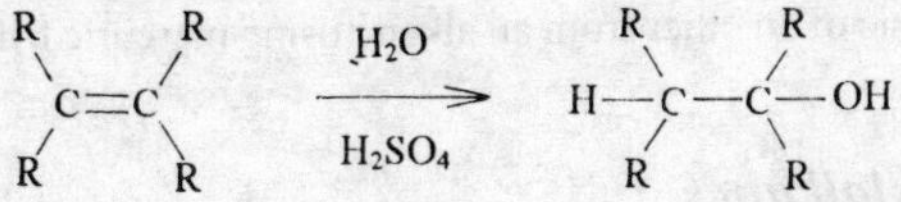

$$R_2C{=}CR_2 \xrightarrow[H_2SO_4]{H_2O} H{-}CR_2{-}CR_2{-}OH$$

Fig. 8.17. Synthesis of an alcohol from an alkene.

Alkenes to Alcohols

Sometimes the reaction conditions used in this reaction are too harsh since heating is involved and rearrangement reactions can occur. A milder method that gives better results is to treat the alkene with mercuric acetate [$Hg(OAc)_2$] then sodium borohydride (Fig. 8.18). The reaction involves electrophilic addition of the mercury reagent to form an intermediate **mercuronium** ion. This reacts with water to give an organomercury intermediate. Reduction with sodium borohydride replaces the mercury substituents with hydrogen and gives the final product (Fig. 8.18).

Alkenes can also be converted to alcohols by hydroboration.

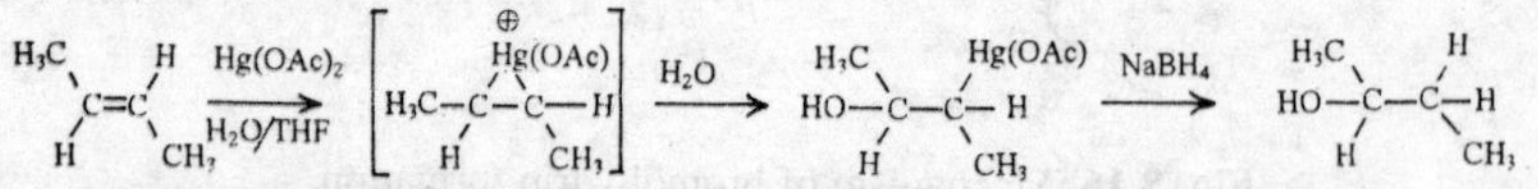

$$H_3C(H)C{=}C(H)CH_3 \xrightarrow[H_2O/THF]{Hg(OAc)_2} \left[H_3C{-}\overset{\overset{\oplus}{Hg(OAc)}}{C(H)}{-}C(H)(CH_3){-}H\right] \xrightarrow{H_2O} HO{-}C(H_3C)(H){-}C(Hg(OAc))(CH_3){-}H \xrightarrow{NaBH_4} HO{-}C(H_3C)(H){-}C(H)(CH_3){-}H$$

Fig. 8.18. Synthesis of an alcohol from an alkene using mercuric acetate.

Alkenes to Ethers

A similar reaction to the mercuric acetate/sodium borohydride synthesis of alcohols allows the conversion of alkenes to ethers. In this case, mercuric trifluoracetate is used (Fig. 8.19).

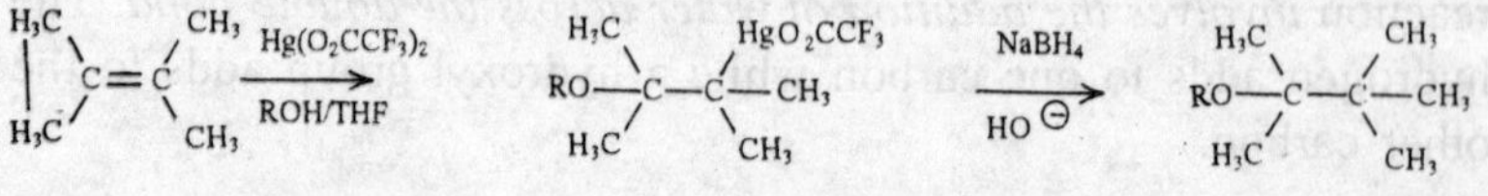

$$(H_3C)_2C{=}C(CH_3)_2 \xrightarrow[ROH/THF]{Hg(O_2CCF_3)_2} RO{-}C(H_3C)_2{-}C(HgO_2CCF_3)(CH_3)_2 \xrightarrow[HO^{\ominus}]{NaBH_4} RO{-}C(H_3C)_2{-}C(CH_3)_2{-}CH_3$$

Fig. 8.19. Synthesis of an ether from an alkene using mercuric trifluoroacetate.

Alkenes to Arylalkanes

The reaction of an aromatic ring such as benzene with an alkene under acid conditions results in the formation of an arylalkane (Fig.

8.20). as far as the alkene is concerned this is another example of electrophilic addition involving the addition of a proton to one end of the double bond and the addition of the aromatic ring to the other. As far as the aromatic ring is concerned this is an example of an electrophilic substitution reaction called the **Friedel-Crafts alkylation.**

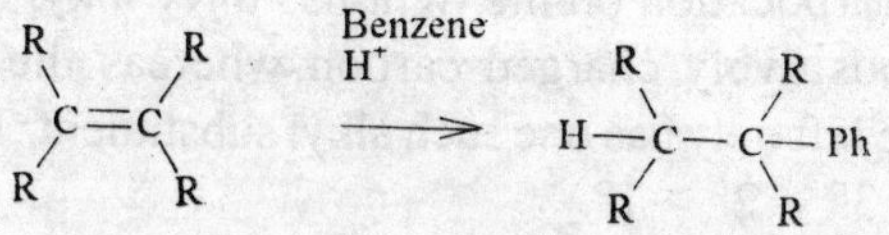

$$R_2C{=}CR_2 \xrightarrow[H^+]{\text{Benzene}} H{-}CR_2{-}CR_2{-}Ph$$

Fig. 8.20. Synthesis of arylalkanes from alkenes.

8.4 Electrophilic Addition to Unsymmetrical Alkenes

Addition of Hydrogen Halides

The reaction of a symmetrical alkene with hydrogen bromide produces the same product irrespective of whether the hydrogen of HBr is added to one end of the double bond or the other. However, this is not the case with unsymmetrical alkenes (Fig. 8.21). In this case, two different products are possible. These are not formed to an equal extent and the *more substituted alkyl halide (II) is preferred.* The reaction proceeds according to **Markovnikov rule** with hydrogen ending up on the least substituted position and the halogen ending up on the most substituted position. *Markovnikov's rule* states that 'in the addition of HX to an alkene, the hydrogen atom adds to the carbon atom that already has the greater number of hydrogen atoms'. This produces the more substituted alkyl halide.

$$(H_3C)_2C{=}CH_2 \xrightarrow{HBr} (H_3C)_2C(H){-}CH_2Br \;(\text{I}) \quad (H_3C)_2C(Br){-}CH_3 \;(\text{II})$$

I: Hydrogen added to left hand end of double bond

II: Hydrogen added to right hand end of double bond

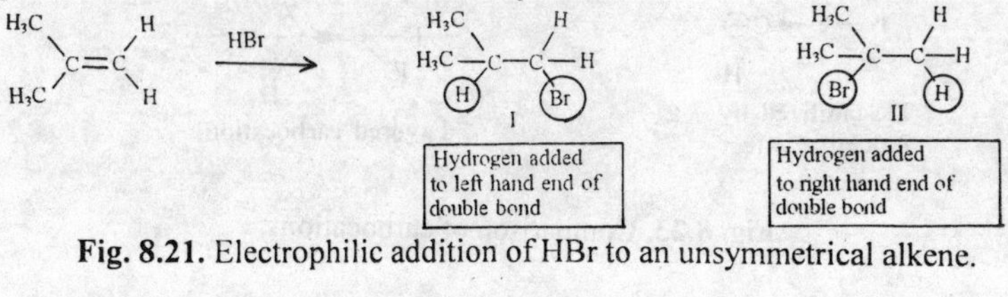

Fig. 8.21. Electrophilic addition of HBr to an unsymmetrical alkene.

Carbocation Stabilities

This reaction can be understood by assuming that in it the carbocation intermediate is formed which leads to product II, which is more stable than the carbocation intermediate leading to product I (Fig. 8.22). It is possible to predict the more stable carbocation by counting the number of alkyl groups attached to the positive center. The more stable carbocation on the right has three alkyl substituents attached to the positively charged carbon whereas the less stable carbocation on the left only has one such alkyl substituent. The stability of carbocation is 3° > 2° > 1°.

However, Markovnikov's rule is not always hold true. For example, the reaction of $CF_3CH = CH_2$ with HBr gives $CF_3CH_2CH_2Br$ rather than $CF_3CHBrCH_3$. Here, *the presence of electron-withdrawing fluorine* substituents has a destabilizing influence on the two possible intermediate carbocations (Fig. 8.23). *The destabilizing effect will be greater for the more substituted carbocation since the carbocation is closer to the fluorine substituents and so the favored carbocation is the least substituted one in this case.*

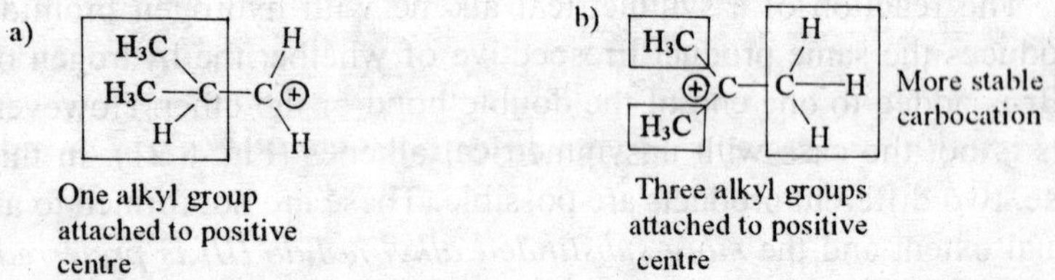

a) H_3C, $H_3C—C—C^{\oplus}$ with H, H, H

One alkyl group attached to positive centre

b) H_3C, H_3C, $^{\oplus}C—C—H$ with H, H — More stable carbocation

Three alkyl groups attached to positive centre

Fig. 8.22. (a) Carbocation leading to product I; (b) carbocation leading to product II.

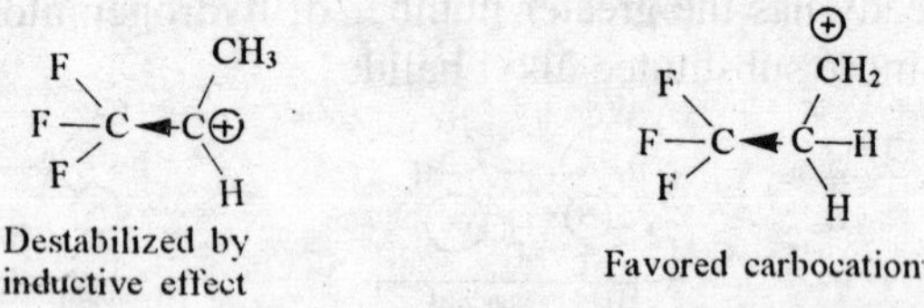

$F_3C \leftarrow C^{\oplus}(CH_3)H$

Destabilized by inductive effect

$F_3C \leftarrow CH(H)CH_2^{\oplus}$

Favored carbocation

Fig. 8.23. Comparison of carbocations.

Addition of Halogens

There is no possibility of different products when a halogen such as bromine or chlorine is added to an unsymmetrical alkene. However, in case water is used as a solvent, the halogen is attached to the least substituted carbon and the hydroxyl group is attached to the more substituted carbon (Fig. 8.24). This can be explained by assuming that the bromonium ion is not symmetrical and that although the positive charge is shared between the bromine and the two carbon atoms, the *positive charge is greater on the more substituted carbon compared with the less substituted carbon.*

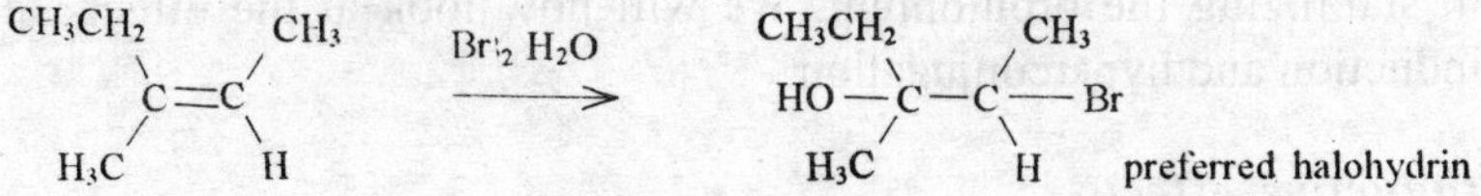

$$(CH_3CH_2)(H_3C)C{=}C(CH_3)(H) \xrightarrow{Br_2,\ H_2O} HO{-}C(CH_3CH_2)(H_3C){-}C(CH_3)(H){-}Br$$

preferred halohydrin

Fig. 8.24. Reaction of 3-methyl-2-pentene with bromine and water.

Addition of Water

With unsymmetrical alkenes, the more substituted alcohol is preferred product (Fig. 8.25).

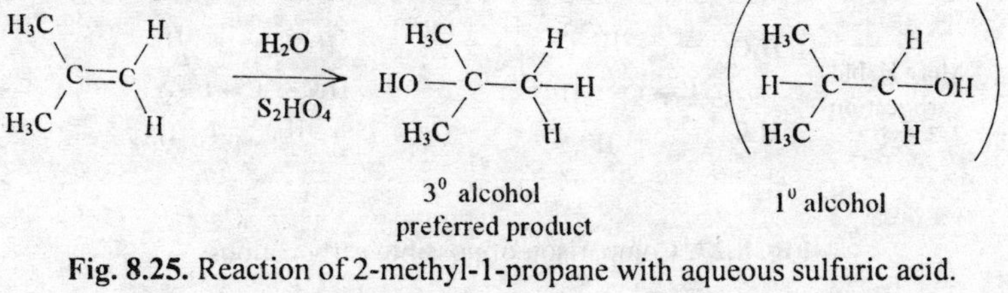

$$(H_3C)_2C{=}CH_2 \xrightarrow[S_2HO_4]{H_2O} HO{-}C(H_3C)_2{-}CH_2{-}H \quad \left(H{-}C(H_3C)_2{-}CH_2{-}OH \right)$$

3^0 alcohol preferred product; 1^0 alcohol

Fig. 8.25. Reaction of 2-methyl-1-propane with aqueous sulfuric acid.

The same holds true for the organomercuric synthesis of alcohols (Fig. 8.26).

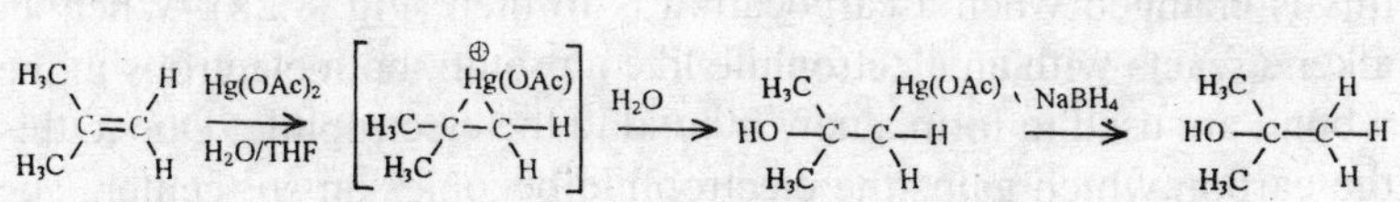

$$(H_3C)_2C{=}CH_2 \xrightarrow[H_2O/THF]{Hg(OAc)_2} \left[H_3C{-}C(H_3C){-}CH_2{-}H \text{ bridged by } \overset{\oplus}{Hg}(OAc) \right] \xrightarrow{H_2O} HO{-}C(H_3C)_2{-}CH(H){-}Hg(OAc) \xrightarrow{NaBH_4} HO{-}C(H_3C)_2{-}CH_2{-}H$$

Fig. 8.26. Organomercuric synthesis of alcohols.

8.5 Carbocation Stabilization

Stabilization

Any positively charge species like carbocations are inherently reactive and unstable. The more unstable they are, the less easily they are formed and the less likely the overall reaction. Any factor that helps to stabilize the positive charge (and by inference the carbocation) will make the reaction more likely. The three ways in which a positive charge can be stabilized are (i) inductive effects, (ii) hyperconjugation, and (iii) delocalization. We have already seen the effects of delocalization in stabilizing the bromonium. We will now look at the effects of induction and hyperconjugation.

Inductive effects

Alkyl groups can donate electrons towards a neighboring positive center and this helps to stabilize the ion since some of the positive charge is partially dispersed over the alkyl group (Fig. 8.27). More the alkyl groups attached, the greater is the electron donating power and the more stable the carbocation.

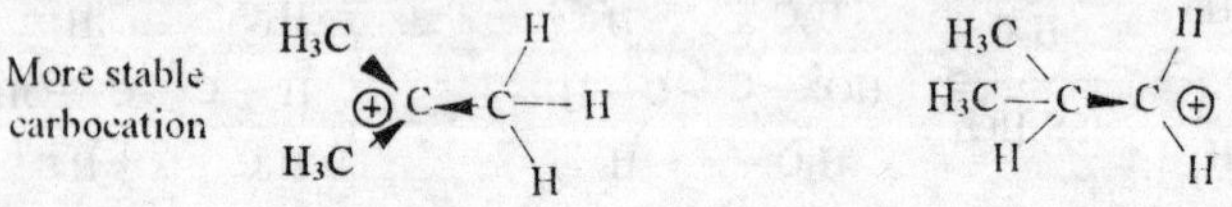

More stable carbocation

Fig. 8.27. Comparison of possible carbocations.

Hyperconjugation

We know that both carbons of an alkene are sp^2 hybridized. But, this is changed when a carbocation is formed (Fig. 8.28). When an alkene reacts with an electrophile like a proton, both electrons in the π bond are used to form a new σ bond to the electrophile. Due to this the carbon which gains the electrophile becomes an sp^3 center. The other carbon containing the positive charge remains as an sp^2 center. Thus, it has three sp^2 hybridized orbitals (used for the three σ bonds still present) and one vacant $2p$ orbital which is not involved in bonding. *Hyperconjugation involves the overlap of the vacant 2p orbital with a neighboring C–H σ-bond orbital* (Fig. 8.29).

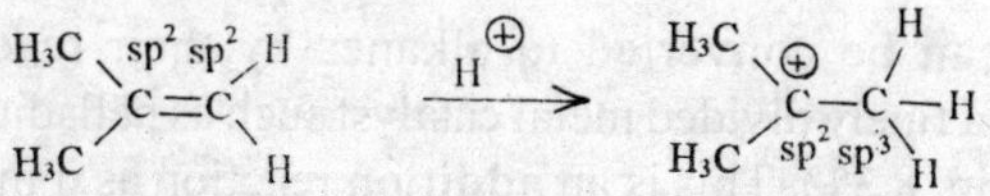

Fig. 8.28. Hybridization of alkene and carbocation.

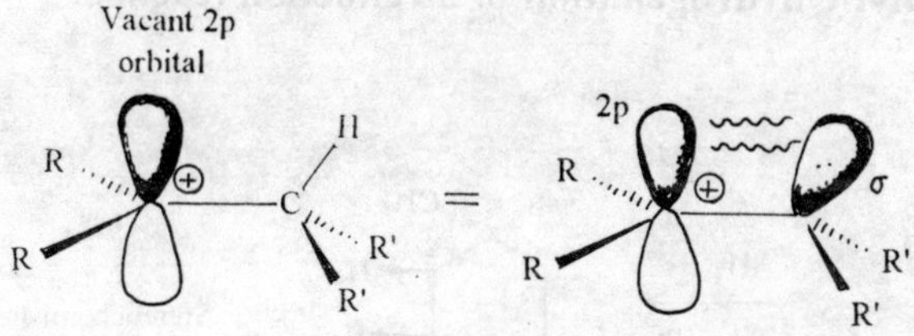

Fig. 8.28. Hyperconjugation.

This interaction means that the $2p$ orbital is not completely vacant as the σ electrons of the C–H bond can spend a small amount of time entering the space occupied by the $2p$ orbital. Thus the C–H bond becomes slightly electron deficient. Due to this, the positive charge is delocalized and hence stabilized. The more alkyl groups attached to the carbocation, the more possibilities there are for hyperconjugation and the more stable the carbocation. For example, the more substituted carbocation (Fig. 8.30a) can be stabilized by hyperconjugation to nine C–H bonds, whereas the less substituted carbocation (Fig. 8.30b) can only be stabilized by hyperconjugation to one C–H bond.

Fig. 8.30. (a) More substituted carbocation; (b) less substituted carbocation.

8.6 Reduction and Oxidation of Alkenes

Alkenes to Alkanes

Alkenes can be converted to alkanes by their reaction with hydrogen over a finely divided metal catalyst such as palladium, nickel, or platinum (Fig. 8.31). This is an **addition** reaction as it involves the addition of hydrogen atoms to each end of the double bond. It is also called a **catalytic hydrogenation** or a **reduction** reaction.

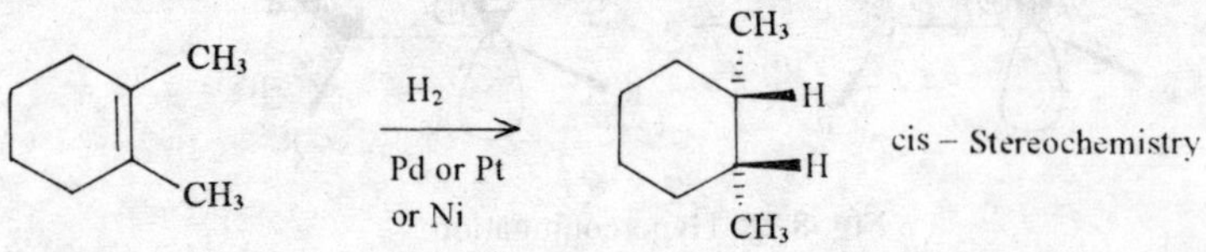

CH3
CH3
H2
Pd or Pt
or Ni
CH3
H
H
CH3
cis – Stereochemistry

Fig. 8.31. Hydrogenation of an alkene.

The catalyst is important since the reaction will not occur at room temperature in its absence. This is because hydrogenation has a high **free energy of activation** (ΔG_1*) (Fig. 8.32). The role of the catalyst is to bind the alkene and the hydrogen to a common surface such that they can react more easily. This results in a much lower energy of activation (ΔG_2*) allowing the reaction to proceed in the much milder conditions. The catalyst itself is unchanged after the reaction and can be used in small quantity.

Potential energy
ΔG1*
ΔG2*
ΔH0
Reaction coordinate

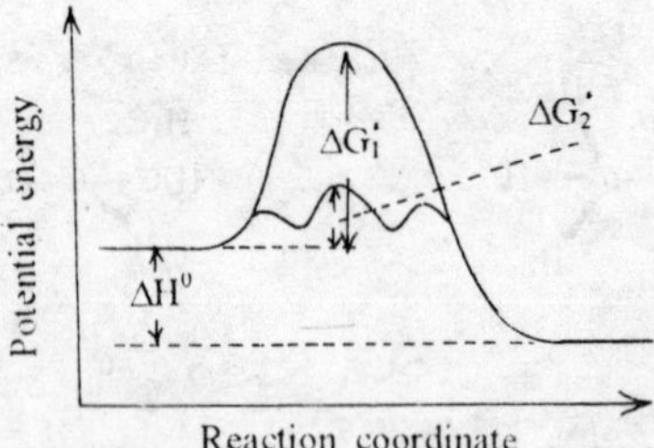

Fig. 8.32. Graph of potential energy versus reaction coordinate for an uncatalyzed and a catalyzed hydrogenation reaction of an alkene.

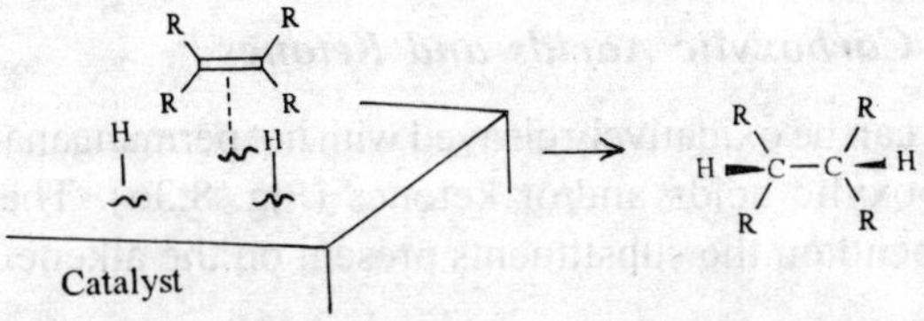

Fig. 8.33. Binding of alkene and hydrogen to catalytic surface.

Both the hydrogen and the alkene are bound to the catalyst surface before the hydrogen atoms are transferred, which means that both hydrogens are added to the same side of the double bond (*see* Fig. 8.33). – ***syn*-addition**. *Note that the hydrogen molecule is split once it has been added to the catalyst.*

Alkenes to Aldehydes and Ketones

Considering that an alkene on oxidation with ozone (Fig. 8.34) that proceeds with the formation of an initial **ozonide** which then rearranges to an isomeric ozonide. This second ozonide is unstable and potentially explosive and so it is not generally isolated. Instead, it is reduced with zinc and water leading to the formation of two separate molecules.

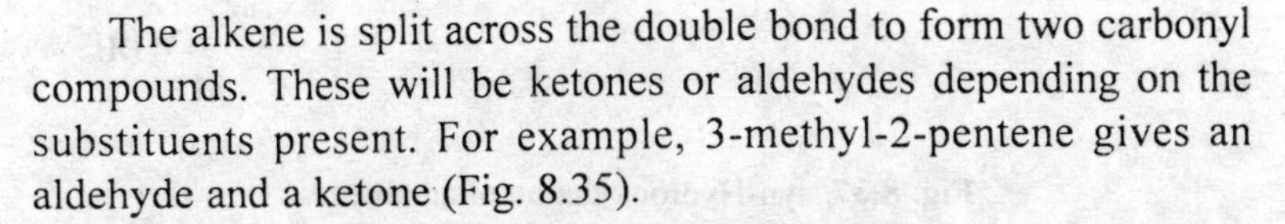

Fig. 8.34. Ozonolysis of an alkene.

The alkene is split across the double bond to form two carbonyl compounds. These will be ketones or aldehydes depending on the substituents present. For example, 3-methyl-2-pentene gives an aldehyde and a ketone (Fig. 8.35).

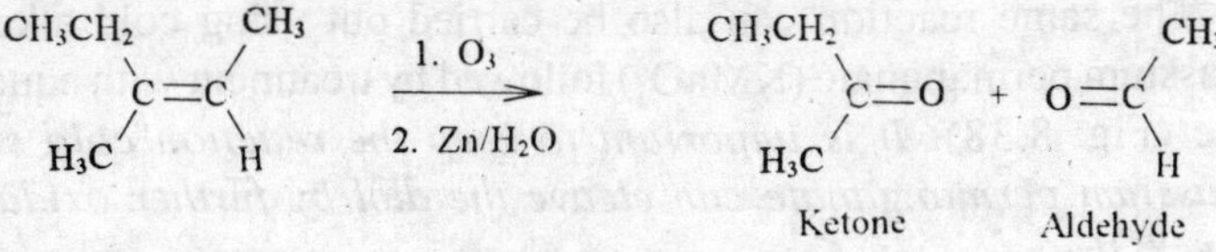

Fig. 8.35. Ozonolysis of 3-methyl-2-pentene.

Alkenes to Carboxylic Aacids and Ketones

Alkenes can be oxidatively cleaved with hot permanganate solution to give carboxylic acids and/or ketones (Fig. 8.36). The products obtained depend on the substituents present on the alkene.

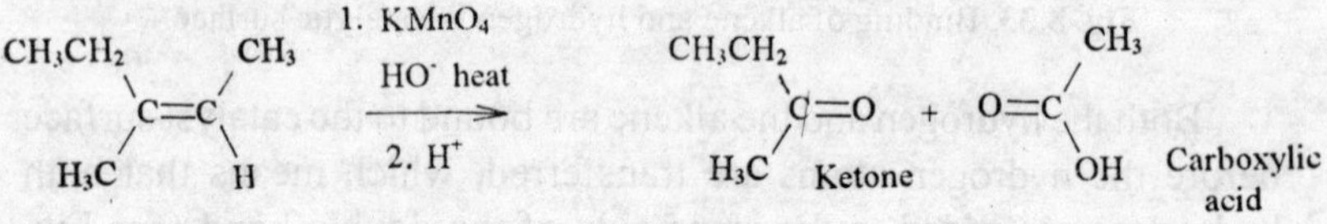

$$\underset{\text{H}_3\text{C} \quad\quad \text{H}}{\overset{\text{CH}_3\text{CH}_2 \quad\quad \text{CH}_3}{\text{C}=\text{C}}} \xrightarrow[\text{2. H}^+]{\text{1. KMnO}_4,\ \text{HO}^-\ \text{heat}} \underset{\text{Ketone}}{\underset{\text{H}_3\text{C}}{\overset{\text{CH}_3\text{CH}_2}{\text{C}=\text{O}}}} + \underset{\text{Carboxylic acid}}{\underset{\text{OH}}{\overset{\text{CH}_3}{\text{O}=\text{C}}}}$$

Fig. 8.36. Oxidative cleavage of 3-methyl-2-pentene.

Alkenes to 1,2-diols

The reaction of alkenes with **osmium tetroxide** (OsO_4) is an example of an oxidation reaction (Fig. 8.37). In this case the alkene is not split. but, a 1,2-diol is obtained which is also called a glycol. The reaction involves the formation of a cyclic intermediate where the osmium reagent is attached to one face of the alkene. On treatment with **sodium bisulfite**, the intermediate is cleaved such that the two oxygen atoms linking the osmium remain attached. Due to this both the alcohols being added to the same side of the double bond i.e. ***syn*-hydroxylation**.

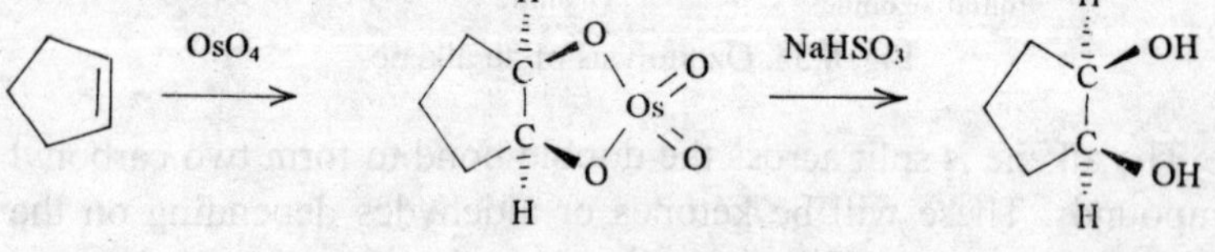

Cyclopentene $\xrightarrow{\text{OsO}_4}$ cyclic osmate ester (Os, O, O) $\xrightarrow{\text{NaHSO}_3}$ cis-cyclopentane-1,2-diol (H, OH / H, OH)

Fig. 8.37. syn-Hydrooxylation of an alkene.

The same reaction can also be carried out using cold alkaline potassium permangnate ($KMnO_4$) followed by treatment with aqueous base (Fig. 8.38). *It is important to keep the reaction cold since potassium permanganate can cleave the diol by further oxidation* (Fig. 8.36).

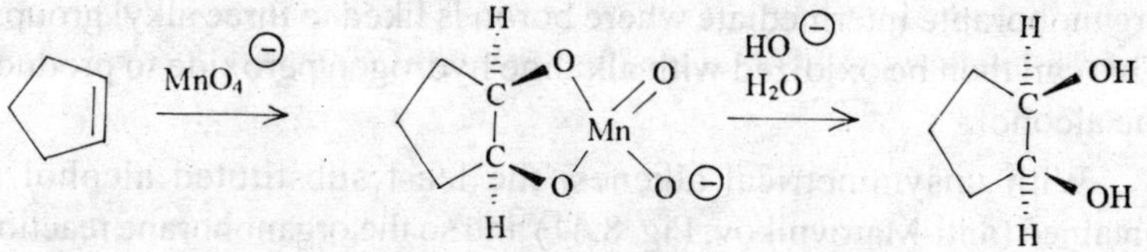

$$\text{cyclopentene} \xrightarrow{MnO_4^{\ominus}} \text{cyclic manganate ester} \xrightarrow[H_2O]{HO^{\ominus}} \textit{cis}\text{-cyclopentane-1,2-diol}$$

Fig. 8.38. syn-Hydroxylation with $KMnO_4$.

The reaction works better with osmium tetroxide. However, this is a highly toxic and expensive reagent and has to be handled with care.

Anti-hydroxylation of the double bond can also be achieved by forming an epoxide, then carrying out an acid-catalyzed hydrolysis.

Alkenes to Expoxides

When an alkene is treated with a **peroxyacid** (RCO_3H) it forms a epoxide (Fig. 8.39). *m*-Chloroperoxybenzoic acid is one of the most commonly used peroxyacids for this reaction. The reaction is unusual because there is no carbocation intermediate, and it involves a one-step process without intermediates.

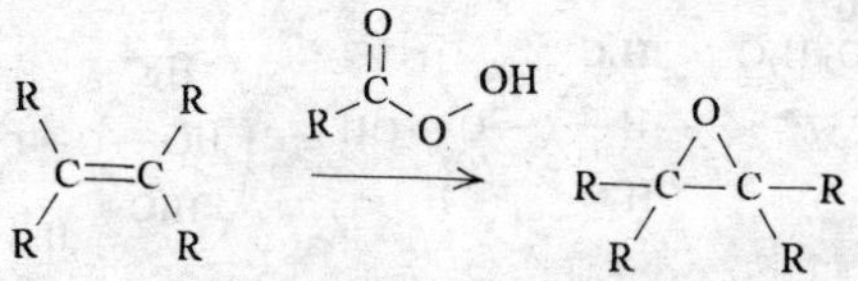

$$R_2C{=}CR_2 \xrightarrow{RC(=O)OOH} R_2C\overset{O}{—}CR_2$$

Fig. 8.39. Expoxidation of an alkene.

8.7 Hydroboration of Alkenes

Reaction

Alcohols can be generated from alkenes by reaction with diborane (B_2H_6 or BH_3), followed by treatment with hydrogen peroxide (Fig. 8.40). The first part of the reaction involves the splitting of a B–H bond in BH_3 with the hydrogen joining one end of the alkene and the boron joining the other. Each of the B–H bonds is split in this way such that each BH_3 molecule reacts with three alkenes to give an organoborante intermediate where boron is liked to three alkyl groups. This can then be oxidized with alkaline hydrogen peroxide to produce the alcohol.

With unsymmetrical alkenes, the least substituted alcohol is obtained (anti-Marovnikov; Fig. 8.41) and so the organoborane reaction is complementary to the electrophilic addition reaction with aqueous acid. Steric factors appear to play a role in controlling this preference with the boron atom preferring to approach the least sterically hindered site. Electronic factors also play a role as described in the mechanism below:

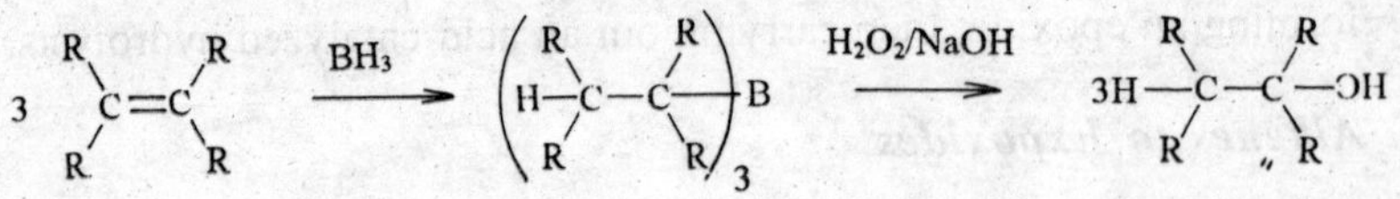

$$3\ R_2C{=}CR_2 \xrightarrow{BH_3} (H{-}CR_2{-}CR_2)_3B \xrightarrow{H_2O_2/NaOH} 3H{-}CR_2{-}CR_2{-}OH$$

Fig. 8.40. Hydroboration of an alkene.

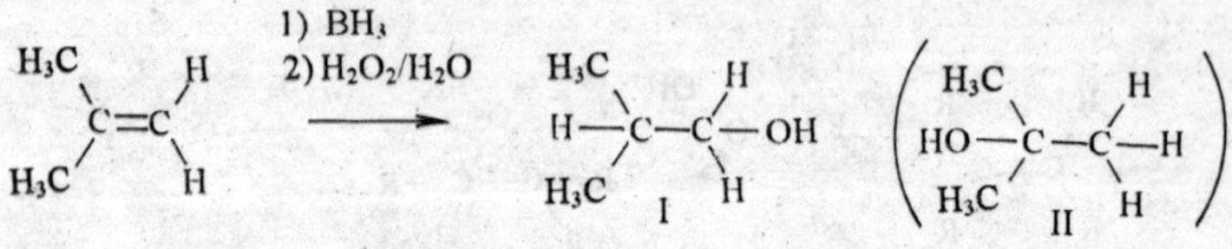

$$(H_3C)_2C{=}CH_2 \xrightarrow[2)\ H_2O_2/H_2O]{1)\ BH_3} \underset{\text{I}}{H{-}C(CH_3)_2{-}CH_2{-}OH} \quad \left(\underset{\text{II}}{HO{-}C(CH_3)_2{-}CH_2{-}H}\right)$$

Fig. 8.41. Hydroboration of 2-methylpropane to give a primary alcohol (I). The tertiary alcohol (II) is not obtained.

Mechanism

The mechanism (Fig. 8.42) involves the alkene π bond interacting with the empty *p* orbital of boron to form a π complex. One of BH_3's hydrogen atom is then transferred to one end of the alkene as boron itself forms a σ bond to the other end. This takes place through a four-centered transition state where the alkene's π bond and the B–H bond are partially broken, and the eventual C–H and C–B bonds are partially formed. There is an imbalance of electrons in the transition state which results in the boron being slightly negative and on of the alkene carbons being slightly positive. The carbon best able to handle this will be the most substituted carbon and so the boron will end up on the least substituted carbon. (Note that boron has six valence electrons and is electrophilic. Therefore, the addition of boron to the least substituted position actually follows Markovnikov's rule.)

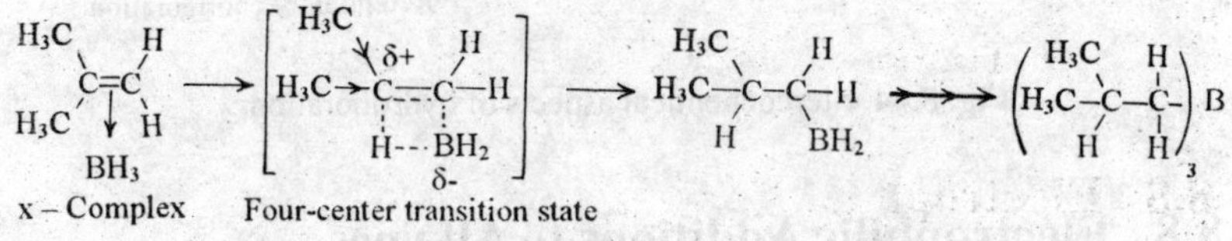

x – Complex Four-center transition state

Fig. 8.42. Mechanism of hydroboration.

Since subsequent oxidation with hydrogen peroxide replaces the boron with a hydroxyl group, the eventual alcohol will be on the least substituted carbon. Furthermore, the addition of the boron and hydrogen atoms takes place such that they are on the same side of the alkenes. This is called ***syn*-addition**. The mechanism of oxidation (Fig. 8.43) involves addition of the hydroperoxide to the electron deficient boron to form an unstable intermediate which then rearranges such that an alkyl group migrates from the boron atom to the neighboring oxygen and expels a hydroxide ion. This process is then repeated for the remaining two hydrogens on boron and the final trialkyl borate $B(OR)_3$ can then be hydrolyzed with water to give three molecules of alcohol plus a borate ion.

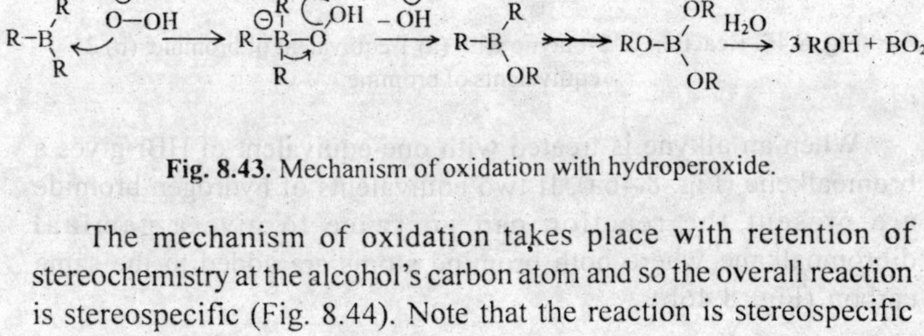

$$R_3B \xrightarrow{^{\ominus}O\text{–}OH} R_3B^{\ominus}\text{–}OOH \xrightarrow{-\,^{\ominus}OH} R_2B\text{–}OR \longrightarrow RO\text{–}B(OR)_2 \xrightarrow{H_2O} 3\,ROH + BO_2^{\,3}$$

Fig. 8.43. Mechanism of oxidation with hydroperoxide.

The mechanism of oxidation takes place with retention of stereochemistry at the alcohol's carbon atom and so the overall reaction is stereospecific (Fig. 8.44). Note that the reaction is stereospecific such that the alcohol group is *trans* to the methyl group in the product. However, it is not enantiospecific and both enantiomers are obtained in equal amounts (a racemate).

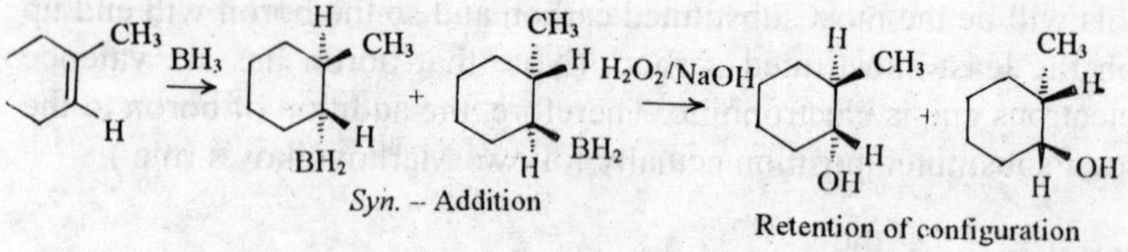

Fig. 8.44. Stereochemical aspects of hydroboration.

8.8 Electrophilic Additions to Alkynes

Additions to symmetrical Alkynes

Alkynes give electrophilic addition reactions with the same reagents as in case of alkenes (e.g. halogens and hydrogen halides). Since there are two π bonds in alkynes, it is possible for the reaction to go once or twice depending on the amount of reagent added. For example, reaction of 2-butyne with one equivalent of bromine gives an (E)-dibromoalkene (Fig. 8.45a). With two equivalents of bromine, the initially formed (E)-dibromoalkene reacts further to give a tetrabromoalkane (Fig. 8.45b).

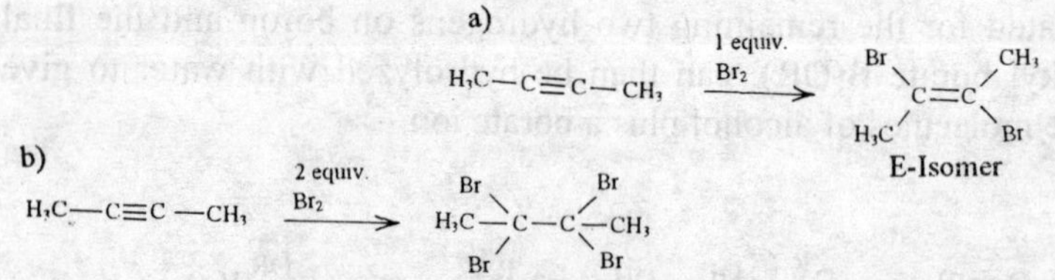

Fig. 8.45. Reaction of 2-butyne with (a) 1 equivalent of bromine; (b) 2 equivalents of bromine.

When an alkyne is treated with one equivalent of HBr gives a bromoalkene (Fig. 8.46a). If two equivalents of hydrogen bromide are present the reaction can go twice to give a **geminal** dibromoalkane where both bromine atoms are added to the same carbon (Fig. 8.46b).

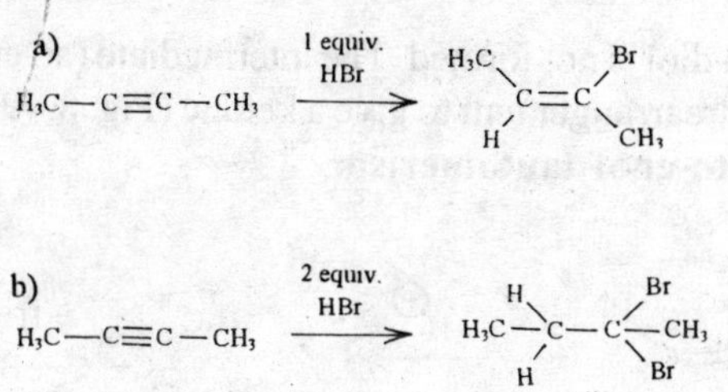

a)

$H_3C-C\equiv C-CH_3 \xrightarrow[\text{HBr}]{\text{1 equiv.}}$ $H_3C(H)C=C(Br)CH_3$

b)

$H_3C-C\equiv C-CH_3 \xrightarrow[\text{HBr}]{\text{2 equiv.}}$ $H_3C-CH_2-CBr_2-CH_3$

Fig. 8.46. Reaction of 2-butyne with (a) 1 equivalent of HBr; (b) 2 equivalents of HBr.

These addition reactions are similar to the addition reactions of alkenes. However, the reaction is much slower for an alkyne, because alkynes are less reactive. Alkynes can be expected to be more nucleophilic, as they are more electron rich in the vicinity of the multiple bond, that is, six electrons in a triple bond as compared to four in a double bond. However, electrophilic addition to an alkyne involves the formation of a **vinylic** carbocation (Fig. 8.47). This carbocation is much less stable than the carbocation intermediate formed during electrophilic addititon to an alkene.

Because this low reactivity, alkynes react slowly wiht aqueous acid and mercuric sulfate has to be added as a catalyst. The product that might be expected from this reaction is a diol (Fig. 8.48).

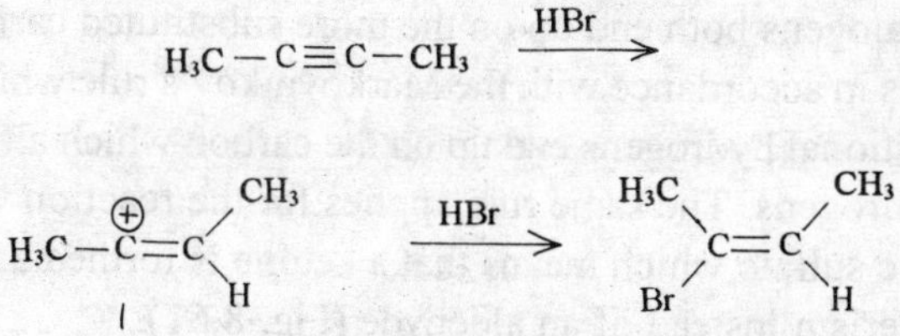

$H_3C-C\equiv C-CH_3 \xrightarrow{\text{HBr}}$

$H_3C-\overset{\oplus}{C}=C(CH_3)H \xrightarrow{\text{HBr}} H_3C(Br)C=C(CH_3)H$

I

Fig. 8.47. Electrophilic addition to an alkyne via a vinylic carbocation (I).

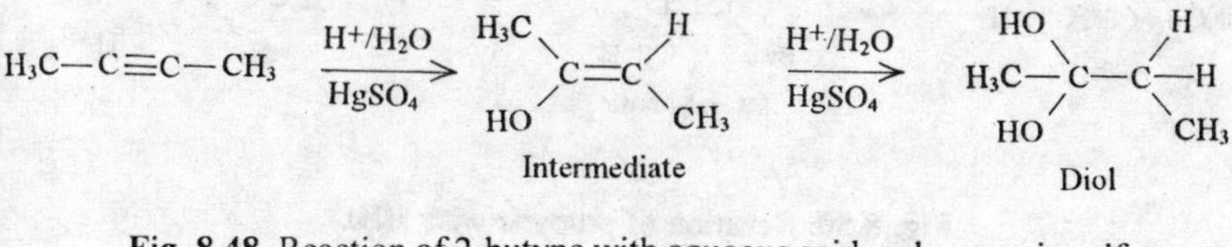

$H_3C-C\equiv C-CH_3 \xrightarrow[\text{HgSO}_4]{\text{H}^+/\text{H}_2\text{O}} H_3C(HO)C=C(H)CH_3 \xrightarrow[\text{HgSO}_4]{\text{H}^+/\text{H}_2\text{O}} H_3C-C(OH)_2-CH_2-CH_3$

Intermediate — Diol

Fig. 8.48. Reaction of 2-butyne with aqueous acid and mercuric sulfate.

Actually, a diol is not formed. The intermediate (an enol) undergoes acid-catalyzed rearrangement to give a ketone (Fig. 8.49). This process is called a **keto-enol tautomerism.**

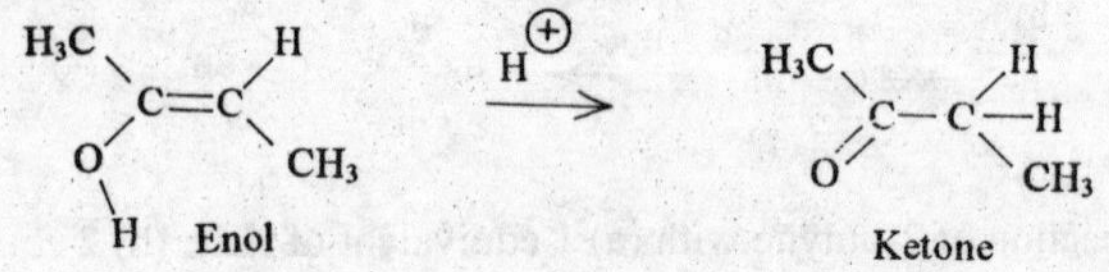

$$\underset{\text{Enol}}{\mathrm{H_3C(HO)C{=}CH(CH_3)}} \xrightarrow{H^{\oplus}} \underset{\text{Ketone}}{\mathrm{H_3C{-}C(=O){-}CH_2(CH_3)}}$$

Fig. 8.49. Keto-enol tautomerism.

Tautomerism is used to describe the rapid interconversion of two different isomeric forms (**tautomers**). In this case the keto and enol tautomers of a ketone. The keto tautomer is by far the dominat species for a ketone and the enol tautomer is generally present in only very small amounts (typically 0.0001%). Therefore, as soon as the enol is formed in the above reaction, it rapidly tautomerized to the keto form and further electrophilic addition does not occur.

Additions to Terminal Alkynes

When a terminal alkyne is treated with an excess of hydrogen halide the halogens both end up on the more substituted carbon (Fig. 8.50). This is in accordance with the Markovnikov's rule which states that the additional hydrogens end up on the carbon which already has the most hydrogens. The same rule applies for the reaction with acid and mercuric sulfate which means that a ketone is formed after keto-enol tautomerism instead of an aldehyde (Fig. 8.51).

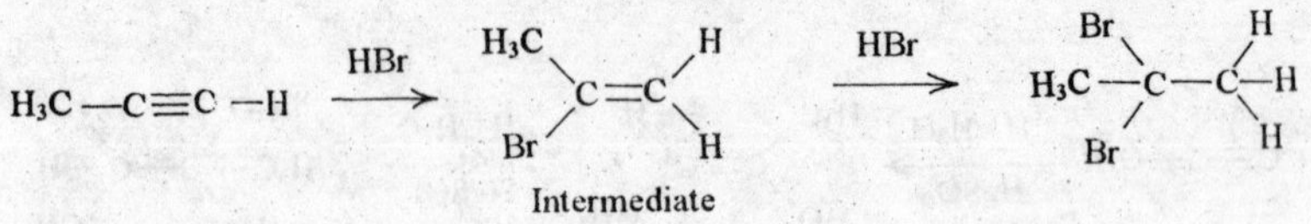

$$\mathrm{H_3C{-}C{\equiv}C{-}H} \xrightarrow{HBr} \underset{\text{Intermediate}}{\mathrm{H_3C(Br)C{=}CH_2}} \xrightarrow{HBr} \mathrm{H_3C{-}CBr_2{-}CH_3}$$

Fig. 8.50. Reaction of propyne with HBr.

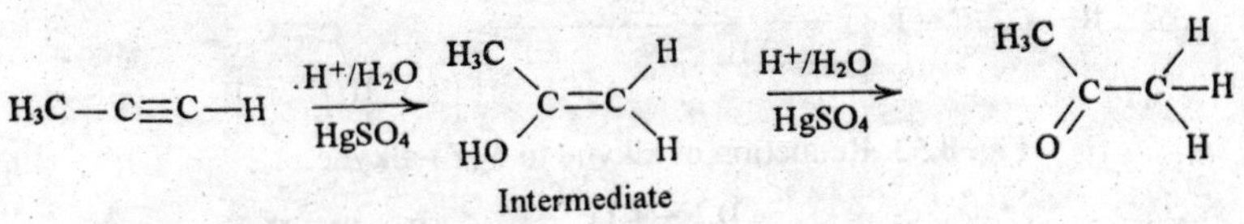

$$H_3C-C\equiv C-H \xrightarrow[HgSO_4]{H^+/H_2O} \underset{\text{Intermediate}}{(H_3C)(HO)C=CH_2} \xrightarrow[HgSO_4]{H^+/H_2O} H_3C-C(=O)-CH_3$$

Fig. 8.51. Reaction of propyne with aqueous acid and mermuric sulfate.

8.9 Reduction of Alkynes

Hydrogenation

Alkynes react with hydrogen gas in the presence of a metal catalyst and the process called *hydrogenation.* It is an example of a **reduction** reaction. With a fully active catalyst like platinum metal, two molecules of hydrogen are added to produce an alkane (Fig. 8.52).

The reaction involves the addition of one molecule of hydrogen to form na alkene intermediate which then reacts with a second molecule of hydrogen to form the alkane. With less active catalysts, it is possible to stop the reaction at the alkene stage. *In particular, (Z)-alkenes can be synthesized from alkynes by reaction with hydrogen gas and Lindlar's catalyst* (Fig. 8.53). This catalyst consists of metallic palladium deposited on calcium carbonate which is then treated with lead acetate and quinoline. The latter treatment 'poisons' the catalyst in such a way that the alkyne reacts with hydrogen to give an alkene, but doe not react further. Since the starting materials are absorbed onto the catalyst surface, both hydrogens are added to the same side of the molecule to produce the (Z) isomer.

The same result can be achieved with nickel boride (Ni_2B) by using the **P-2 catalyst**.

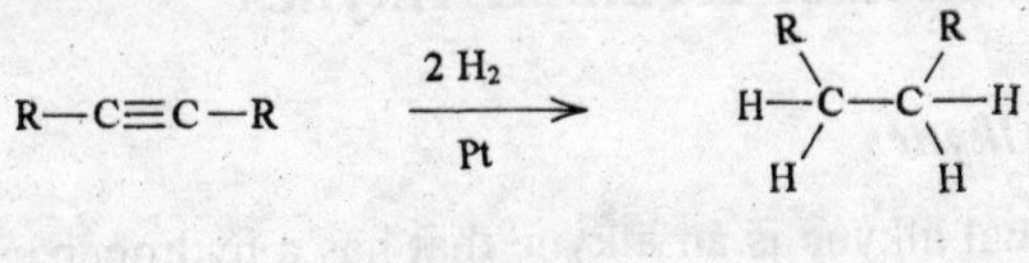

$$R-C\equiv C-R \xrightarrow[Pt]{2\,H_2} H-C(R)(H)-C(R)(H)-H$$

Fig. 8.52. Reduction of an alkyne to an alkene.

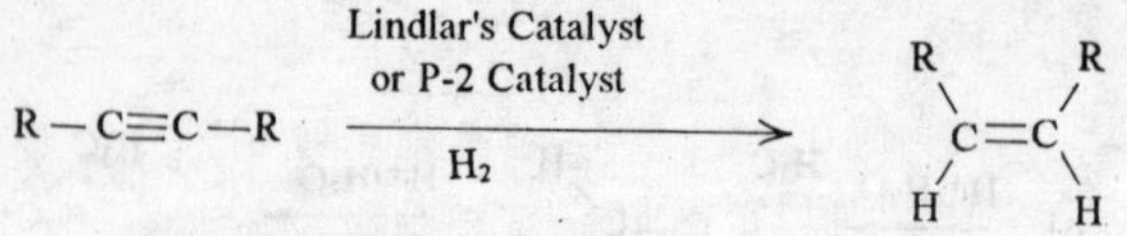

$$R-C\equiv C-R \xrightarrow[H_2]{\text{Lindlar's Catalyst or P-2 Catalyst}} \underset{H\quad\quad H}{\overset{R\quad\quad R}{C=C}}$$

Fig. 8.53. Reduction of alkyne to a (Z)-alkyne.

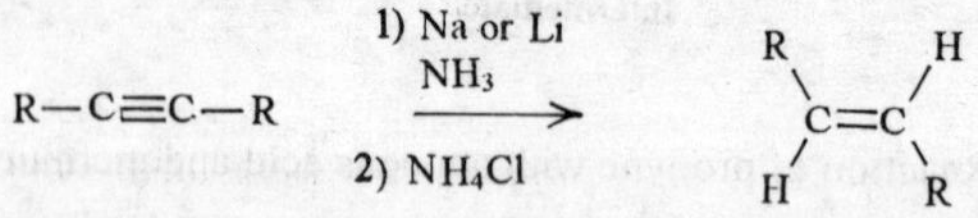

$$R-C\equiv C-R \xrightarrow[2)\ NH_4Cl]{1)\ \text{Na or Li}\ NH_3} \underset{H\quad\quad R}{\overset{R\quad\quad H}{C=C}}$$

Fig. 8.54. Reduction of an alkyne to a E-*alkene.*

Dissolving Metal Reduction

Reduction of an alkyne to an (E)-alkene can be achieved by treating the alkyne with lithium or sodium metal in ammonia at low temperatures (Fig. 8.54). This is called **dissolving metal reduction.**

In this reaction, the alkali metal donates its valence electron to the alkyne to produce a radical anion (Fig. 8.55). This removes a proton from ammonia to produce a *vinylic radical* that receives an electron from a second alkali metal to produce a *trans-vinylic anion.* This anion then removes a proton from a second molecule of ammonia and forms the *trans-* or (E)-alkene. Only half curly arrows are used in the mechanism because this is a **radical reaction** *involving the movement of single electrons.*

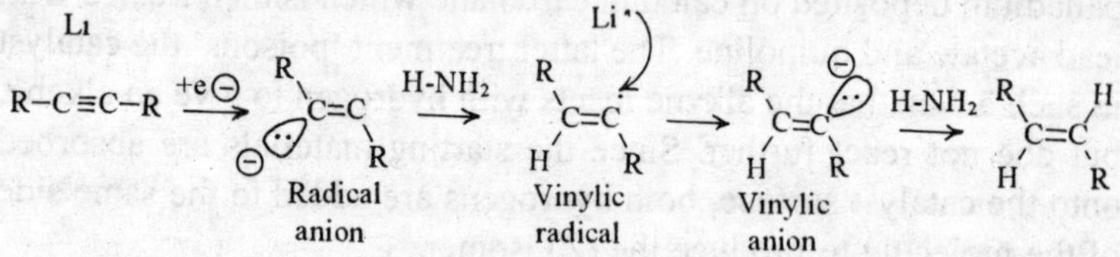

Li
R–C≡C–R →(+e⊖) Radical anion →(H-NH₂) Vinylic radical →(Li•) Vinylic anion →(H-NH₂) (E)-alkene

Fig. 8.55. Mechanism for the dissolving metal reduction of an alkyne.

8.10 Alkylation of Terminal Alkynes

Terminal Alkynes

A terminal alkyne is an alkyne that has a hydrogen substituents (Fig. 8.56). This hydrogen substituent is acidic and can be removed with strong base (e.g. sodium amide $NaNH_2$) to produce an **alkynide** (Fig. 8.57). This is an acid-base reaction.

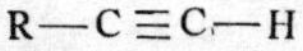

$$R—C \equiv C—H$$

Fig. 8.56. Terminal alkyne.

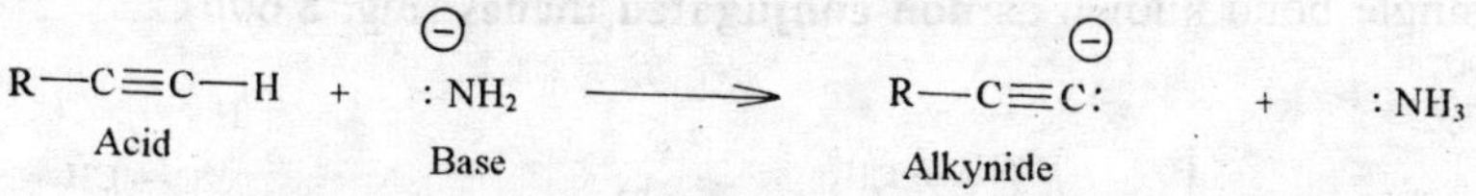

$$\underset{\text{Acid}}{R—C \equiv C—H} + \underset{\text{Base}}{:\overset{\ominus}{N}H_2} \longrightarrow \underset{\text{Alkynide}}{R—C \equiv \overset{\ominus}{C}:} + :NH_3$$

Fig. 8.57. Reaction of a terminal alkyne with a strong base.

Alkylation

Once the alkynide is formed, it can be treated with an alkyl halide to form more complex alkynes (Fig. 8.58). This reaction is called an alkylation and is an example of *nucleophilic substitution.*

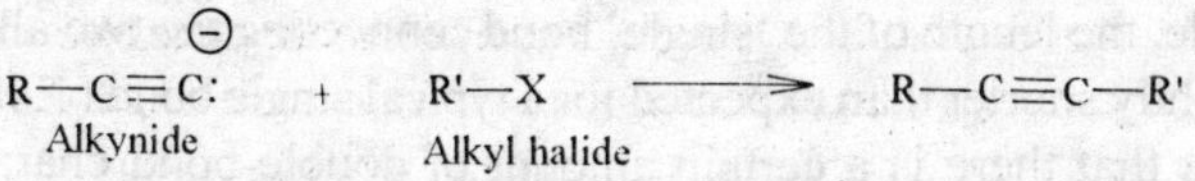

$$\underset{\text{Alkynide}}{R—C \equiv \overset{\ominus}{C}:} + \underset{\text{Alkyl halide}}{R'—X} \longrightarrow R—C \equiv C—R'$$

Fig. 8.58. Reaction of an alkynide ion with an alkyl halide.

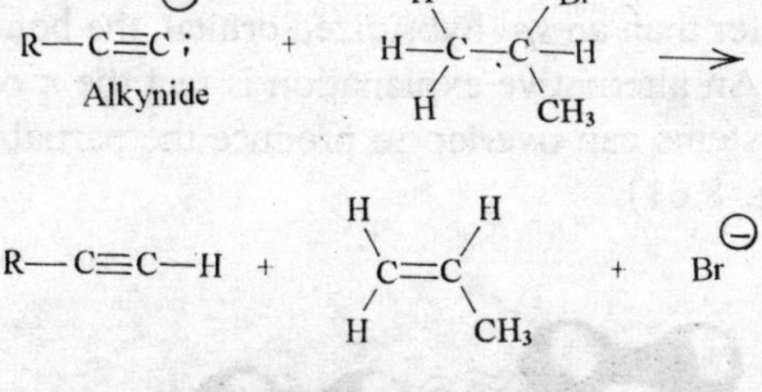

$$\underset{\text{Alkynide}}{R—C \equiv \overset{\ominus}{C}:} + H—\underset{H}{\overset{H}{C}}—\underset{CH_3}{\overset{Br}{C}}—H \longrightarrow$$

$$R—C \equiv C—H + \underset{H}{\overset{H}{C}}=\underset{CH_3}{\overset{H}{C}} + Br^{\ominus}$$

Fig. 8.59. Elimination of HBr.

This reaction works best with primary alkyl halides. When secondary or tertiary alkyl halides are used, the alkynide reacts like a base and this results in elimination of hydrogen halide from the alkyl halide to produce an alkene (Fig. 8.59).

8.11 Conjugated Dienes

Structure

A **conjugate diene** is made up of two alkene units separated by a single bond (Fig. 8.60a). Dienes are separated by more than one single bond known as **non-conjugated dienes** (Fig. 8.60b).

```
                                                                   H
                                                                    \
a)        H     CH2— CH3                b)                           C—CH3
           \   /                                                    //
  H         C=C                          H        CH2—C
   \       /   \                          \      /     \
    C=C         H                          C=C          H
   /   \                                  /   \
H3C     H                              H3C     H
```

Fig. 8.60. (a) Conjugated diene; (b) nonconjugated diene.

Bonding

A conjugated diene does not behave like two isolated alkenes. For example, the length of the 'single' bond connecting the two alkene units is slightly shorter than expected for a typical single bond (1.48Å). This shows that there is a certain amount of double-bond character present in this bond. Two sp^2 hybridized carbons rather than two sp^3 hybridized carbons. Therefore, an sp^2 hybridized orbital from each carbon is used for the single bond. Since this hybridized orbital has more s-character than an sp^3 hybridized orbital, the bond is expected to be shorter. An alternative explanation is that the π orbitals of the two alkene systems can overlap to produce the partial double-bond character (Fig. 8.61).

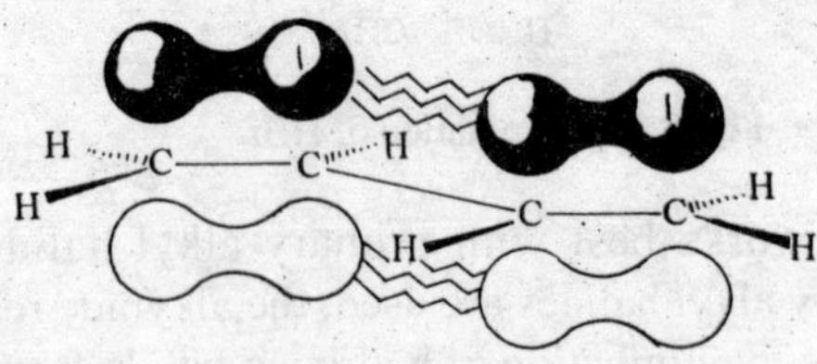

Fig. 8.61. π-Orbital overlap.

Electrophilic Addition

The reactions of a conjugate diene reflect the fact that a conjugated diene should be viewed as a functional group in its own right, rather than as two separate alkenes. Electrophilic addition to a conjugated diene results in a mixture of two possible products arising from **1,2-addition** and **1,4 addition** (Fig. 8.62).

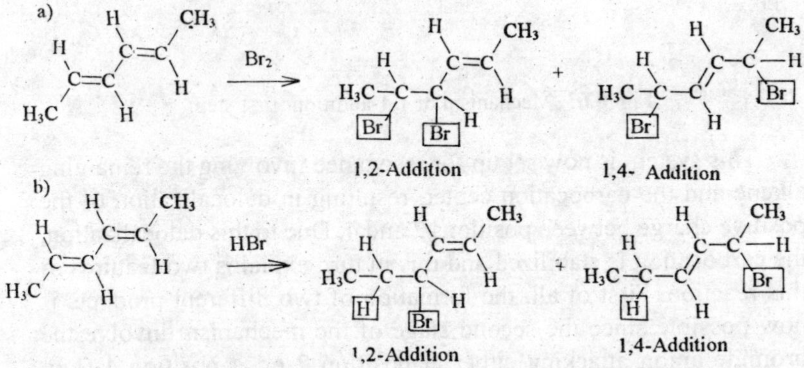

a) Br_2 → 1,2-Addition + 1,4- Addition

b) HBr → 1,2-Addition + 1,4-Addition

Fig. 8.62. Electrophilic addition to a conjugated diene of (a) bromine and (b) HBr.

In 1,2-addition, new atoms have been added to each end of one of the alkene units. This is the normal electrophilic addition of an alkene with which we are familiar. In 1,4-addition, new atoms have been added to each end of the entire diene system. Furthermore, the double bond remaining has shifted position (isomerized) to the 2,3-position.

The mechanism of 1,4-addition starts off in the same way as a normal electrophilic addition. We shall consider the reaction of a conjugated diene with hydrogen bromide as an example (Fig. 8.63). One of the alkene units of the diene uses its π electrons to form a bond to the electrophilic hydrogen of HBr. The H–Br bond breaks at the same time to produce a bromide ion. The intermediate carbocation produced has a double bond next to the carbocation center and is called an **allylic carbocation**.

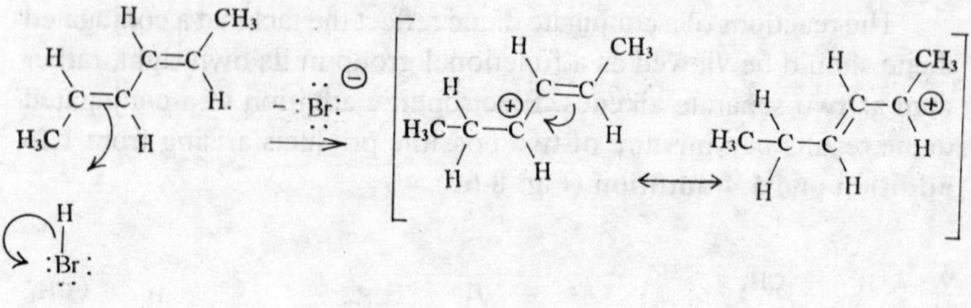

Fig. 8.63. Mechanism of 1,4-addition–first step.

This system is now set up for resonance involving the remaining alkene and the carbocation center, resulting in delocalization of the positive charge between positions 2 and 4. Due to this delocalization, the carbocation is stabilized and this in turn explains two features of this reaction. First of all, the formation of two different products is now possible since the second stage of the mechanism involve the bromide anion attacking either at position 2 or at position 4 (Fig. 8.64). Secondly, it explains why the alternative 1,2-addition product is not formed (Fig. 8.65). The intermediate carbocation required for this 1,2-addition cannot be stabilized by resonance. Therefore, the reaction proceeds through the allyiic carbocation instead.

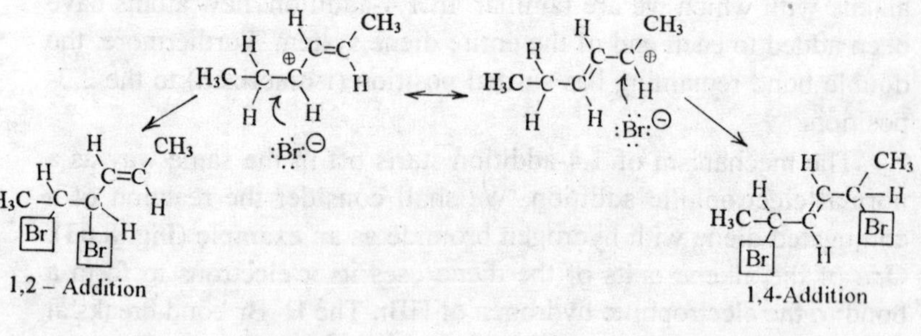

1,2 – Addition

1,4-Addition

Fig. 8.64. Mechanism of 1,2- and 1,4-addition – second step.

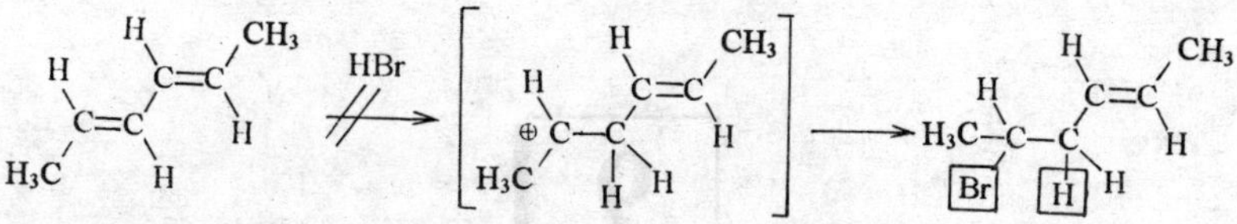

Fig. 8.65. Unfavoured reaction mechanism.

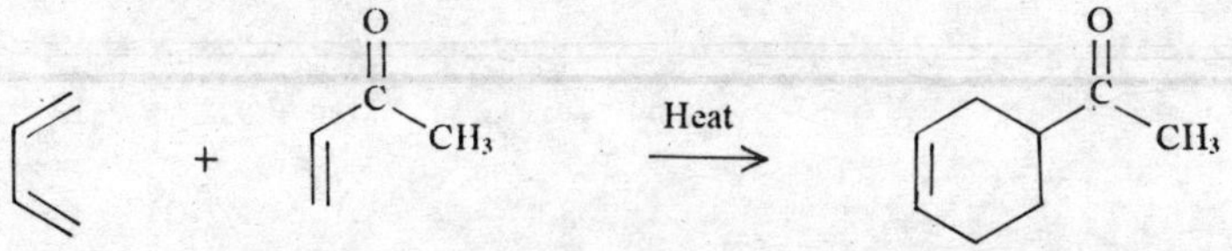

Fig. 8.66. Diels-Alder cycloaddition.

Diels-Alder Cycloaddition

The diels-Alder cycloaddition reaction is an important reaction by which six-numbered rings can be synthesized. It involves a *conjugated diene* and an *alkene* (Fig. 8.66). The alkene is called a **dienophile** (diene-lover) and generally has an electron-withdrawing group linked to it (e.g. a carbonyl group or a nitrile). The *mechanism is concerted* with new bonds being formed simultaneously as old bonds are being broken (Fig. 8.67). No intermediates are involved.

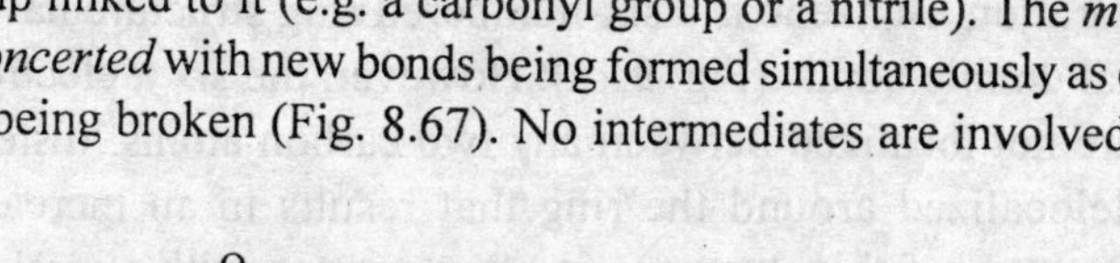

Fig. 8.67. Mechanism of Diels-Alder cycloaddition.

9

Aromatic Chemistry

9.1 Aromaticity

Definition

Originally term **aromatic** was used for benzene-like structures because of the distinctive aroma of such compounds. Aromatic compounds undergo certain distinctive reactions that set them apart from other functional groups. They are *highly unsaturated compounds*, but unlike alkenes and alkynes, they are *relatively unreactive* and tend to undergo reactions that involve a retention of their unsaturation. Benzene is a six-numbered ring structure having three formal double bonds (Fig. 9.1a). However, the six π electrons involved are not localized between any two carbon atoms. Instead, they are delocalized around the ring that results in an increased stability. Because of this, benzene is often written with a circle in the center of the ring to signify the delocalization of the six π electrons (Fig. 9.1b). Reactions that disrupt this delocalization are not favored as it means a loss of stability, so benzene undergoes reactions in which the aromatic ring system is retained. All six carbon *atoms* in benzene are sp^2 hybridized, and the molecule itself is cyclic and planar. The planarity is essential if the $2p$ atomic orbitals on each carbon atom are to overlap and result in delocalization.

Fig. 9.1. Representations of benzene.

Hückel Rule

An **aromatic molecule** must be *cyclic* and *planar with* sp^2 *hybridized atoms* (i.e. conjugated), but it must also obey the **HÜckel rule**. It states, that *the ring system must have* $4n + 2$ π *electrons* where $n = 1, 2, 3,$ *etc*. Therefore, ring systems which have 6, 10, 14, ... π electrons are aromatic. Benzene fits the Hückel rule as it has six π electrons. **Cyclooctatetraene** has eight π electrons and *does not obey the Hückel rule*. Although all the carbon atoms in the ring are sp^2 hybridized, **cyclooctateraene reacts like a conjugated alkene.** It is not planar, the π electrons are not delocalized and the molecule is made up of alternating single and double bonds (Fig. 9.2a). However, the 18-membered cyclic system (Fig. 9.2b) obeys the Hückel rule ($n = 4$) and is a planar molecule with aromatic properties and a delocalized π system.

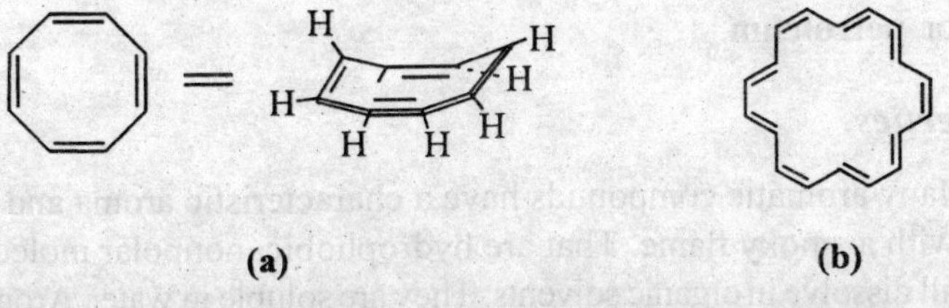

(a) (b)

Fig. 9.2. (a) Cyclooctatetraene; (b) 18-membered aromatic ring.

We can also possible to get aromatic ring. The *cyclopentadienyl anion* and the *cycloheptatrienyl cation* are both **aromatic** (Fig. 9.3). Both are cyclic and planar, containing six π electrons, and all the atoms in the ring are sp^2 hybridized.

Bicyclic and polycyclic systems can also be aromatic (Fig. 9.4).

Fig. 9.3. (a) Cyclopentadienyl anion; (b) cycloheptatrienyl cation.

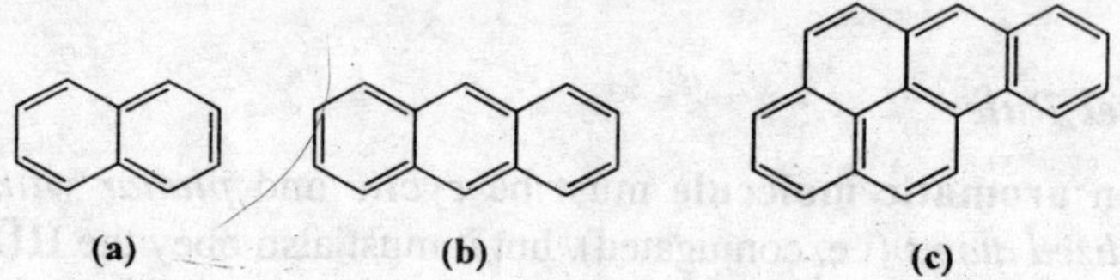

Fig. 9.4. (a) Naphthalene; (b) anthracene; (c) benzo[a]pyrene.

9.2 Preparation and Properties of Aromatic Compounds

Preparation

It is difficult to synthesize aromatic compounds in the laboratory from scratch and most aromatic compounds are prepared from benzene or other simple aromatic compounds (e.g. toluene and naphthalene). These in turn are isolated from natural sources like **coal** or **petroleum**.

Properties

Many aromatic compounds have a characteristic aroma and they burn with a smoky flame. That are hydrophobic, nonpolar molecules and will dissolve in organic solvents. They are soluble in water. Aromatic molecules can interact with each other through *intermolecular bonding by van der Waals interactions* (Fig. 9.5a). However, induced dipole interactions are also possible with alkyl ammonium ions or metal ions where the positive charge of the cation induces a dipole in the aromatic ring such that the face of the ring is slightly negative and the edges are slightly positive (Fig. 9.5b). This results in the cation being sandwiched between two aromatic rings.

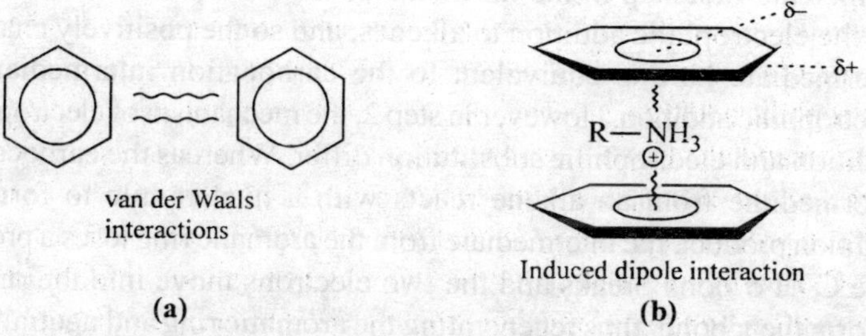

Fig. 9.5. Intermolecular bonding involving aromatic rings.

Aromatic compounds are generally stable and they do not react like alkenes. They prefer to undergo reactions in which the stable aromatic ring is retained. The most common type of reaction for aromatic rings is **electrophilic substitution**, but **reduction** is also possible.

9.3 Electrophilic Substitutions of Benzene

Definition

Aromatic rings undergo **electrophilic substitution**, for example the bromination of benzene (Fig. 9.6). The reaction involves an electrophile (Br^+) replacing another electrophile (H^+) with the aromatic ring remaining intact. Therefore, one electrophile replaces another and the reaction is called an **electrophilic substitution**.

$$C_6H_5\text{–}H + Br^{\oplus} \longrightarrow C_6H_5\text{–}Br + H^{\oplus}$$

Fig. 9.6. Electrophilic substitution of benzene.

Mechanism

In the mechanism of electrophilic substitution reactions (Fig. 9.7) the aromatic ring acts as a nucleophile and it provides two of its π electrons to form a bond to Br^+. The aromatic ring now has lost one of its formal double bonds which results in a positively charged carbon atom. This first step in the mechanism is the same as that described for the electrophilic addition to alkenes, and so the positively charged intermediate here is equivalent to the carbocation intermediate in electrophilic addition. However in step 2, the mechanism of electrophilic addition and electrophilic substitution differ. Whereas the carbocation intermediate from an alkene reacts with a nucleophile to form an addition product, the intermediate from the aromatic ring loses a proton. The C–H σ bond breaks and the two electrons move into the ring to reform the π bond, thus regenerating the aromatic ring and neutralizing the positive charge on the carbon. This is the mechanism undergone in all electrophilic substitutions. The only difference is the nature of the electrophile (Fig. 9.8).

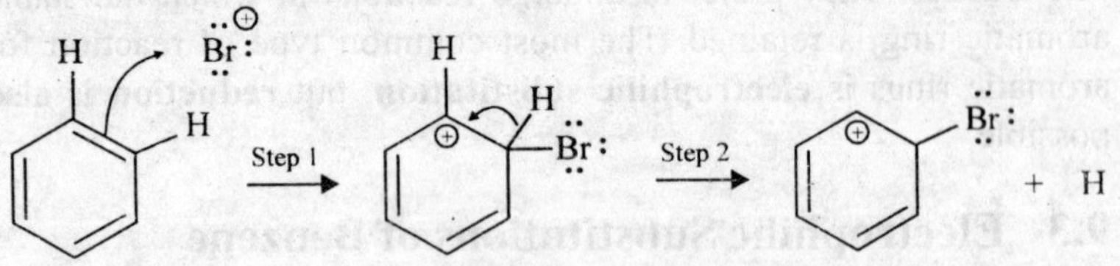

Step 1 Step 2 Intermediate

Fig. 9.7. Mechanism of electrophilic substitution.

$Br^{\oplus}$ $O=\overset{\oplus}{N}=O$ $O_2S=O$ $R_2\overset{\oplus}{C}-R$ $R-\overset{\oplus}{C}=O$

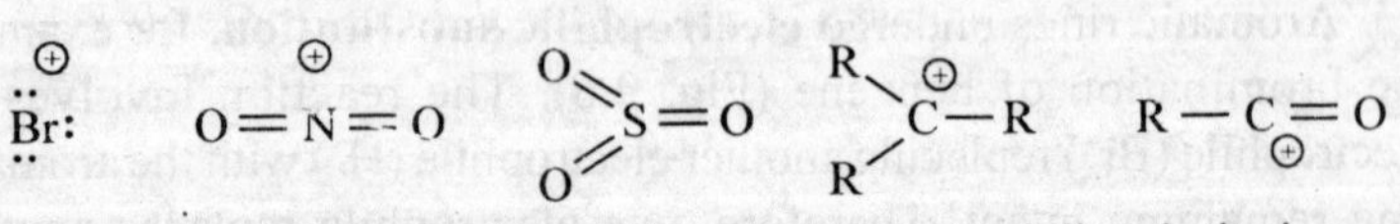

Fig. 9.8. Examples of electrophiles used in electrophilic substitution.

Intermediate Stabilization

The **rate-determining step** in the electrophilic substitution is the *formation of the positively charged intermediate*, and so the rate of the reaction is determined by the energy level of the transition state leading to that intermediate. The transition state resembles the intermediate in character and so any factor stabilizing the intermediate also stabilizes the transition state and lowers the activation energy needed for the reaction. Therefore, electrophilic substitution is more likely to occur if the positively charged intermediate can be stabilized. Stabilization is possible if the positive charge can be spread amongst different atoms–i.e. by **delocalization**. The process by which this can occurs is called **resonance** (Fig. 9.9).

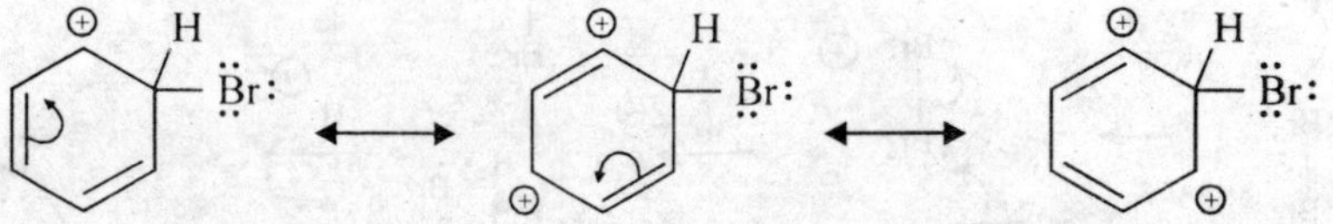

Fig. 9.9. Resonance stabilization of the charged intermediate.

The **resonance** process involves two π electrons shifting their position round the ring to provide the 'top' carbon with a fourth bond and thus neutralize its positive charge. In the process, another carbon in the ring is left short of bonds and gains the positive charge. This process can be repeated such that the positive charge is spread to a third carbon. The structures drawn in Fig. 9.9 are called **resonance structures** (Canonical forms).

Halogenation

The stable aromatic ring means that aromatic compounds are less reactive than alkenes to electrophiles. For example, an alkene will react with Br_2, but an aromatic ring will not. Therefore, we have to activate the aromatic ring (i.e. make it a better nucleophile) or activate the Br_2 (i.e. make it a better electrophile) if we want a reaction to take place. The electron-donating substituents on an aromatic ring increase the nucleophilicity of the aromatic ring. A Br_2 molecule can be activated to make it a better electrophile by adding a Lewis acid like $FeCl_3$, $FeBr_3$, or $AlCl_3$ to the reaction medium. These compounds contain a central atom (iron or aluminum) that is strongly electrophilic and does not have a full valence shell of electrons. Due to this, the central atom can accept a lone pair of electrons, even from a weakly nucleophilic atom like a halogen. For example (Fig. 9.10) bromine uses a lone pair of electrons to form a bond to the Fe atom in $FeBr_3$ and becomes positively charged. Bromine is now activated to act as an electrophile and will react more easily with a nucleophile (the aromatic ring) by the normal mechanism for electrophilic substitution.

An aromatic ring can be chlorinated in a similar manner, making use of Cl_2 in the presence of $FeCl_3$.

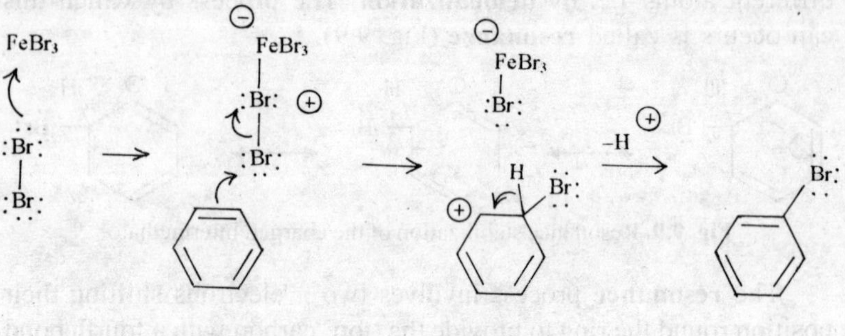

Fig. 9.10. Mechanism by which a Lewis acid activates bromine towards electrophilic substitution.

Friedel-Crafts Alkylation and Acylation

Friedel-Crafts alkylation and acylation (Fig. 9.11) are two other examples of *electrophilic substitution requiring the presence of a Lewis acid*. These are particularly important as they allow the synthesis of larger organic molecules by adding alkyl (R) or acyl (RCO) side chains to an aromatic ring.

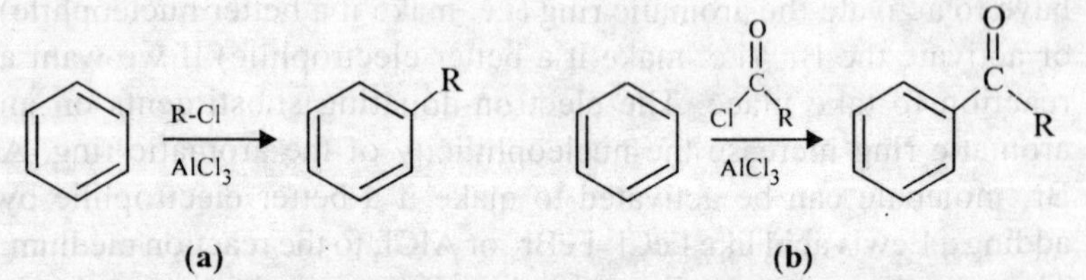

Fig. 9.11. (a) Friedel-Crafts alkylation; (b) Friedel-Crafts acylation.

The example of Friedel-Crafts alkylation is the reaction of benzene with 2-chloropropane (Fig. 9.12). The Lewis acid ($AlCl_3$) promotes the formation of the carbocation needed for the reaction and does so by accepting a lone pair of electrons from chlorine to form an unstable intermediate that fragments to give a carbocation and $AlCl_4^-$ (Fig. 9.13). Once the carbocation is formed it reacts as an electrophile with the aromatic ring by the electrophilic substitution mechanism already described (Fig. 9.14).

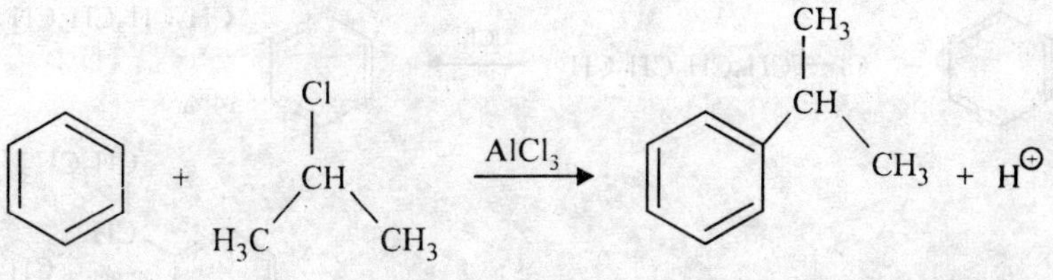

Fig. 9.12. Friedel-Crafts reaction of benzene with 2-chloropropane.

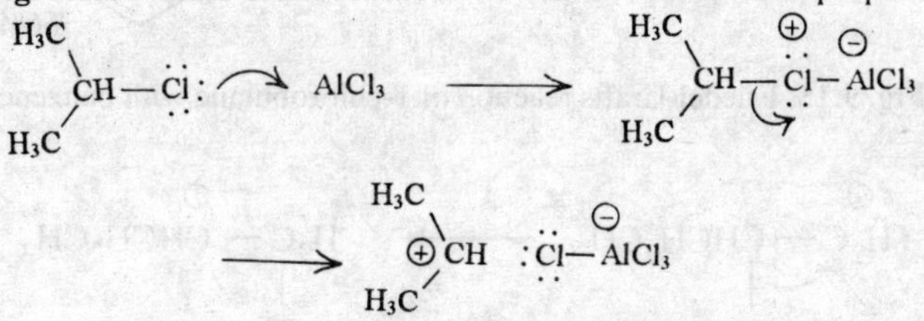

Fig. 9.13. Mechanism of carbocation formation.

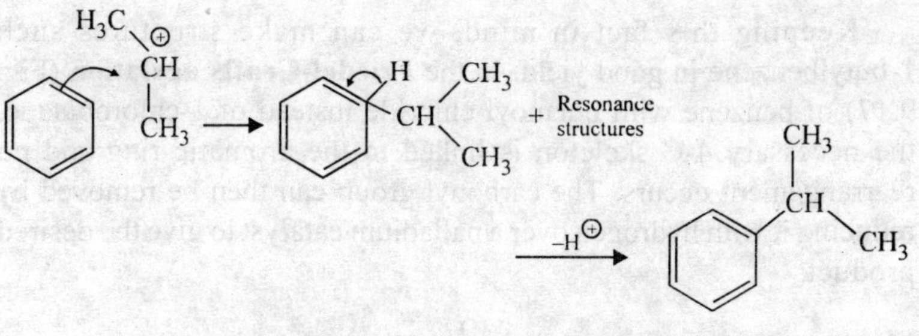

Fig. 9.14. Mechanism for the Friedel-Crafts alkylation.

Following are the limitations to the Friedel-Crafts alkylation. For example, the reaction of 1-chlorobutane with benzene gives two products with only 34% of the desired product (Fig. 9.15). This is because of the fact that the primary carbocation that is generated can rearrange to a more stable secondary carbocation where a hydrogen (and the two sigma electrons making up the C–H bond) 'shift' across to the neighboring carbon atom (Fig. 9.16). This is called a **hydride shift** and it occurs because the secondary carbocation is more stable than the primary carbocation. Such rearrangements limit the type of alkylations that can be carried out by the Friedel-Crafts reaction.

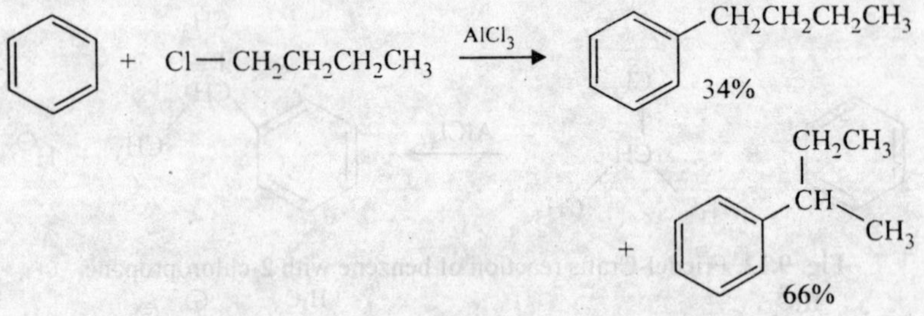

$$C_6H_6 + Cl-CH_2CH_2CH_2CH_3 \xrightarrow{AlCl_3} C_6H_5CH_2CH_2CH_2CH_3\ (34\%) + C_6H_5CH(CH_3)CH_2CH_3\ (66\%)$$

Fig. 9.15. Friedel-Crafts reaction of 1-chlorobutane with benzene.

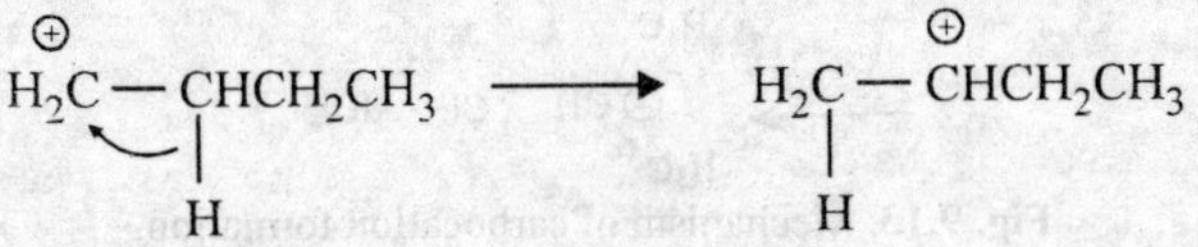

$$H_2\overset{\oplus}{C}-CH(H)CH_2CH_3 \longrightarrow H_2C(H)-\overset{\oplus}{C}HCH_2CH_3$$

Fig. 9.16. Hydride shift.

Keeping this fact in mind, we can make structures such 1-butylbenzene in good yield. In the **Friedel-Crafts acylation** (Fig. 9.17) of benzene with butanoyl chloride instead of 1-chlorobutane, the necessary 4-C skeleton is linked to the aromatic ring and no rearrangement occurs. The carbonyl group can then be removed by reducing it with hydrogen over a palladium catalyst to give the desired product.

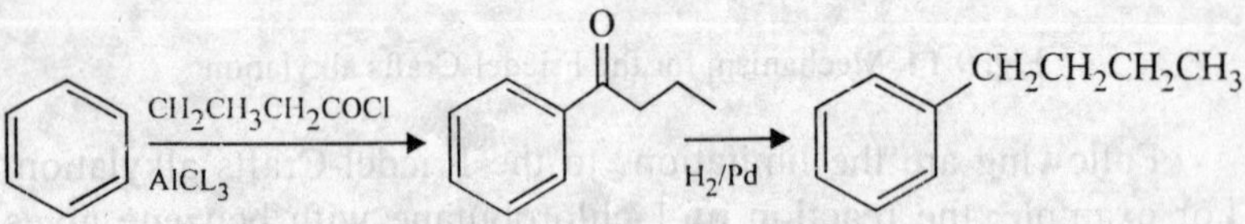

$$C_6H_6 \xrightarrow[AlCl_3]{CH_3CH_2CH_2COCl} C_6H_5COCH_2CH_2CH_3 \xrightarrow{H_2/Pd} C_6H_5CH_2CH_2CH_2CH_3$$

Fig. 9.17. Synthesis of 1-butylbenzene by Friedel-Crafts acylation and reduction.

The mechanism of the Friedel-Crafts acylation is the same as the Friedel-Crafts alkylation. It involves an **acylium** ion instead of a carbocation. Like Friedel-Crafts alkylation, a Lewis acid is needed to generate the acylium ion $(R-C \equiv O)^+$, but unlike a carbocation the acylium ion does not rearrange since there is resonance stabilization from the oxygen (Fig. 9.18).

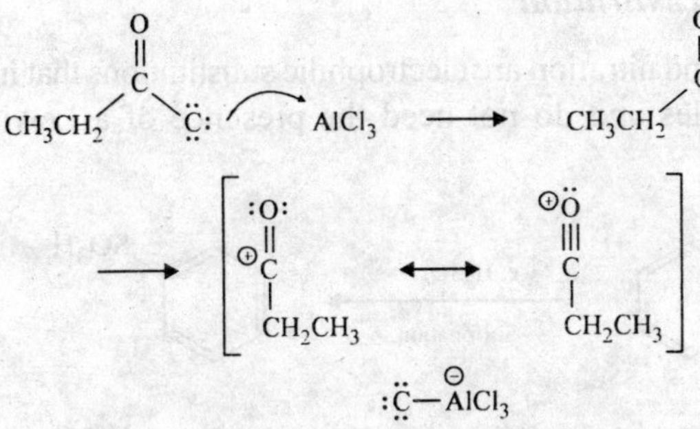

$$CH_3CH_2-C(=O)-\ddot{C}l: + AlCl_3 \longrightarrow CH_3CH_2-C(=O)-\overset{\oplus}{C}l-\overset{\ominus}{A}lCl_3$$

$$\longrightarrow \left[{}^{\oplus}C(=\ddot{O}:)-CH_2CH_3 \longleftrightarrow {}^{\oplus}\ddot{O}\equiv C-CH_2CH_3 \right] \quad :\ddot{C}l-\overset{\ominus}{A}lCl_3$$

Fig. 9.18. Generation of the acylium ion.

Friedel-Crafts alkylations can also be done using alkenes instead of alkyl halides. A Lewis acid is not needed, but a mineral acid is required. Treatment of the alkene with the acid leads to a carbocation that can then react with an aromatic ring by the electrophilic substitution mechanism already described (Fig. 9.19). For an alkene, this is another example of electrophilic addition where a proton is attached to one end of the double bond and a phenyl group is added to the other.

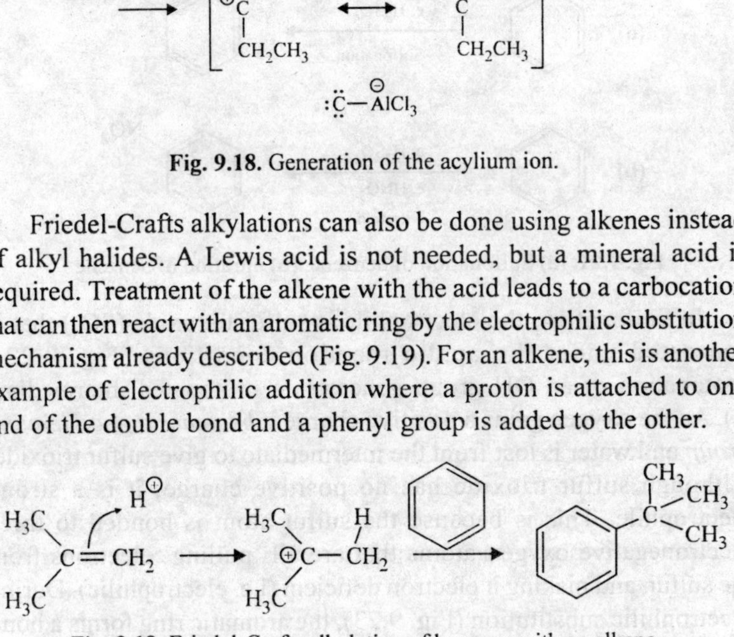

$$(H_3C)_2C=CH_2 \xrightarrow{H^{\oplus}} (H_3C)_2\overset{\oplus}{C}-CH_2H \xrightarrow{C_6H_6} C_6H_5-C(CH_3)_3$$

Fig. 9.19. Friedel-Crafts alkylation of benzene with an alkene.

Friedel-Crafts reactions can also be done with alcohols in the presence of mineral acid. The acid leads to the elimination of water from the alcohol resulting in the formation of an alkene that can then be converted to a carbocation (Fig. 9.20).

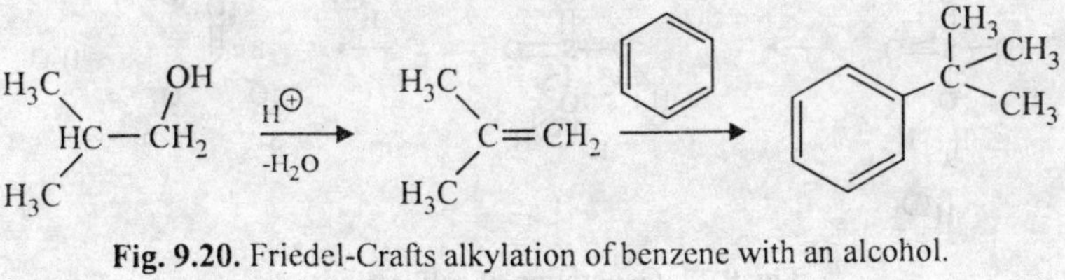

$$(H_3C)_2HC-CH_2OH \xrightarrow[-H_2O]{H^{\oplus}} (H_3C)_2C=CH_2 \xrightarrow{C_6H_6} C_6H_5-C(CH_3)_3$$

Fig. 9.20. Friedel-Crafts alkylation of benzene with an alcohol.

Sulfonation and Nitration

Sulfonation and nitration are electrophilic substitutions that involve strong electrophiles and do not need the presence of a Lewis acid (Fig. 9.21).

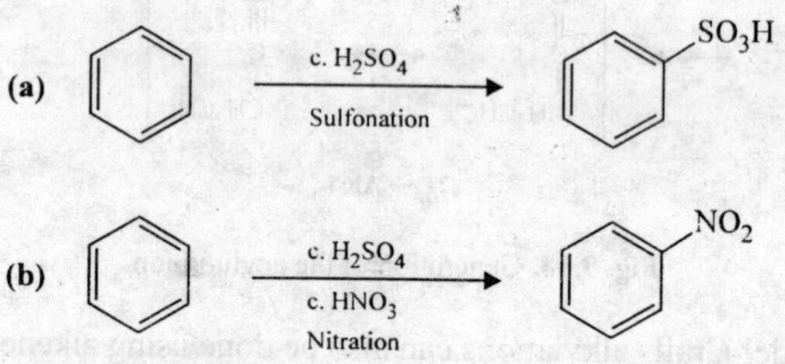

(a) benzene →(c. H_2SO_4, Sulfonation)→ benzene-SO_3H

(b) benzene →(c. H_2SO_4, c. HNO_3, Nitration)→ benzene-NO_2

Fig. 9.21. (a) Sulfonation of benzene; (b) nitration of benzene.

In Sulfonation, the electrophile is sulfur tetroxide (SO_3) that is generated under the acidic reaction conditions (Fig. 9.22). Protonation of an OH group generates a protonated intermediate (I). As the oxygen gains a positive charge it *becomes a good leaving group* and water is lost from the intermediate to give sulfur trioxide. Although sulfur trioxide has no positive charge, it is a strong electrophile. This is because the sulfur atom is bonded to three electronegative oxygen atoms that are all 'pulling' electrons from the sulfur, and making it electron deficient (i.e. electrophilic). During electrophilic substitution (Fig. 9.23), the aromatic ring forms a bond to sulfur and one of the π bonds between sulfur and oxygen is broken. Both electrons move to the more electronegative oxygen top form a third lone pair and produce a negative charge on that oxygen. This finally gets neutralized when the third lone pair of electrons is used to form a bond to a proton.

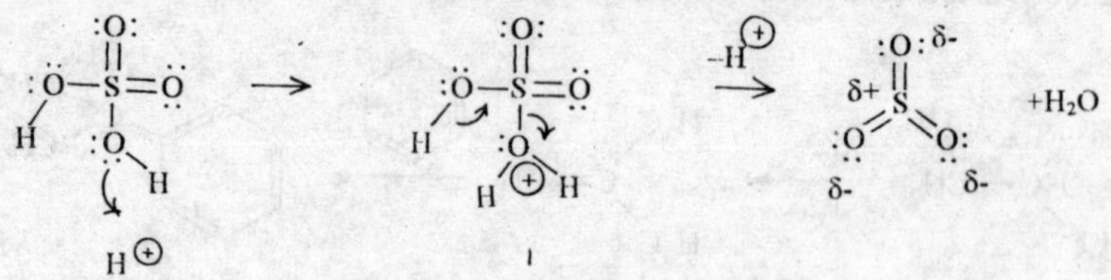

H_2SO_4 + $H^{\oplus}$ → I (protonated H_2SO_4, $-OH_2^{\oplus}$) →($-H^{\oplus}$)→ SO_3 (δ+ on S, δ- on each O) + H_2O

Fig. 9.22. Generation of sulfur trioxide.

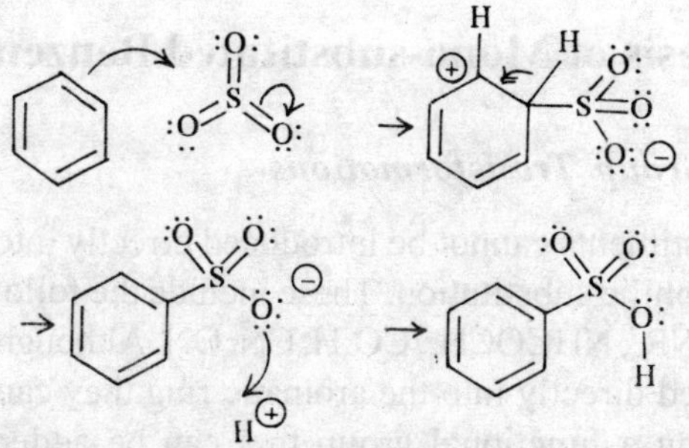

Fig. 9.23. Sulfonation of benzene.

In nitration, sulfuric acid acts as an *acid catalyst* for the formation of a nitronium ion (NO_2^+) that is generated from nitric acid by a similar mechanism to that used in the generation of sulfur trioxide from sulfuric acid (Fig. 9.24).

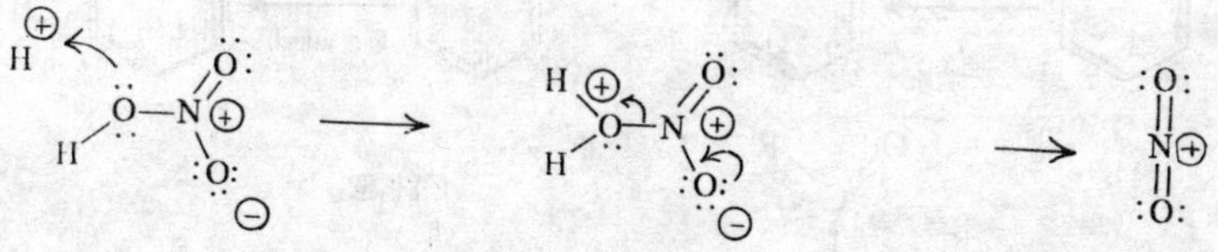

Fig. 9.24. Generation of the nitronium ion.

The mechanism for the nitration of benzene is similar to sulfonation (Fig. 9.25). As the aromatic ring forms a bond to the electrophilic nitrogen atom, a π bond between N and O breaks and both electrons move onto the oxygen atom. Unlike sulfonation, this oxygen keeps its negative charge and does not pick up a proton. This is because it acts as a counterion to the neighboring positive charge on nitrogen.

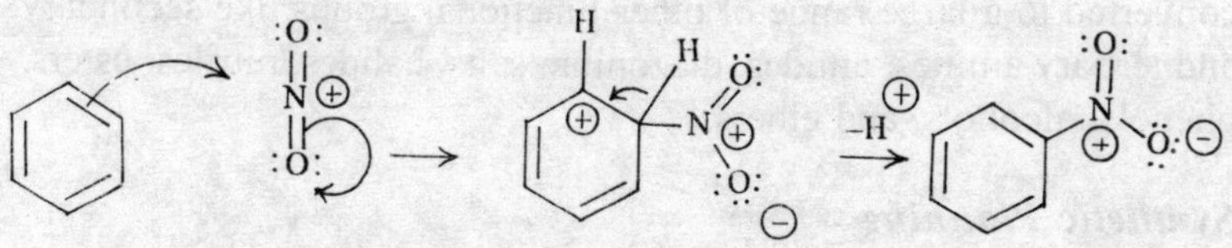

Fig. 9.25. Nitration of benzene.

9.4 Synthesis of Mono-substituted Benzenes

Functional Group Transformations

Some substituents cannot be introduced directly into an aromatic ring by electrophilic substitution. These include the following groups: $-NH_2$, $-NHR$, NR_2, $NHCOCH_3$, CO_2H, CN, OH. Although these groups cannot be added directly into the aromatic ring they can be obtained by transforming a functional group that can be added directly by electrophilic substitution. Some of the most important transformations are shown (Fig. 9.26).

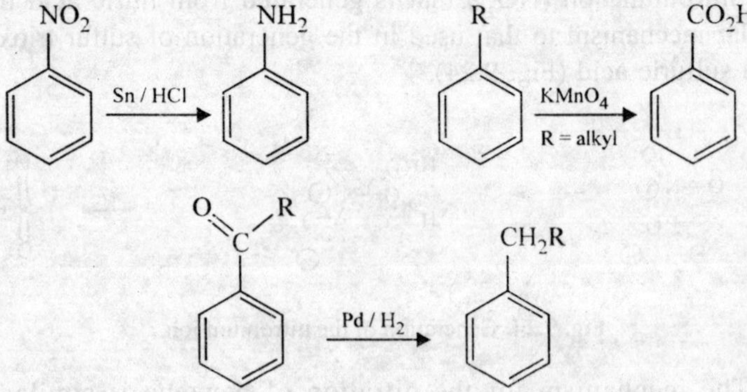

NO_2 → (Sn / HCl) → NH_2 R → ($KMnO_4$, R = alkyl) → CO_2H

RCO → (Pd / H_2) → CH_2R

Fig. 9.26. Functional group transformations of importance in aromatic chemistry.

Nitro ($-NO_2$), alkyl (-R) and acyl (RCO-) groups can readily be added by electrophilic substitution and can then be converted to amino, carboxylic acid, and alkyl groups respectively. Once the amino and carboxylic acid groups have been obtained, they can be further converted to a large range of other functional groups like secondary and tertiary amines, amides, diazonium salts, halides, nitriles, esters, phenols, alcohols, and ethers.

Synthetic Planning

In planning, the synthesis of an aromatic compound, it is best to work backwards from the products and to ask what it could have been synthesized from–a process known as **retrosynthesis.** To illustrate this, consider the synthesis of an aromatic ester (Fig. 9.27). An ester functional group cannot be attached directly by electrophilic substitution, so the synthesis must involve various steps. The ester can be prepared from an acid chloride which can be synthesized from a carboxylic acid. Alternatively, the ester can able made directly from the carboxylic acid by treating it with an alcohol and an acid catalyst. Either way, benzoic acid is required to synthesize the ester. Carboxylic acids cannot be added directly to aromatic rings either, so we have to look for a different functional group that can be added directly, then transformed to a carboxylic acid. A carboxylic acid group can be obtained from the oxidation of a methyl group. Methyl groups can be added directly by *Friedel-Crafts alkylation*. Therefore a possible synthetic route would be as shown in Fig. 9.27.

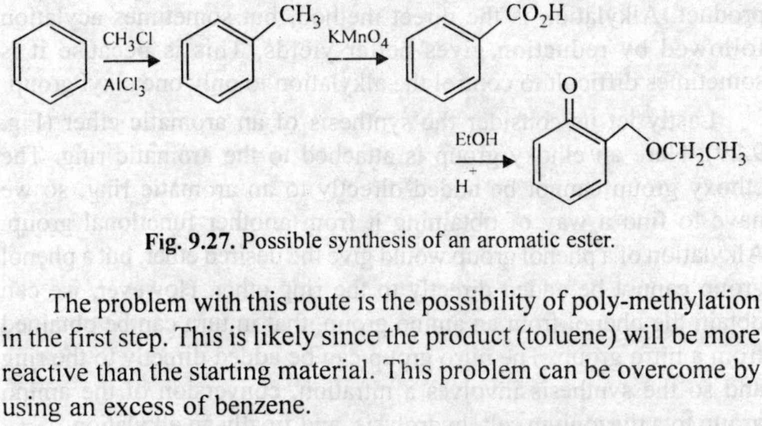

CH_3Cl, $AlCl_3$ → CH_3 ; $KMnO_4$ → CO_2H ; EtOH, H^+ → O, OCH_2CH_3

Fig. 9.27. Possible synthesis of an aromatic ester.

The problem with this route is the possibility of poly-methylation in the first step. This is likely since the product (toluene) will be more reactive than the starting material. This problem can be overcome by using an excess of benzene.

Again consider the synthesis of an aromatic amine (Fig. 9.28). The alkylamine group cannot be applied to an aromatic ring directly and so must be obtained by modifying another functional group. The alkyamine group could be obtained by alkylation of an amino group (NH_2). An amino group cannot be directly applied to an aromatic ring either. However, an amino group could be obtained by reduction of a nitro group. A nitro group can be applied directly to an aromatic ring. Thus, the overall synthesis would be nitration followed by reduction, followed by alkylation.

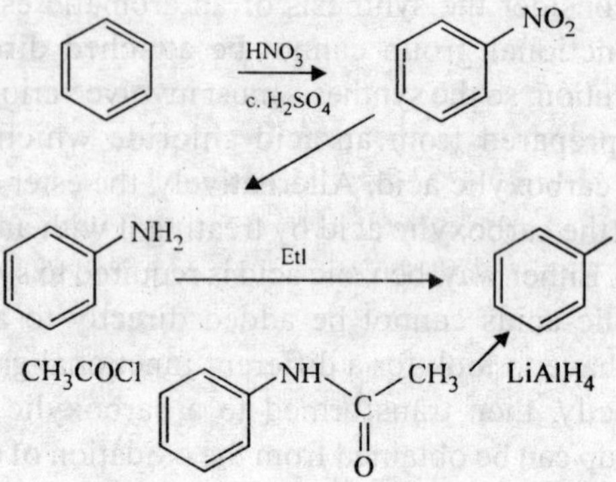

NO_2

HNO_3 / c. H_2SO_4

NH_2 — EtI → $NHCH_2CH_3$

CH_3COCl — NH–C(=O)–CH_3 — $LiAlH_4$

Fig. 9.28. Possible synthetic of an aromatic amine.

There are two methods of converting aniline ($PhNH_2$) to the final product. Alkylation is the direct method, but sometimes acylation followed by reduction gives better yields. This is because it is sometimes difficult to control the alkylation to only one alkyl group.

Lastly let us consider the synthesis of an aromatic ether (Fig. 9.29). Here an ethoxy group is attached to the aromatic ring. The ethoxy group cannot be added directly to an aromatic ring, so we have to find a way of obtaining it from another functional group. Alkylation of a phenol group would give the desired ether, but a phenol group cannot be added directly to the ring ether. However, we can obtain the phenol from an amino group, that in turn can be obtained from a nitro group. The nitro group can be added directly to the ring and so the synthesis involves a nitration, conversion of the amino group to a diazonium salt, hydrolysis, and finally an alkylation.

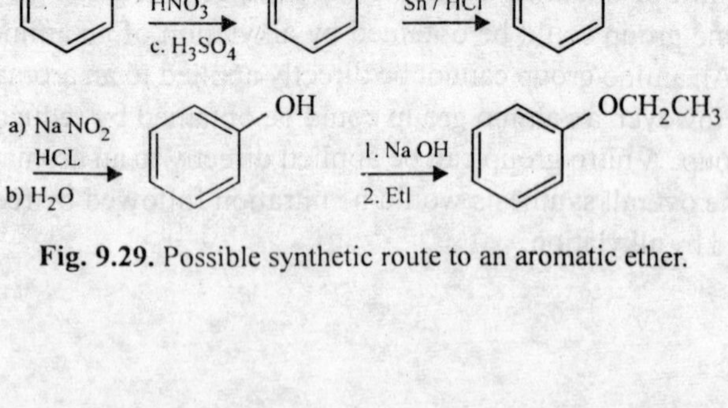

HNO_3 / c. H_2SO_4 → NO_2 — Sn / HCl → NH_2

a) Na NO_2 / HCL, b) H_2O → OH — 1. Na OH, 2. EtI → OCH_2CH_3

Fig. 9.29. Possible synthetic route to an aromatic ether.

9.5 Electrophilic Substitutions of Mono-substituted Aromatic Rings

Ortho, Meta and Para Substitution

An aromatic compounds containing a substituent can undergo electrophilic substitution at three different positions relative to the substituent. For example in the bromination of toluene (Fig. 9.30), three different products are possible depending on where the bromine enters the ring. These products have the same molecular formula and are therefore **constitutional isomers**. The aromatic ring is said to be disubstituted and the three possible isomers are referred to as being *ortho, meta,* and *para*. The mechanisms leading to these three isomers are shown in Fig. 9.31.

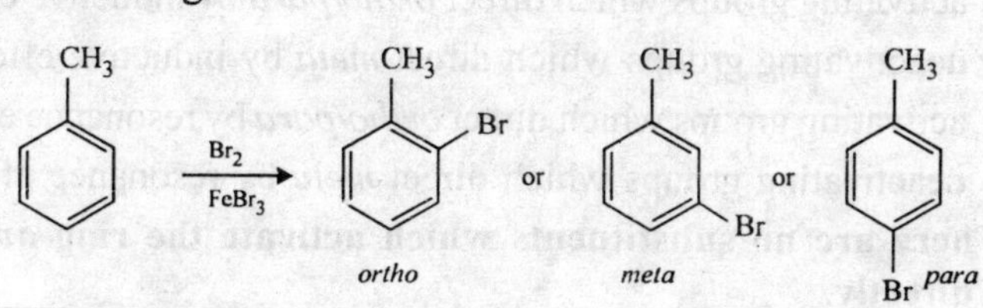

Fig. 9.30. ortho, meta and para isomers of bromotoluene.

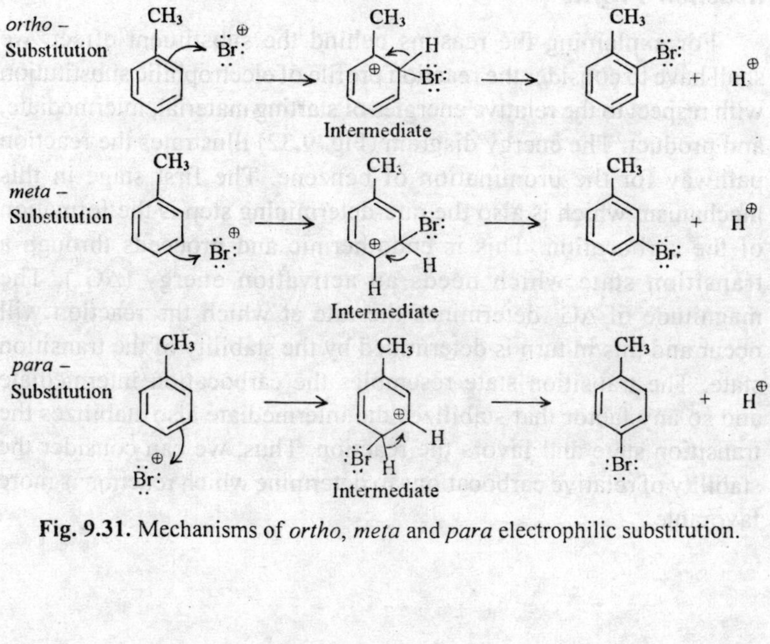

Fig. 9.31. Mechanisms of *ortho*, *meta* and *para* electrophilic substitution.

Substituent Effect

Out of the three possible isomers arising from the bromination of toluene; only two (the *ortho* and *para*) are formed in significant quantity. Moreover, the bromination of toluene goes at a faster rate than the bromination of benzene. This is so because the methyl substituent can affect the rate and the position of further substitution. A substituent can either activate or deactivate the aromatic ring towards electrophilic substitution and does so through inductive or resonance effects. A substituent can also direct the next substitution so that it goes mainly *ortho/para* or mainly *meta*.

The substituents can be classified into four groups depending on the effect they have on the rate and the position of substitution:-

(i) activating groups which direct *ortho/para* by inductive effects;

(ii) deactivating groups which direct *meta* by inductive effects;

(iii) activating groups which direct *ortho/para* by resonance effects;

(iv) deactivating groups which direct *meta* by resonance effects.

There are no substituents which activate the ring and are *meta* directly.

Reaction Profile

For explaining the reasons behind the substituent effect, we shall have to consider the reaction profile of electrophilic substitution with respect to the relative energies of starting material, intermediate, and product. The energy diagram (Fig. 9.32) illustrates the reaction pathway for the bromination of benzene. The first stage in this mechanism which is also the rate-determining step is the formation of the carbocation. This is endothermic and proceeds through a transition state which needs an activation energy ($\Delta G^{\#}$). The magnitude of $\Delta G^{\#}$ determines the rate at which the reaction will occur and this in turn is determined by the stability of the transition state. The transition state resembles the carbocation intermediate and so any factor that stabilizes the intermediate also stabilizes the transition state and favors the reaction. Thus, we can consider the stability of relative carbocations to determine which reaction is more favorable.

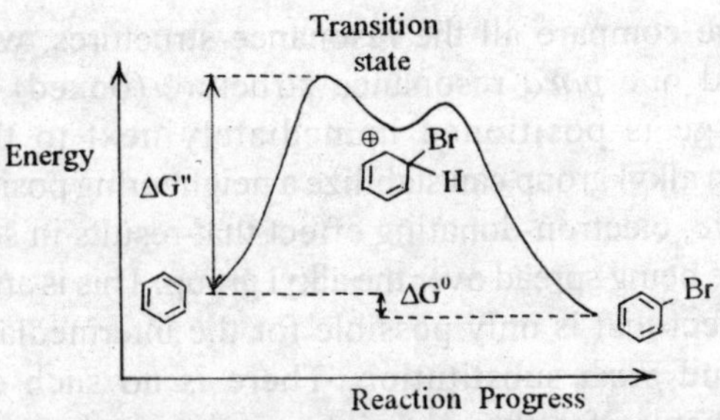

Fig. 9.32. Energy diagram for electrophilic substitution.

Activating Groups—Inductive o/p Directing

A methyl substituent is an inductive activating group and let us consider again the bromination of toluene. To explain the directing properties of the methyl group, we look more closely at the mechanisms involved in generating the *ortho*, *meta,* and *para* isomers (Fig. 9.31). The preferred reaction pathway will be the one that goes through the most stable intermideiate. Since a meth1 group directs *ortho* and *para,* the intermediate involved in these reaction pathways are more stable than the intermediates involved in *meta* substitution. The relevant intermediates and their resonance structures are shown in Fig. 9.33.

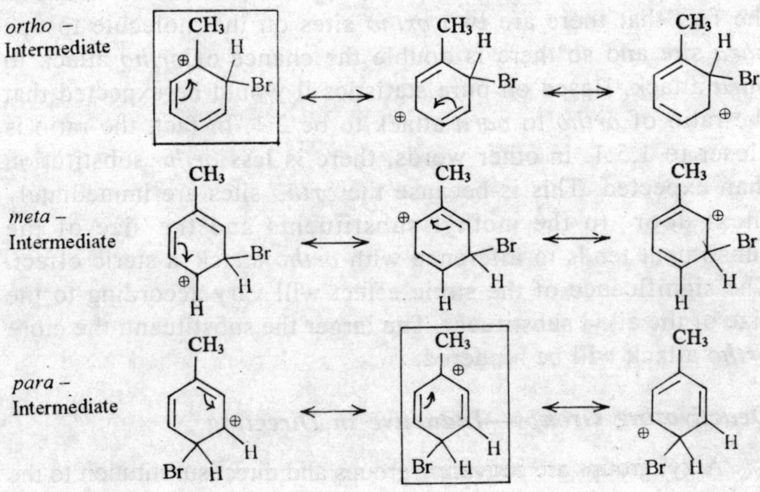

Fig. 9.33. Intermediates for *ortho, meta,* and *para* substitution.

Now if we compare all the resonance structures, we can spot one *ortho* and one *para* resonance structure (boxed) where the positive charge is positioned immediately next to the methyl substituent. An alkyl group can stabilize a neighboring positive charge by an inductive, electron-donating effect that results in some of the positive charge being spread over the alkyl group. This is an additional stabilizing effect that is only possible for the intermediates arising from *ortho* and *para* substitution. There is no such equivalent resonance structure for the *meta* intermediate and so that means that the *ortho* and *para* intermediates experience an increased stability over the *meta*, that results in a preference for these two substitution pathways.

Similarly, toluene will be more reactive than benzene. The electron-donating effect of the methyl group into the aromatic ring makes the ring inherently more nucleophilic and more reactive to electrophiles, as well as providing extra stabilization of the reaction intermediate. Thus, alkyl groups are **activating groups** and are ***ortho, para directing***.

Consider the nitration of toluene (Fig. 9.34). The amount of *meta* substitution is very small as expected and there is a preference for the *ortho* and *para* products. The formation of more *ortho* substitution compared to *para* substitution is due to the fact that there are two *ortho* sites on the molecule to one *para* site and so there is double the chance of *ortho* attack to *para* attack. Based on pure statistics it would be expected that the ratio of *ortho* to *para* attack to be 2:1. In fact, the ratio is closer to 1.5:1. in other words, there is **less** *ortho* substitution than expected. This is because the *ortho* sites are immediately 'next door' to the methyl substituents and the size of the substituent tends to inference with *ortho* attack–a steric effect. The significance of the steric effect will vary according to the size of the alkyl substituent. The larger the substituent, the more *ortho* attack will be hindered.

Deactivating Groups—Inductive in Directing

Alkyl groups are activating groups and direct substitution to the *ortho, para* positions. Electron withdrawing substitutions (Fig. 9.35) have the opposite effect. They deactivate the ring, make the ring less nucleophilic and less likely to react with an electrophile. The electron-withdrawing effects also destabilizes the reaction intermediate and makes the reaction more difficult. This destabilization is more pronounced in the intermediates arising from *ortho/para* attack and so *meta* attack is favored.

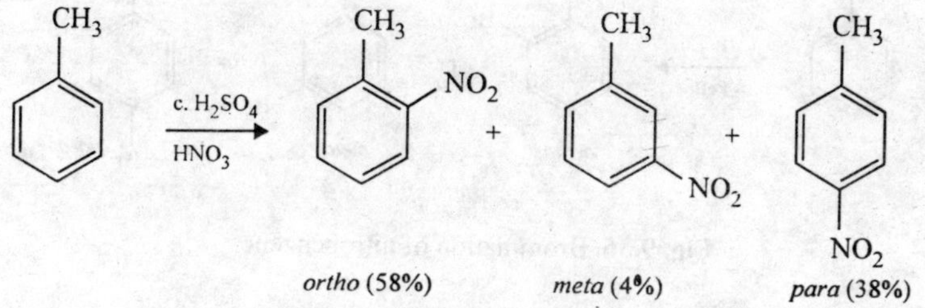

ortho (58%) *meta* (4%) *para* (38%)

Fig. 9.34. Nitration of toluene.

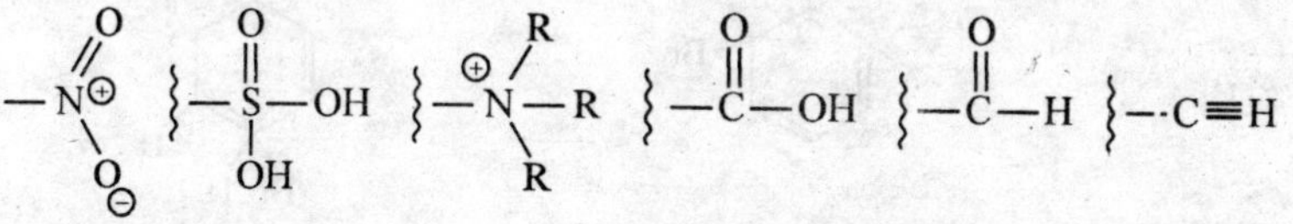

Fig. 9.35. Examples of electron-withdrawing groups.

All of these groups possess a positively charged atom or an electron deficient atom (i.e. an electrophilic center) directly attached to the aromatic ring. Since this atom is electron deficient, it has an electron-withdrawing effect on the ring.

Deactivating groups make electrophilic substitution more difficult but the reaction will proceed under more forcing reaction conditions. However, substitution is now directed to the *meta* position. This can be explained by comparing all the possible resonance structures arising form *ortho, meta* and *para* attack. For example, consider the bromination of nitrotoluene (Fig. 9.36). Of all the possible resonance structures arising from *ortho, meta,* and *para* attack, there are two specific resonance structures (arising from *ortho* and *para* attack) in which the positive charge is placed directly next to the electron-withdrawing nitro group (Fig. 9.37). Due to this, these resonance structures are generally destabilized. This does not take place with any of the resonance structures arising from *meta* attack and so *meta* attack is favored.

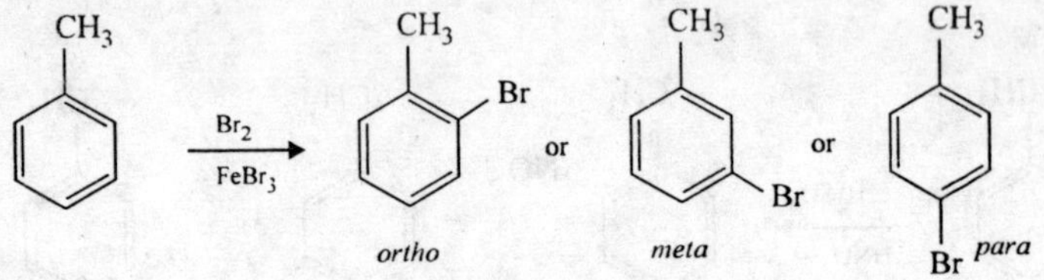

Fig. 9.36. Bromination of nitrobenzene.

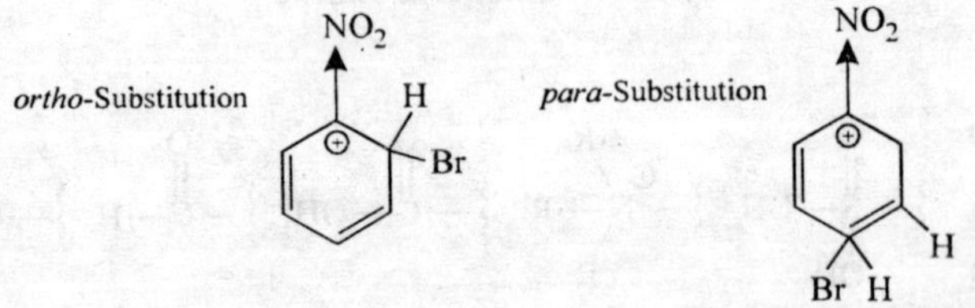

Fig. 9.37. Destabilizing resonance structures for the intermediate arising from *ortho* and *para* substitution.

Activating Groups—Resonance o/p Directing

Phenol group activates the aromatic ring by resonance effects and it directs substitution to the *ortho* and *para* positions. In phenol, an electronegative oxygen atom is next to the aromatic ring. As oxygen is electronegative so it should have an electron-withdrawing inductive effect and it might be expected to *deactivate the ring.* The fact that the phenolic group is a powerful activating group is because of the fact that *oxygen is electron rich and can also act as a nucleophile, feeding electrons into the ring through a resonance process.* For example, consider the nitration of phenol (Fig. 9.38).

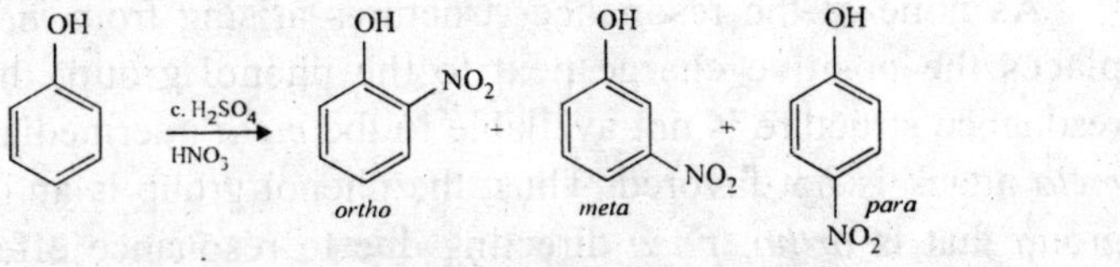

Fig. 9.38. Nitration of phenol.

There are three resonance structures for the intermediate formed in each form of electrophilic substitution, but there are two crucial ones to consider (Fig. 9.39), arising from *ortho* and *para* substitution. These resonance structures have the positive charge next to the OH substituent. If oxygen only had an inductive effect, these resonance structures would be highly unstable. However, oxygen can act as a nucleophile and so can use one of its lone pairs of electrons to form a new π bond to the neighboring electrophilic center (Fig. 9.40). This results in a fourth resonance structure where the positive charge is moved out of the ring and onto the oxygen atom. Delocalizing the charge like this further stabilizes it and makes the reaction proceed more easily.

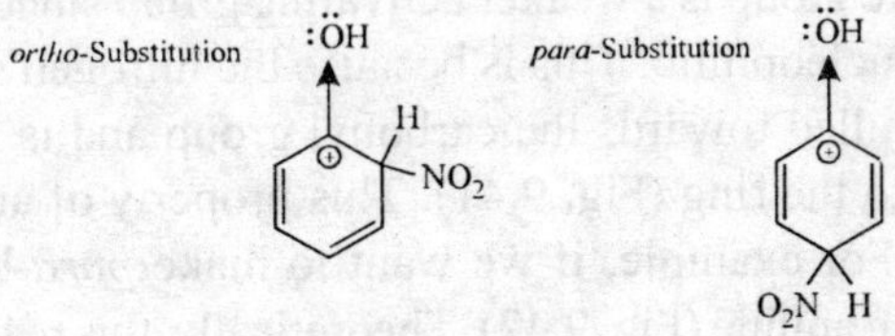

Fig. 9.39. Resonance structures for the intermediates arising from *ortho* and *para* substitution.

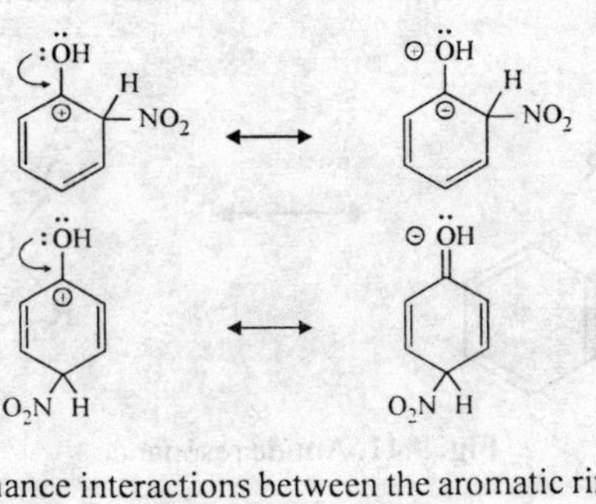

Fig. 9.40. Resonance interactions between the aromatic ring and oxygen.

As none of the resonance structures arising from *meta* attack places the positive charge next to the phenol group, this fourth resonance structure is not available to the *meta* intermediate and so *meta* attack is not favored. Thus, the phenol group is an activating group that is *ortho, para* directing due to resonance effects. This resonance effect is more important than any inductive effect that the oxygen might have.

The same is true for the following substituents: alkoxy (–OR), esters (–OCOR), amines ($-NH_2$, –NHR, $-NR_2$) and amides (–NHCOR). In all these cases, there is either a nitrogen or an oxygen next to the ring. Both these atoms are nucleophilic and have lone pairs of electrons which can be used to form an extra bond to the ring. The ease with which the group can do this depends on the nucleophilicity of the attached atom and how well it can cope with a positive charge.

Nitrogen is more nucleophilic than oxygen since it is better able to cope with the resulting positive charge. Therefore amine substituents are stronger activating groups than ethers. On the other hand, an amide group is a weaker activating group since the nitrogen atom is less nucleophilic. This is because the nitrogen's lone pair of electrons is pulled towards the carbonyl group and is less likely to form a bond to the ring (Fig. 9.41). This property of amides can be quite useful. For example, if we want to make *para*-bromoaniline by brominating aniline (Fig. 9.42). Theoretically, this reaction scheme should give the desired product. In practice, the NH_2 group is such a strong activating group that the final bromination goes three times to give the tri-bromianted product rather than the mono-brominated product.

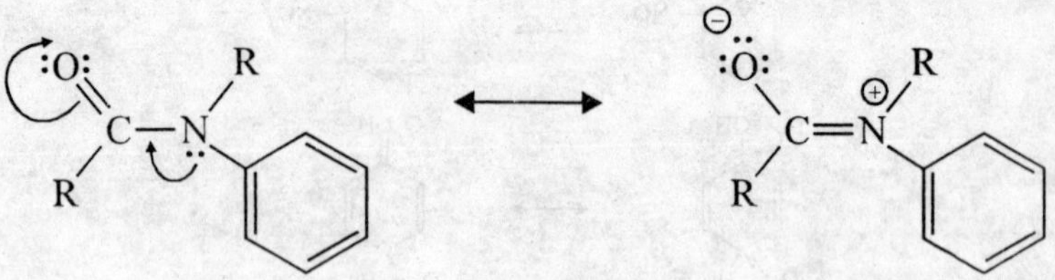

Fig. 9.41. Amide resonance.

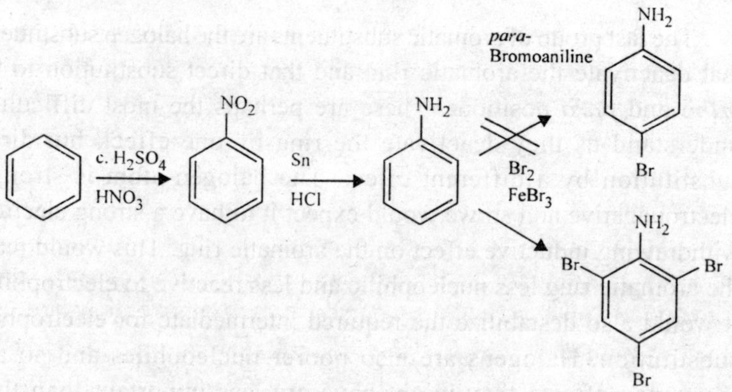

Fig. 9.42. Bromination of aniline.

To lower the activation of the amino group, we can convert it to the less activating amide group (Fig. 9.43). The bromination then only goes once. We also find that the bromination reaction is more selective for the *para* position than for the *ortho* position. This is because the amide group is bulkier than the NH_2 group and tends to shield the *ortho* positions from attack. Once the bromination has been completed the amide can be converted back to the amino group by hydrolysis.

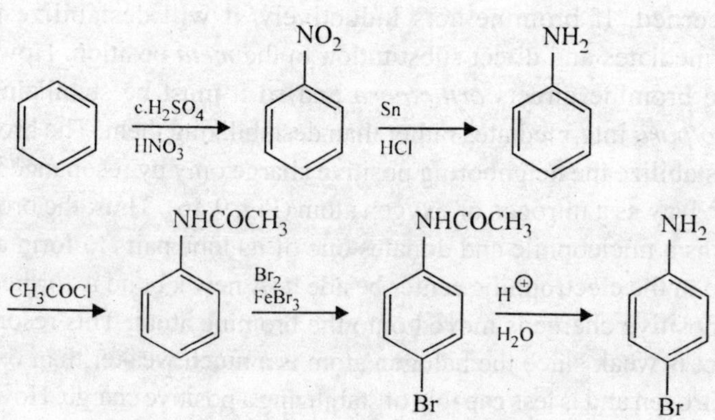

Fig. 9.43. Synthesis of *para*-bromoaniline.

Deactivating Groups—Resonance o/p Directing

The last group of aromatic substituents are the halogen substituents that deactivate the aromatic ring and that direct substitution to the *ortho* and *para* positions. These are perhaps the most difficult to understand as they deactivate the ring by one effect, but direct substitution by a different effect. The halogen atom is strongly electronegative and so we would expect it to have a strong electron-withdrawing inductive effect on the aromatic ring. This would make the aromatic ring less nucleophilic and less reactive to electrophiles. It would also destabilize the required intermediate for electrophilic substitution. Halogens are also poorer nucleophiles and so any resonance effects they might have are less important than their conductive effects.

However, if halogen atoms are deactivating the ring due to inductive effects, they should not direct substitution to the *meta* position like other electron-withdrawing groups? Consider the nitration of bromobenzene. There are three resonance structures for each of the three intermediates leading to these products, but the crucial ones to consider are those which position a positive charge next to the substituent. These occur with *ortho* and *para* substitution, but not *meta* substitution (Fig. 9.45). These are the crucial resonance structures as far as the directing properties of the substituents is concerned. If bromine acts inductively, it will destabilize these intermediates and direct substitution to the *meta* position. However, since bromine directs *ortho/para* and so it must be stabilizing the *ortho/para* intermediates rather than destabilizing them. The bromine can stabilize the neighboring positive charge only by resonance in the same way as a nitrogen or oxygen atom (Fig. 9.46). Thus, the bromine acts as a nucleophile and donates one of its lone pairs to form a new bond to the electrophilic center beside it. A new π bond is formed and the positive charge is moved onto the bromine atom. This resonance effect is weak since the halogen atom is a much weaker than oxygen or nitrogen and is less capable of stabilizing a positive charge. However, it is significant enough to direct substitution to the *ortho* and *para* positions.

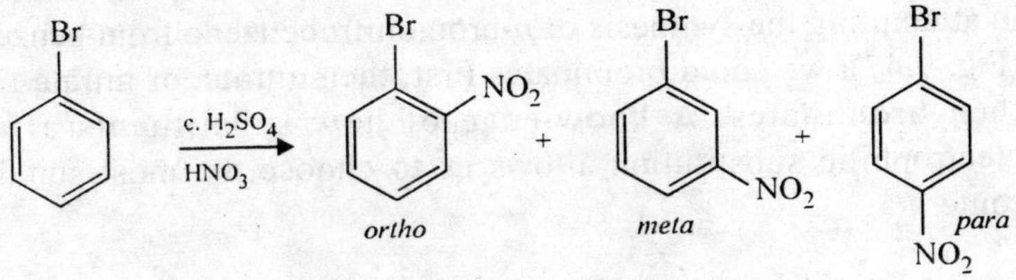

Fig. 9.44. Nitration of bromobenzene.

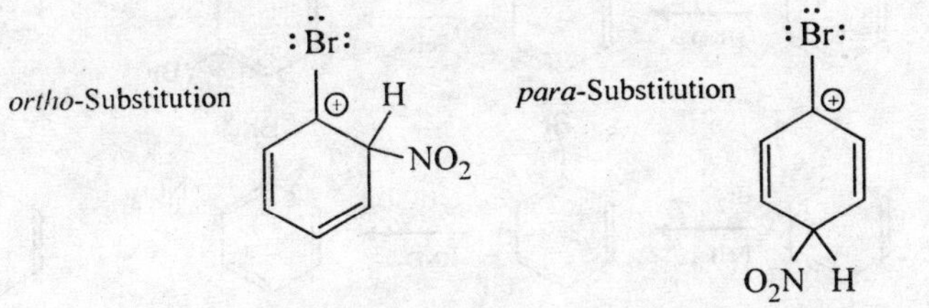

Fig. 9.45. Crucial resonance structures for *ortho* and *para* substitution.

In case of halogen substituents, the inductive effect is more important than the resonance effect in deactivating the ring. However, once electrophilic substitution **does** occur, resonance effects are more important than inductive effects in directing substitution.

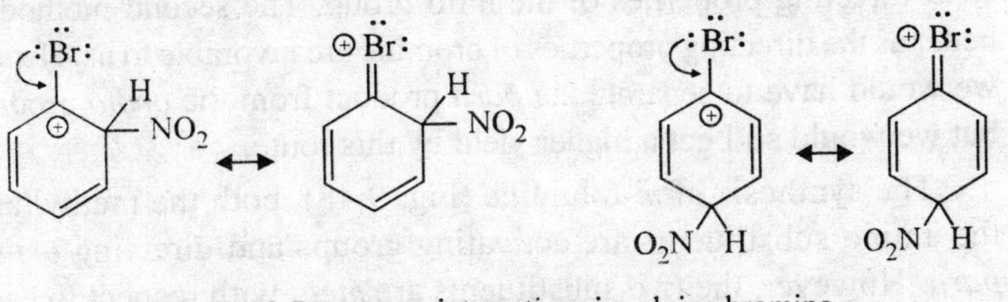

Fig. 9.46. Resonance interactions involving bromine.

9.6 Synthesis of Di- and Tri- Substituted Benzenes

Di-substituted Benzenes

To fully understand how substituents direct further substitution is important in planning the synthesis of a di-substituted aromatic compound. For example, there are two choices which can be made in attempting the synthesis of *p*-bromonitrobenzene from benzene (Fig. 9.47). we could brominates first, then nitrate, or nitrate first than brominates. A knowledge of how substituents affect electrophilic substitution allows us to choose the most suitable route.

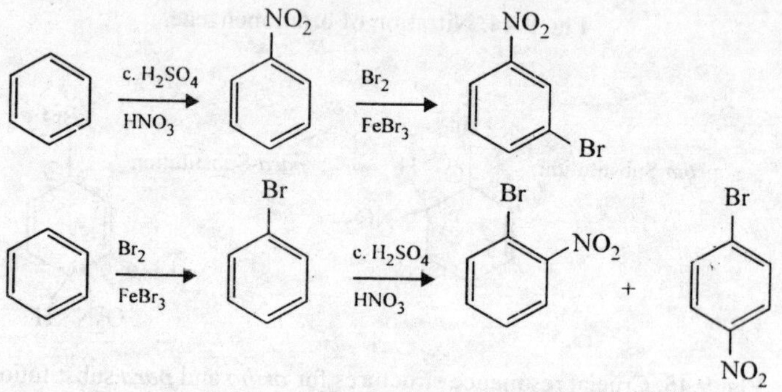

NO_2 NO_2
c. H_2SO_4 / HNO_3 → ; Br_2 / $FeBr_3$ → Br
Br Br Br
Br_2 / $FeBr_3$ → ; c. H_2SO_4 / HNO_3 → NO_2 + NO_2

Fig. 9.47. Synthetic planning to di-substituted benzenes.

In the first method, nitrating first then brominating would give predominantly the *meta* isomer of the final product because of the *meta* directing properties of the nitro group. The second method is better as the directing properties of bromine are favorable to us. Hence, we would have to separate the *para* product from the *ortho* product but we would still get a higher yield by this route.

The synthesis of *m*-toluidine (Fig. 9.48), both the methyl and the amino substituents are activating groups and directing *ortho/para.* However, the two substituents are *meta* with respect to each other. Thus to get *meta* substitution we should introduce a substituent other than the methyl or nitro group that will direct the second substitution to the *meta* position. Moreover, once this has been achieved, the *meta* directing substituent has to be converted to one of the desired substituents. The nitro group is ideal for this as it directs *meta* and can then be converted to the required amino group.

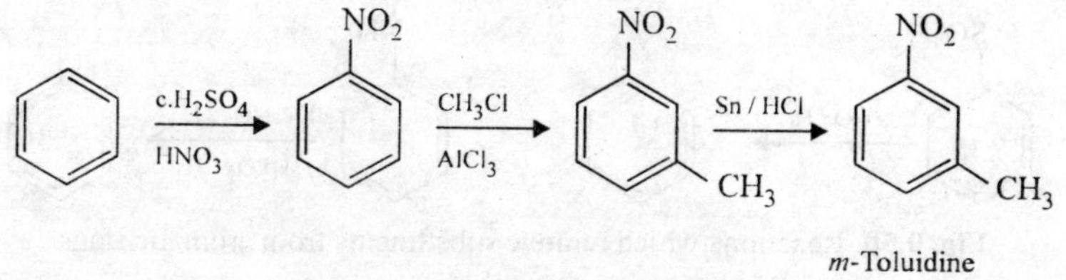

Fig. 9.48. Synthesis of m-toluidine.

The same strategy can be used for a large range of *meta*-disubstituted aromatic rings where both substituents are *ortho/para* directing since the nitro group can be transformed to an amino group which can then be transformed to a large range of different functional groups. Another difficult situation is where there are two *meta*-directing substituents at *ortho* or *para* positions with respect to each other, for example, *p*-nitrobenzoic acid (Fig. 9.49). in this case, a methyl substituent is added which is o/p directing. Nitration is then carried out and the *para* isomer is separated from any *ortho* isomer which might be formed. The methyl group can then be oxidized to the desired carboxylic acid.

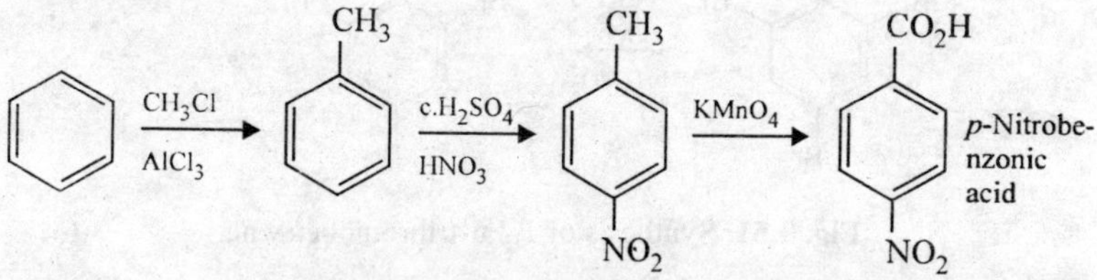

Fig. 9.49. Synthesis of *para*-nitrobenzoic acid.

Larger alkyl groups could be used to increase the ratio of *para* to *ortho* substituents since they can all be oxidized down to the carboxylic acid.

Removable Substituents

At times it is useful to have a substituent present that can direct or block a particular substitution, and that can then be removed once the desired substituents have been added. The reactions in Fig. 9.50 are used to remove substituents from aromatic rings.

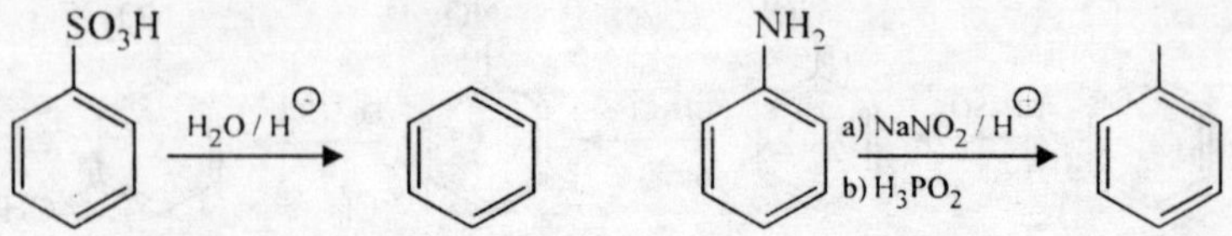

Fig. 9.50. Reactions which remove substituents from aromatic rings.

Consider the synthesis of 1,3,5-tribromobenzene (Fig. 9.51). This structure cannot be made directly from benzene by bromination. The bromine atoms are in the *meta* positions with respect to each other, but bromine atoms direct *ortho/para*. Moreover, bromine is a deactivating group and so it would be difficult to introduce three such groups directly to benzene.

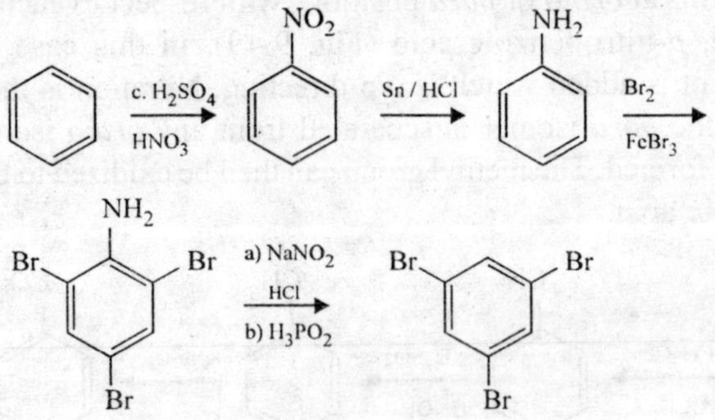

Fig. 9.51. Synthesis of 1,3,5-tribromobenzene.

To overcome this we use a strong activating group that will direct *ortho/para* and that can then be removed at the end of the synthesis. The amino group is ideal for this and the complete synthesis is shown in Fig. 9.51.

For the synthesis of *ortho*-bromotoluene we use a sulfonic acid. *o*-Bromotoluene could be synthesized by bromination of toluene or by Friedel-Crafts alkylation of bromobenzene (Fig. 9.52). However, the reaction would also give the *para*-substitution product and this is more likely if the electrophile is hindered from approaching the *ortho* position by unfavorable steric interactions. Alternatively we can substitute a group at the *para* position before carrying out the bromination. This group would then act as a blocking group at the *para* position and would force the bromination to occur *ortho* to the methyl group. If the blocking group could then be removed, the desired product would be obtained. The sulfonic acid group is specially useful in this respect since it can be easily removed once the synthesis is over (Fig. 9.53).

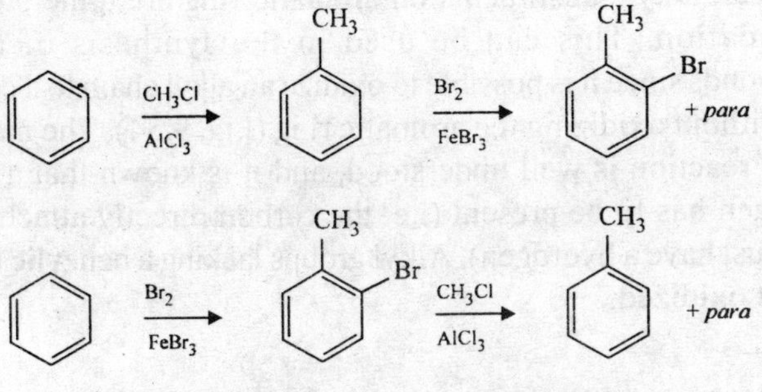

Fig. 9.52. Possible synthetic routes to ortho-bromotoluene.

The sulfonation of toluene could theoretically occur at the *ortho* position as also at the *para* position. However, the SO_3 electrophile is bulky and so the latter position is preferred for steric reasons. Once the sulfonic acid group is present, both it and the methyl group direct bromination to the same position (*ortho* to the methyl group = *meta* to the sulfonic acid group).

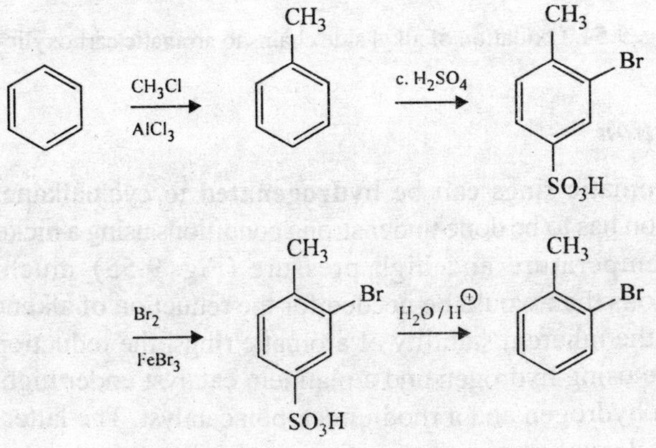

Fig. 9.53. Synthesis of ortho-bromotoluene.

9.7 Oxidation and Reduction

Oxidation

Aromatic rings are quite stable to oxidation and are resistant to oxidizing agents like potassium permanganate or sodium dichromate. However, alkyl substituents on aromatic ring are quite susceptible to oxidation. This can be used in the synthesis of aromatic compounds since it is possible to oxidize an alkyl chain to a carboxylic acid without oxidizing the aromatic ring (Fig. 9.54). The mechanism of this reaction is well understood, and it is known that a **benzylic** hydrogen has to be present (i.e. the carbon directly attached to the ring must have a hydrogen). Alkyl groups lacking a benzylic hydrogen are not oxidized.

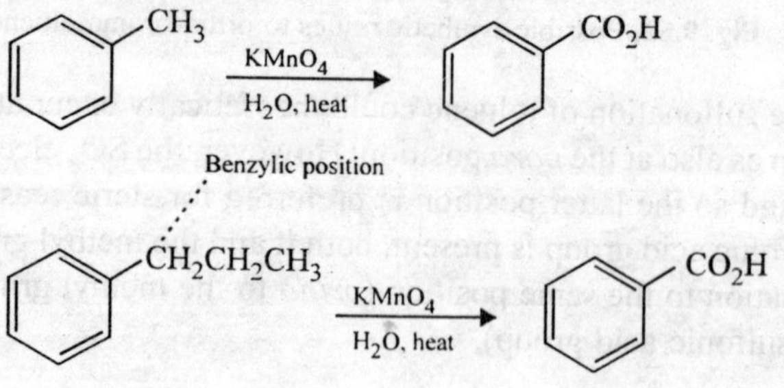

CH_3 — $KMnO_4$ / H_2O, heat → CO_2H

Benzylic position

$CH_2CH_2CH_3$ — $KMnO_4$ / H_2O, heat → CO_2H

Fig. 9.54. Oxidation of alkyl side chains to aromatic carboxylic acids.

Reduction

Aromatic rings can be **hydrogenated** to cycloalkanes, but the reduction has to be done under strong conditions using a nickel catalyst, high temperature and high pressure (Fig. 9.55)–much stronger conditions than would be needed for the reduction of alkenes. This is due to the inherent stability of aromatic rings, the reduction can also be done using hydrogen and a platinum catalyst under high pressure, or with hydrogen and a rhodium/carbon catalyst. The latter is a more powerful catalyst and the reaction can be done at room temperature and at atmospheric pressure.

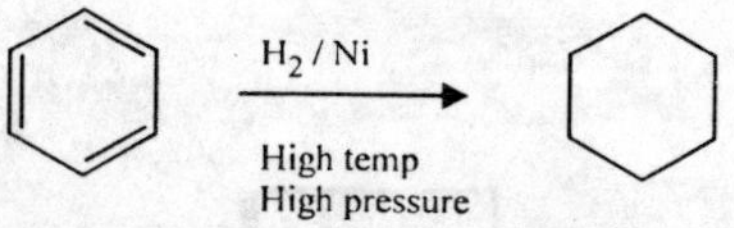

H$_2$ / Ni
High temp
High pressure

Fig. 9.55. Reduction of benzene to cyclohexane.

The resistance of the aromatic ring to reduction is useful as it is possible to reduce functional groups that might be arranged to the ring without reducing the aromatic ring itself. For example, the carbonyl group of an aromatic ketone can be reduced with hydrogen over a palladium catalyst without affecting the aromatic ring (Fig. 9.56). This permits the synthesis of primary alkylbenzenes that cannot be synthesized directly by the Friedel-Crafts alkylation. It must be noted that the aromatic ring makes the ketone group more reactive to reduction than would normally be the case. Aliphatic ketones would not be reduced under these conditions. Nitro groups can also be reduced to amino groups under these conditions without affecting the aromatic ring.

O
C
CH_3
Pd / H$_2$
CH_3

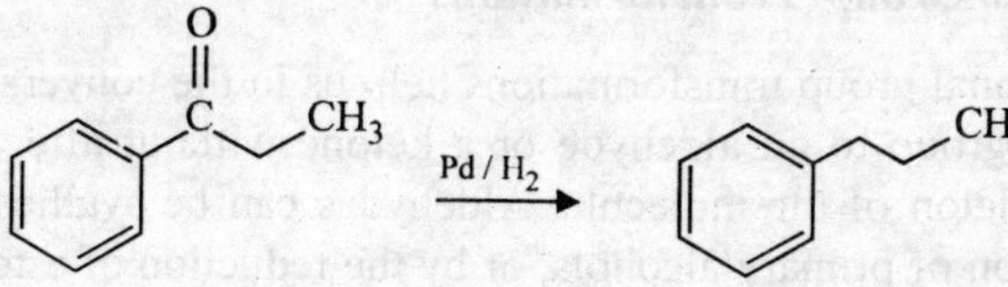

Fig. 9.56. Reduction of an aromatic ketone.

10

Aldehydes and Ketones

10.1 Methods of Preparation

Functional Group Transformations

Functional group transformations help us in the conversion of a functional group to an aldehyde or a ketone without affecting the carbon skeleton of the molecule. Aldehydes can be synthesized by the oxidation of primary alcohols, or by the reduction of esters, acid chlorides, or nitriles. Since nitriles can be obtained from alkyl halides, this a way of adding an aldehyde unit (CHO) to an alkyl halide (Fig. 10.1).

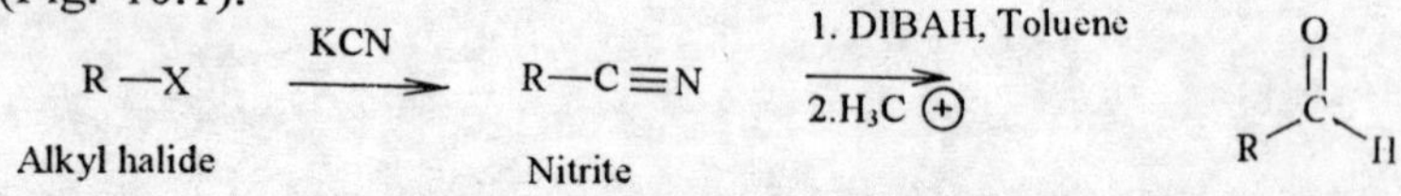

$$\underset{\text{Alkyl halide}}{R{-}X} \xrightarrow{\text{KCN}} \underset{\text{Nitrite}}{R{-}C{\equiv}N} \xrightarrow[\text{2.}H_3C^{\oplus}]{\text{1. DIBAH, Toluene}} R{-}\overset{O}{\overset{\|}{C}}{-}H$$

Fig. 10.1. Synthesis of an aldehyde from an alkyl halide with 1C chain extension.

Ketones can be synthesized by the oxidation of secondary alcohols. Methyl ketones can be synthesized from terminal alkynes.

C–C Bond Formation

Reactions that result in the formation of ketones by carbon–carbon bond formation are extremely important as they can be used to synthesize complex carbon skeleton from simple starting materials. Ketones can be synthesized from the reaction of acid chlorides with organocuprate reagents, or from the reaction of nitriles with a Grignard or organolithium reagent. Aromatic ketones can be synthesized by the Friedel-Crafts acylation of an aromatic ring.

C–C Bond Cleavage

Aldehydes and ketones can be obtained from the ozonolysis of suitably substituted alkenes.

10.2 Properties of Aldehydes and Ketones

Carbonyl Group

Both aldehydes and ketones have a carbonyl group (C=O). The substituents attached to the carbonyl group determine whether it is an aldehyde or a ketone, and whether it is aliphatic or aromatic.

The geometry of the carbonyl group is planar with bond angles of 120° (Fig. 10.2). The carbon and oxygen atoms of the carbonyl group are sp^2 hybridized and the double bond between the atoms consists of a strong σ bond and a weaker π bond. The carbonyl bond is shorter than a C–O single bond (1.22 Å vs. 1.43 Å) and is also stronger since two bonds are present as opposed to one (732 kJ mol^{-1} vs. 385 kJ mol^{-1}). The carbonyl group is more reactive than a C–O single bond because of the relatively weak π bond.

Fig. 10.2. Geometry of the carbonyl group.

The carbonyl group is polarized in such a way that the oxygen is slightly negative and the carbon is slightly positive. Both the polarity of the carbonyl group and the presence of the weak π bond explain much of the chemistry and the physical properties of aldehydes and ketones. The polarity of the bond also means that the carbonyl group has a resultant dipole moment.

Properties

Because of polar nature of the carbonyl group, aldehydes and ketones have higher boiling points than alkanes of similar molecular weight. However, hydrogen bonding is not possible between carbonyl groups and so aldehydes and ketones have lower bonding points than alcohols or carboxylic acids.

Low molecular weight aldehydes and ketones (e.g. formaldehyde and acetone) are soluble in water. This is because the oxygen of the carbonyl group can participate in intermolecular hydrogen bonding with water molecules (Fig. 10.3).

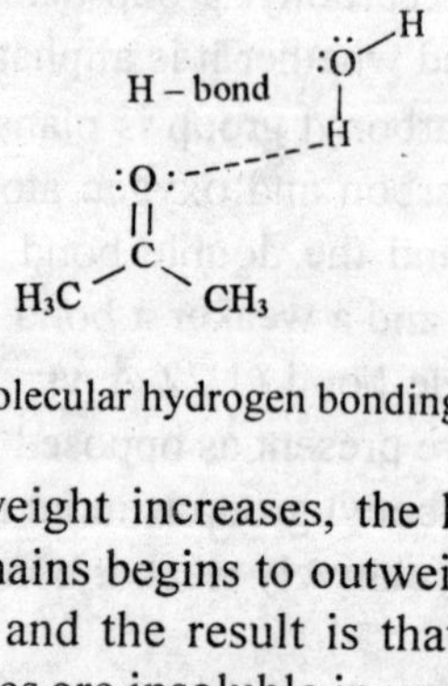

H – bond

Fig. 10.3. Intermolecular hydrogen bonding of a ketone with water.

As molecular weight increases, the hydrophobic character of the attached alkyl chains begins to outweigh the water solubility of the carbonyl group and the result is that large molecular weight aldehydes and ketones are insoluble in water. Aromatic ketones and aldehydes are insoluble in water because of the hydrophobic aromatic ring.

Nucleophilic and Electrophilic Centers

Because of the polarity of the carbonyl group, aldehydes and ketones have a nucleophilic oxygen center and an electrophilic carbon center as shown for propanal (Fig. 10.4). Therefore, nucleophiles react with aldehydes and ketones at the carbon center, and electrophiles react at the oxygen center.

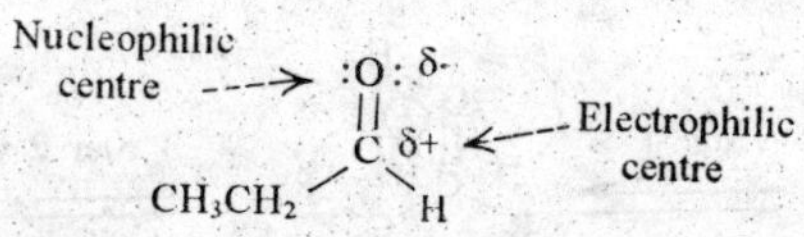

Nucleophilic centre

Electrophilic centre

Fig. 10.4. Nucleophilic and electrophilic centers of the carbonyl group.

Keto-enol Tautomerism

Ketones that have hydrogen atoms on their α-carbon (the carbon next to the carbonyl group) are in rapid equilibrium with an isomeric structure known as an **enol** in which the α-hydrogen ends up on the oxygen instead of the carbon. The two isomeric forms are known as **tautomers** and the process of equilibrium is known as **tautomerism** (Fig. 10.5). Generally the equilibrium favors the keto tautomer and the enol tautomer may only be present in very small quantities.

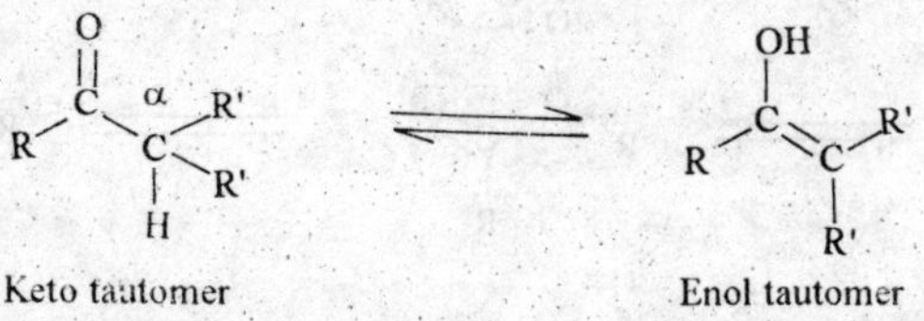

Keto tautomer

Enol tautomer

Fig. 10.5. Keto-enol tautomerism.

The tautomerism mechanism is catalyzed by acid or base. When catalyzed by acid (Fig. 10.6), the carbonyl group acts a nucleophiles with the oxygen using a lone pair of electrons to form a bonds to a proton. This results in the carbonyl oxygen gaining a positive charge that activates the carbonyl group to attack by weak nucleophiles (Step 1). The weak nucleophiles is a water molecule that removes the α-proton from the ketone, resulting in the formation of a new C = C double bond and cleavage of the carbonyl π bond. The enol tautomer is formed thus neutralizing the unfavorable positive charge on the oxygen (Step 2).

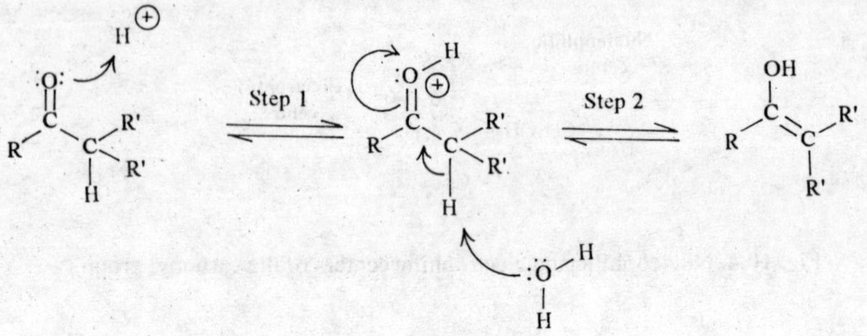

Fig. 10.6. Acid-catalyzed mechanism for keto-enol tautomerism.

Under basic condition (Fig. 10.7), an enolate ion is formed, which then reacts with water to form the enol.

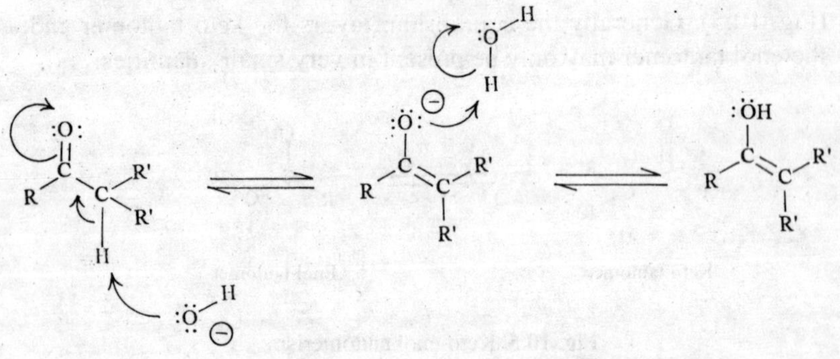

Fig. 10.7. Base-catalyzed mechanism for keto-enol tautomerism.

10.3 Nucleophilic Addition

Definition

The nucleophilic addition involves the addition of a nucleophilic to a molecule. This is a distinctive reaction for ketones and aldehydes and the nucleophile will add to electrophilic carbon atom of the carbonyl group. The nucleophile can be a negatively charged ion like cyanide or hydride, or it can be a neutral molecule like water or alcohol.

Overview

Generally, addition of charged nucleophiles results in the formation of a charged intermediate (Fig. 10.8). The reaction stops at this stage and acid has to be added to complete the reaction.

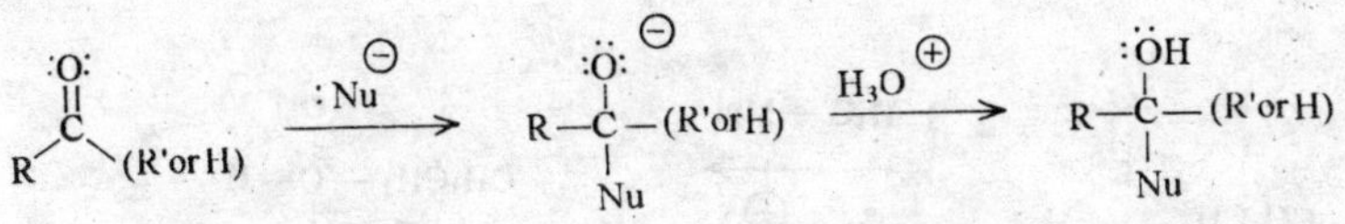

$$R-C(=O)-(R'\text{ or }H) \xrightarrow{:Nu^{\ominus}} R-C(O^{\ominus})(Nu)-(R'\text{ or }H) \xrightarrow{H_3O^{\oplus}} R-C(OH)(Nu)-(R'\text{ or }H)$$

Fig. 10.8. Nucleophilic addition to a carbonyl group.

Neutral nucleophiles where nitrogen or oxygen is the nucleophilic center are relatively weak nucleophiles, and an acid catalyst is generally needed. After nucleophilic addition has taken place, further reactions may occur leading to structures such as imines, enamines, acetals, and ketals (Fig. 10.9).

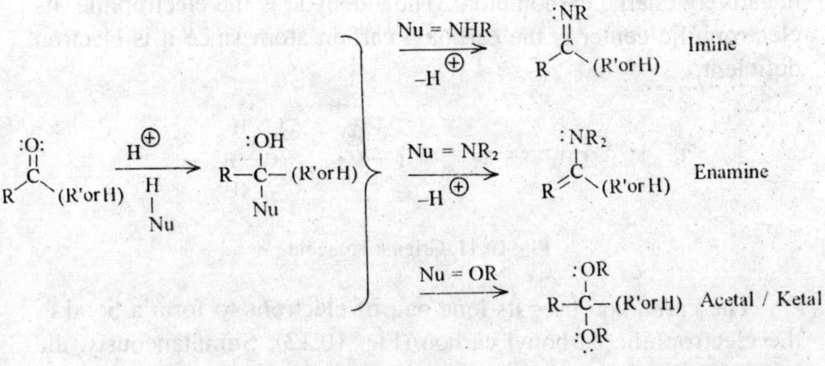

$$R-C(=O)-(R'\text{ or }H) \xrightarrow[H-Nu]{H^{\oplus}} R-C(OH)(Nu)-(R'\text{ or }H)$$

$$\xrightarrow[-H^{\oplus}]{Nu = NHR} R-C(=NR)-(R'\text{ or }H) \quad \text{Imine}$$

$$\xrightarrow[-H^{\oplus}]{Nu = NR_2} R{=}C(NR_2)-(R'\text{ or }H) \quad \text{Enamine}$$

$$\xrightarrow{Nu = OR} R-C(OR)(OR)-(R'\text{ or }H) \quad \text{Acetal / Ketal}$$

Fig. 10.9. Synthesis of imines, enamines, acetals, and ketals.

10.4 Nucleophilic Addition—Charged Nucleophiles

Carbon Addition

Carbanions are highly reactive and they do not occur in isolation. However there are two reagents that can supply the equivalent of a carbanion. These are **Grignard reagent** and **Organolithium reagents**. Let us first look at the reaction of Grignard reagent with aldehydes and ketones (Fig 10.10).

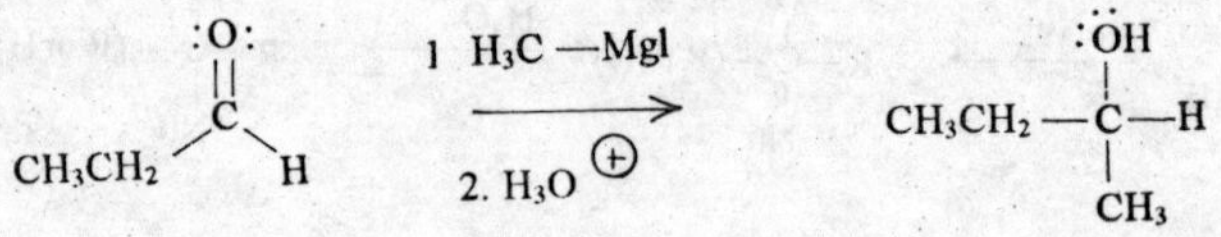

$$CH_3CH_2-\overset{\overset{:O:}{\|}}{C}-H \xrightarrow[2.\ H_3O^{\oplus}]{1\ H_3C-MgI} CH_3CH_2-\underset{\underset{CH_3}{|}}{\overset{\overset{:\ddot{O}H}{|}}{C}}-H$$

Fig. 10.10. Grignard reaction.

The Grignard reagent in this reaction is *methyl magnesium iodide* (CH_3MgI) and it is the source of a methyl carbanion (Fig. 10.11). Actually, the methyl carbanion is never present as a separate ion, but the reaction proceeds as if it were present. The methyl carbanion is the nucleophile in this reaction and the nucleophilic center is the negatively charge carbon atom. The aldehyde is the electrophile. Its electrophilic center is the carbonyl carbon atom since it is electron deficient.

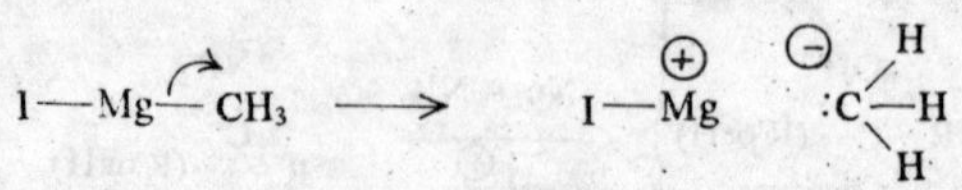

$$I-Mg-CH_3 \longrightarrow I-\overset{\oplus}{Mg} \quad :\overset{\ominus}{C}H_3$$

Fig. 10.11. Grignard reagent.

The carbanion uses its lone pair of electrons to form a bond to the electrophilic carbonyl carbon (Fig. 10.12). Simultaneously, the relatively weak π bond of the carbonyl group breaks and both electrons move to the oxygen to give it a third lone pair of electron and a negative charge (Step 1). The reaction stops at this stage, because the negatively charged oxygen is complexed with magnesium that acts as a counterion (not shown). Aqueous acid is now added to provide an electrophile in the shape of a proton. The intermediate is negatively charged and can acts as a nucleophile/base. A lone pair electrons on the negatively charged oxygen is used to form a bond to the proton and the final product is obtained (Step 2).

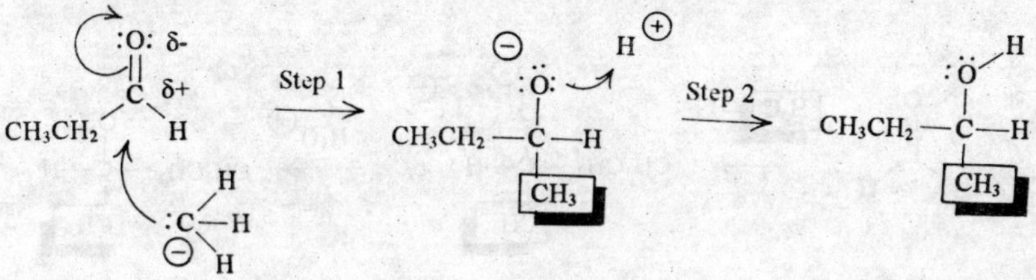

Fig. 10.12. Mechanism for the nucleophilic addition of a Grignard reagent.

The reaction of aldehydes and ketones with Grignard reagents is a useful method of synthesizing primary, secondary, and tertiary alcohols (Fig. 10.13). Primary alcohols can be obtained from formaldehydes, secondary alcohols can be obtained from aldehydes, and tertiary alcohols can be obtained from ketones. The reaction involves the formation of a carbon-carbon bond and so this is an important way of building up complex organic structures from simple starting materials.

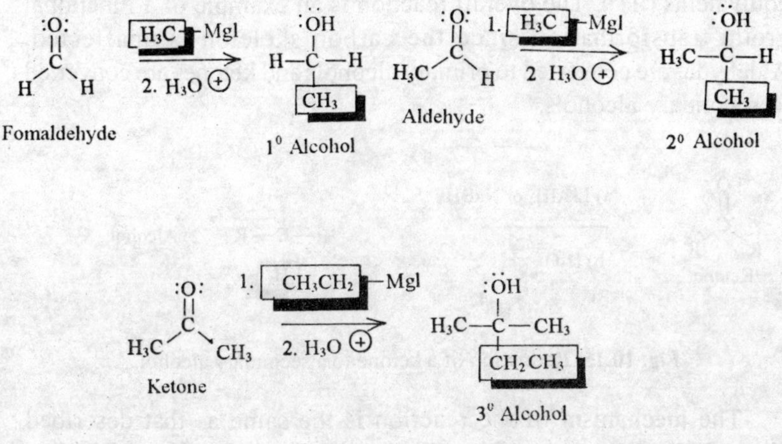

Fig. 10.13. Synthesis of primary, secondary, and tertiary alcohols by the Grignard reaction.

The Grignard reagent itself is synthesized from an alkyl halide and a large variety of reagents are possible.

Organolithium reagents like CH_3Li can also be used to provide the nucleophilic carbanion and the reaction mechanism is eaxtly the same as that described for the Grignard reaction (Fig. 10.14.)

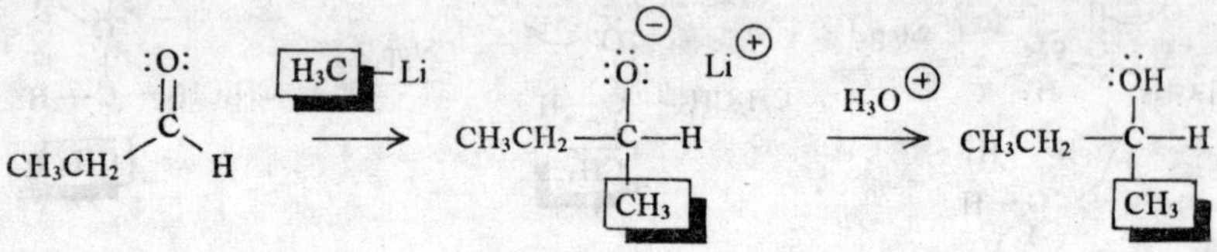

$$CH_3CH_2\text{–}C(=O)\text{–}H \xrightarrow{H_3C\text{–}Li} CH_3CH_2\text{–}C(O^{-}Li^{+})(CH_3)\text{–}H \xrightarrow{H_3O^{+}} CH_3CH_2\text{–}C(OH)(CH_3)\text{–}H$$

Fig. 10.14. Nucleophilic addition with an organolithium reagent.

Hydride Addition

Reducing agents like sodium borohydride ($NaBH_4$) and lithium aluminum hydride ($LiAlH_4$) react with aldehydes and ketones as if they are providing a hydride ion ($:H^-$; Fig. 10.15). This species is not present as such and the reaction mechanism is more complex. We can explain the reaction by considering these reagents as hydride equivalents ($:H^-$). The overall reaction is an example of a functional group transformation since the carbon skeleton is unaffected. Aldehydes are converted to primary alcohols and ketones are converted to secondary alcohols.

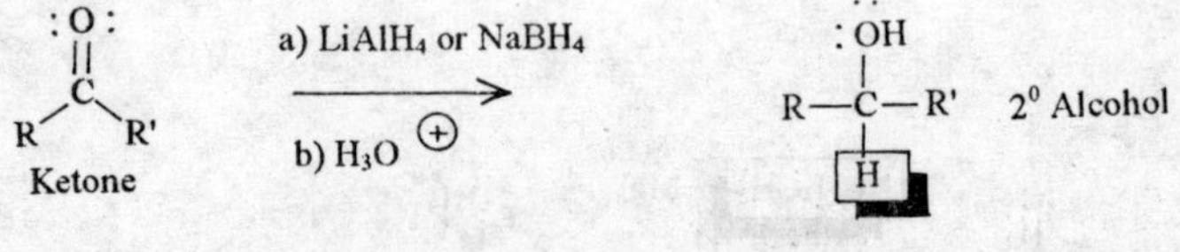

$$\underset{\text{Ketone}}{R\text{–}C(=O)\text{–}R'} \xrightarrow[\text{b) } H_3O^{+}]{\text{a) } LiAlH_4 \text{ or } NaBH_4} R\text{–}C(OH)(H)\text{–}R' \quad 2^0 \text{ Alcohol}$$

Fig. 10.15. Reduction of a ketone to a secondary alcohol.

The mechanism of the reaction is the same as that described above for the Grignard reaction (Fig 10.16). The hydride ion equivalent adds to the carbonyl group and a negatively charged intermediate is obtained that is complexed as a lithium salt (Step 1). Subsequent treatment with acid gives the final product (Step 2). The mechanism is actually more complex that this because the hydride ion is too reactive to exist in isolation.

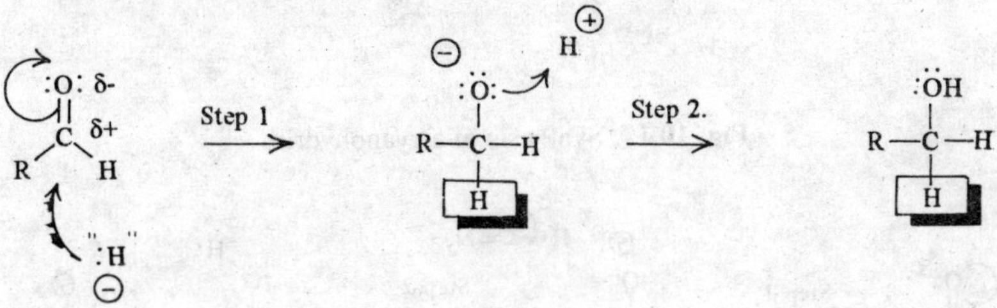

Fig. 10.16. Mechanism for the reaction of a ketone with $LiAlH_4$ or $NaBH_4$.

Cyanide Addition

Nucleophilic addition of a cyanide ion to an aldehyde or ketone forms **cyanohydrin** (Fig 10.17). In this reaction, there is a catalyst amount of potassium cyanide present and this supplies the attacking nucleophile in the form of the cyanide ion (CN^-). The nucleophilic center of the nitrile group is the carbon atom as this is the atom with the negative charge. The carbon atom uses its alone pair of electrons to form a new bond to the electrophilic carbon of the carbonyl group (Fig 10.18). As this new bond forms, the relatively weak π bond of the carbonyl group breaks and the two electrons making up that bond move onto the oxygen to give it a third lone pair of electrons and a negative charge (Step 1). The intermediate formed can now act as a nucleophile/base as it is negatively charged and so it reacts with the acidic hydrogen of HCN. A lone pair of electrons from oxygen is used to form a bond to the acidic proton and the H–CN σ bond is broken the same time such that these electrons move onto the neighboring carbon to give it a lone pair of electrons and a negative charge (Step 2). The products are the cyanohydrin and the cyanide ion. A cyanide ion started the reaction and a cyanide ion is regenerated. Therefore, only a catalyst amount of cyanide ion is needed to start the reaction and once the reaction has occurred, a cyanide ion is regenerated to continue the reaction with another molecule of ketone.

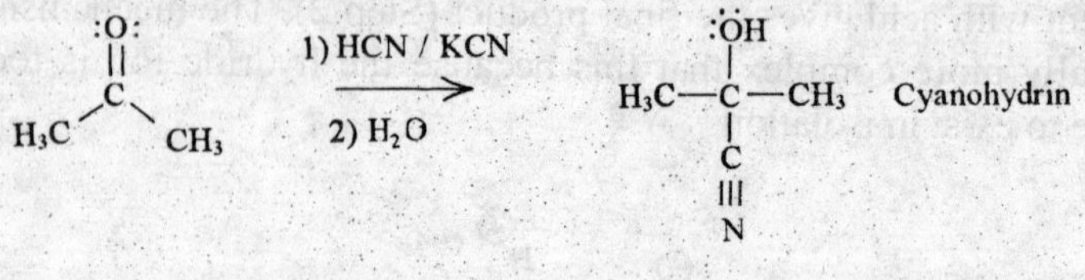

Fig. 10.17. Synthesis of a cyanohydrin.

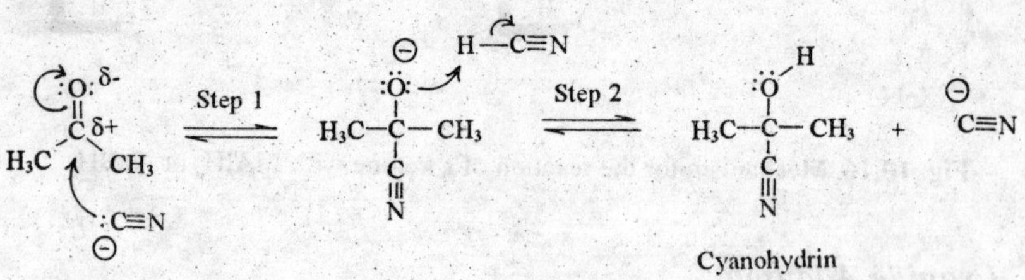

Fig. 10.18. Mechanism for the formation of a cyanohydrin.

Cyanohydrins are useful in synthesis as the cyanide group can be converted to an amine or to a carboxylic acid (Fig. 10.19).

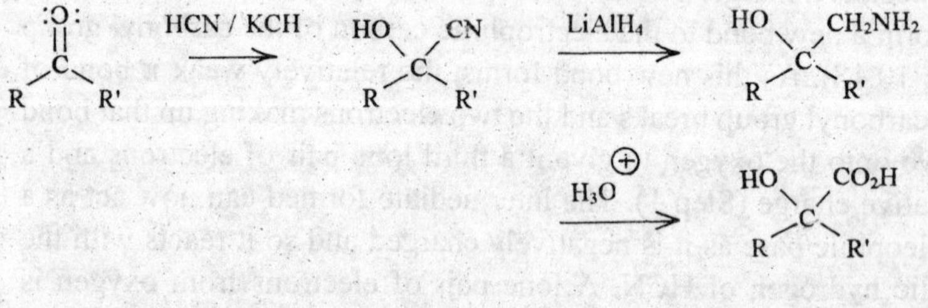

Fig. 10.19. Further reactions of cyanohydrins.

Bisulfite Addition

The reaction of an aldehyde or a methyl ketone with sodium bisulfite ($NaHSO_3$) involves nucleophilic addition of a bisulfite ion ($^{-}$:SO_3H) to the carbonyl group to give a water soluble salt (Fig. 10.20.).

The bisulfite ion is a relatively weak nucleophile compared to other charged nucleophiles and so only the most reactive carbonyl compounds will react. Larger ketones do not react because larger alkyl groups hinder attack. The reaction is reversible and so it is a useful method of separating aldehydes and methyl ketones from other ketones or from other organic molecules. This is generally done during an experimental work up where the products of the reaction are dissolved in a water immiscible organic solvent. Aqueous sodium bisulfite is then added and the mixture is shaken thoroughly in a separating funnel. Once the layers have separated, any aldehydes and methyl ketones will have undergone nucleophilic addition with the bisulfite solution and will be dissolved in the aqueous layer as the water soluble salt. The layers can then be separated. If the aldehydes or methyl ketone is desired, it can be recovered by adding acid or base to the aqueous layer which reverses the reaction and regenerates the carbonyl compound.

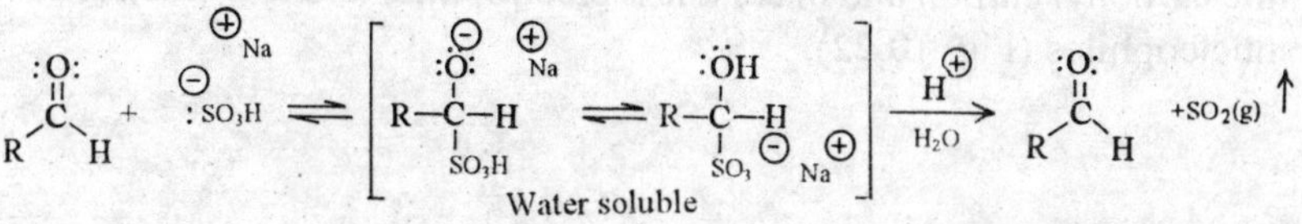

$$RCHO + Na^{\oplus}\ {}^{\ominus}\!:SO_3H \rightleftharpoons \left[R{-}CH(O^{\ominus})(SO_3H)\ Na^{\oplus} \rightleftharpoons R{-}CH(OH)(SO_3^{\ominus})\ Na^{\oplus} \right] \xrightarrow[H_2O]{H^{\oplus}} RCHO + SO_2(g)\uparrow$$

Water soluble

Fig. 10.20. Reaction of the bisulfite ion with an aldehyde.

Aldol Reaction

Another nucleophilic addition involving a charged nucleophile is the Aldol reaction. This involves the nucleophilic addition of enolate ions to aldehydes and ketones to form β-hydroxycarbonyl compounds (Fig. 10.21).

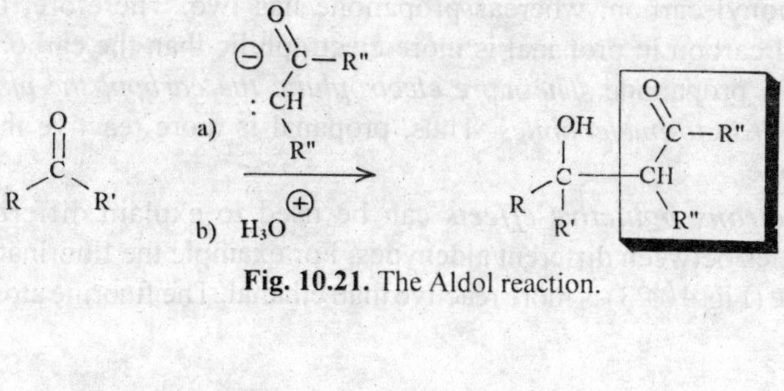

$$RCOR' \xrightarrow[b)\ H_3O^{\oplus}]{a)\ {}^{\ominus}:CH(R'')C(=O)R''} R(R')C(OH){-}CH(R''){-}C(=O)R''$$

Fig. 10.21. The Aldol reaction.

10.5 Electronics and Steric Effects

Reactivity

It is a fact that *aldehydes are more reactive to nucleophiles than ketones*. To explain this difference in reactivity, we will consider the two factors (i.e. electronic and steric effects).

Electronic Factors

The carbonyl carbon in aldehydes is more electrophilic as compared to that in ketones because of the substituents attached to the carbonyl carbon. A ketone has two alkyl groups attached whereas the aldehyde has only one. The carbonyl carbon is electron deficient and electrophilic because the neighboring oxygen has a greater share of the electrons in the double bond. However, neighboring alkyl groups have an inductive effect by which they push electrons density towards the carbonyl carbon and make it less electrophilic and less reactive to nucleophiles (Fig. 10.22).

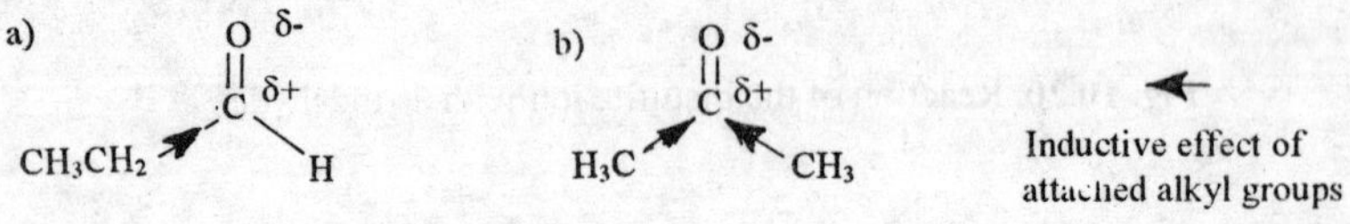

Fig. 10.22. Inductive effect in (a) propanal; (b) propanone.

In case of propanal there is one alkyl group feeding electrons into the carbonyl carbon, whereas propanone has two. Therefore, the carbonyl carbon in propanal is more electrophilic than the carbonyl carbon in propanone. *The more electrophilic the carbon, the more reactive it is to nucleophiles*. Thus, propanal is more reactive than propanone.

Electrons inductive effects can be used to explain differing reactivities between different aldehydes. For example the fluorinated aldehyde (Fig. 10.23) is more reactive than ethanal. The fluorine atoms are electronegative and have an electron-withdrawing effect on the neighboring carbon, making it electron deficient. This in turn has an inductive effect on the neighboring carbonyl carbon. Since electrons are being withdrawn, the electrophilicity of the carbonyl carbon is increased, making it more reactive to nucleophiles.

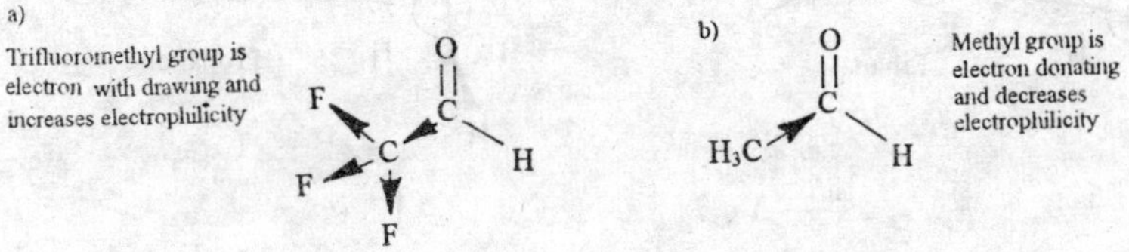

a) Trifluoromethyl group is electron with drawing and increases electrophilicity

b) Methyl group is electron donating and decreases electrophilicity

Fig. 10.23. Inductive effect of (a) trifluoroethanal; (b) ethanal.

Steric Factors

Steric factors also play an important role in the reactivity of aldehydes and ketones. We may look at the relative ease with which the attacking nucleophile can approach the carbonyl carbon or consider how steric factors influence the stability of the transition state leading to the final product.

Let us consider the relative ease with which a nucleophile can approach the carbonyl carbon of an aldehyde and a ketone. Let us consider the bonding and the shape of these functional groups (Fig. 10.24). Both molecules have planar carbonyl group. The atoms that are in the plane are circled in white. A nucleophile will approach the carbonyl group from above or below the plane. The diagram (Fig. 10.24) shows a nucleophile attacking from above. In it the hydrogen atoms on the neighboring methyl groups are not in the plane of the carbonyl group and so these atoms can hinder the approach of a nucleophile and thus hinder the reaction. This effect will be more significant for a ketone where there are alkyl groups on either side of the carbonyl group. An aldehyde has only one alkyl group attached and so the carbonyl group is more accessible to nucleophilic attack.

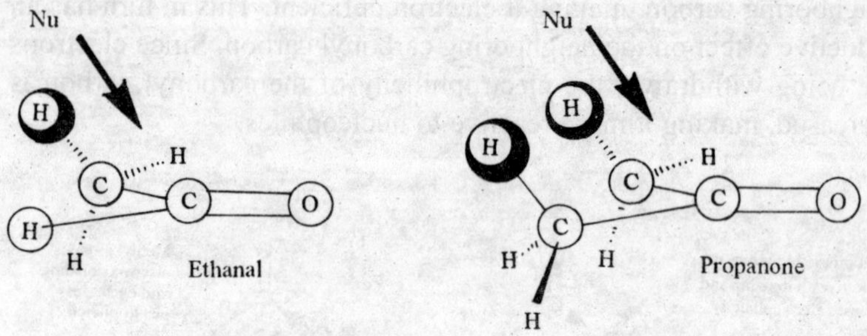

Fig. 10.24. Steric factors.

To understand how steric factors affect the stability of the transition state leading to the final product. Let us consider the reactions of propanone and propanal with HCN to give cyanohydrin products (Fig. 10.25).

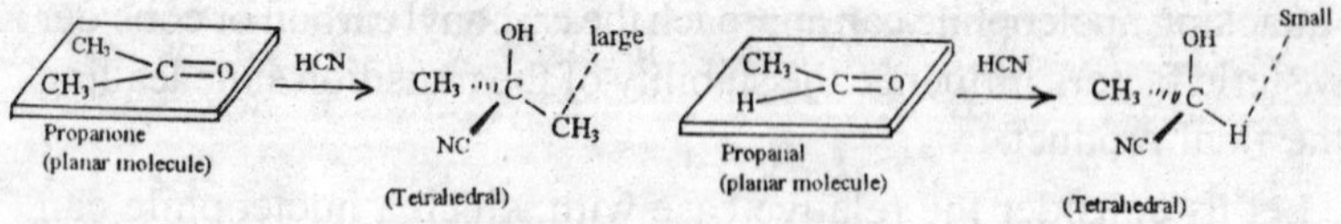

Fig. 10.25. Reactions of propanone and propanal with HCN.

Both propanone and propanal are planar molecules. The cyanohydrin products are tetrahedral. Thus, the reaction leads to a marked difference in shape between the starting carbonyl compound and the cyanohydrin product. There is also a marked difference in the space available to the substituents attached to the reaction site i.e. the carbonyl carbon. The tetrahedral molecule is more crowded because there are four substituents crowded round a central carbon, whereas in the planar starting material, there are only three substituents attached to the carbonyl carbon. The crowding in the tetrahedral product arising from the ketone will be greater than that arising from the aldehyde because one of the substituents from the aldehyde is a small hydrogen atom.

The ease with which nucleophilic addition occurs depends on the ease with which the transition state is formed. In nucleophilic addition, the transition state is considered to resemble the tetrahedral product more than it does the planar starting material. Therefore, any factor that affects the stability of the product will also affect the stability of the transition state. As crowding has a destabilizing effect so the reaction of propanone should be more difficult than the reaction of propanal. Therefore, ketones in general will be less reactive than aldehydes.

The bigger the alkyl groups, the larger the steric effect. For example, 3-pentanone is less reactive than propanone and fails to react with the weak bisulfite nucleophile whereas propanone does (Fig. 10.26).

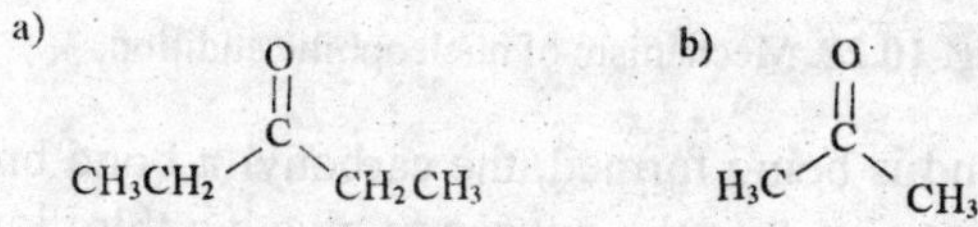

Fig. 10.26. (a) 3-Pentanone; (b) propanone.

10.6 Nucleophilic Addition—Nitrogen Nucleophiles

Imine Formation

The reaction of primary amines with aldehydes and ketones do not give the products expected from nucleophilic addition alone. This is because of the further reaction taking place once nucleophilic addition occurs e.g. consider the reaction of acetaldehyde (ethanal) with a primary amine methylamine (Fig. 10.27). The product contains the methylamine skeleton, but there is no alcohol group and there is a double bond between the carbon and the nitrogen. This product is known as **imine** or a **Schiff base**.

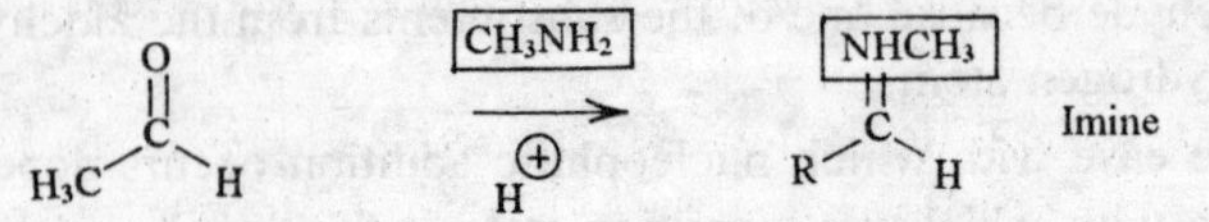

Fig. 10.27. Reaction of ethanal with methylamine.

The first stage of the mechanism (Fig. 10.28) is a normal nucleophilic addition. The amine acts as the nucleophile and the nitrogen atom is the nucleophilic center. The nitrogen uses its lone pair of electrons to form a bond to the electrophilic carbonyl carbon.

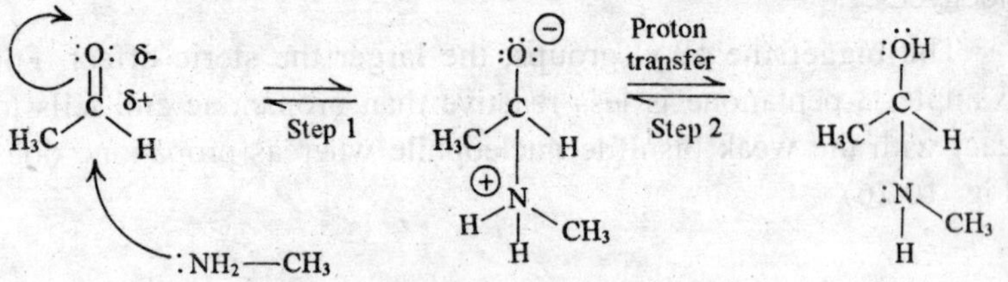

Fig. 10.28. Mechanism of nucleophilic addition.

As this bond is being formed, the carbonyl π bond breaks with both electrons moving onto the oxygen to give it a third lone pair of electrons and a negative charge. The nitrogen also gains a positive charge, but both these charges can be neutralized by the transfer of a proton from the nitrogen to the oxygen (Step 2). The oxygen uses up one of its lone pair to form the new O–H bond and the electrons in the N–H bond end up on the nitrogen as a lone pair. An acid catalyst is present, but is not needed for this part of the mechanism because nitrogen is a good nucleophile and although the amine is neutral, it is sufficiently nucleophilic to attack the carbonyl group without the need for acid catalyst. The intermediate obtained is the structure one would expect from nucleophilic addition alone, but the reaction does not stop there. The oxygen atom is now protonated by the acid catalyst and gains a positive charge (Fig. 10.29, Step 3). Since oxygen is electronegative, a positive charge is not favored and so there is a strong drive to neutralize the charge. This can be done if the bond to carbon breaks and the oxygen leaves as part of a water molecule *Therefore, protonation has turned the oxygen into a good leaving group*. The nitrogen helps the departure of the water by using its lone pair of electrons to form a π bond to the neighboring carbon atom and a positive charged intermediate is formed (Step 4). The water now acts as a nucleophile and removes a proton from the nitrogen such that the nitrogen's lone pair is restored and the positive charge is neutralized (Step 5).

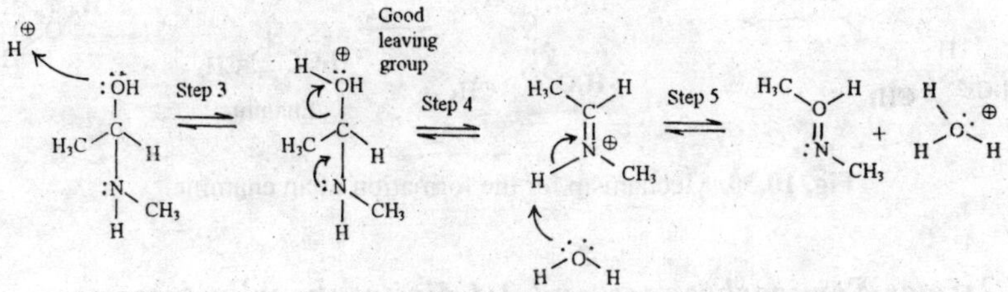

Good leaving group

Step 3 Step 4 Step 5

Fig. 10.29. Mechanism for the elimination of water.

Overall, a molecule of water has been lost in the second part of the mechanism. Acid catalyst is important in creating a good leaving group. If protonation did not take place, the leaving group would have to be hydroxide ion that is a more reactive molecule and a poorer leaving group.

Although acid catalyst is important to the reaction mechanism, too much acid may hinder the reaction. This is because a high acid concentration leads to protonation of the amine, and prevents it from acting as a nucleophile.

Enamine Formation

The reaction of carbonyl compounds with secondary amines cannot give imines as there is no NH proton to be lost in the final step of the mechanism. However, there is another way in which the positive charge on the nitrogen can be neutralized. It involves loss of a proton from a neighboring carbon atom (Fig. 10.30). Water acts as a base to remove the proton and the electrons that make up the C–H σ bond are used to form a new π bond to the neighboring carbon. This forces the existing π bond between carbon and nitrogen to break such that both the π electrons end up on the nitrogen atom as a lone pair, thus neutralizing the charge. The final structure is called an **enamine** and is a useful product for organic synthesis.

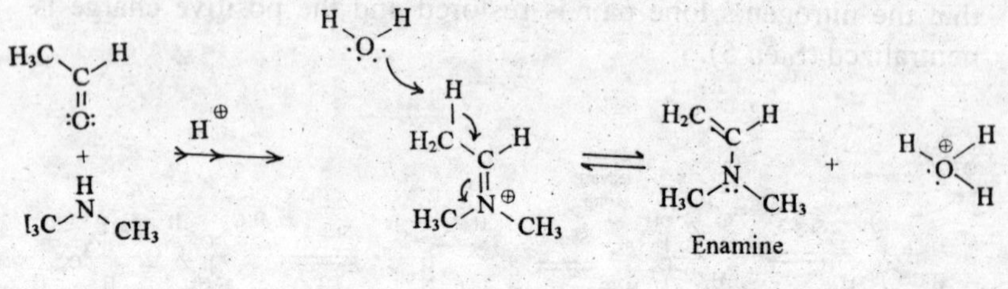

Fig. 10.30. Mechanism for the formation of an enamine.

Oximes Semicarbazones and 2,4-dinitrophenylhydrazones

The reaction of aldehydes and ketones with hydroxylamine (NH_2OH), semicarbazide ($NH_2NHCONH_2$) and 2,4-dinitropheylhydrazine occurs by the same mechanism as describe for primary amines to give Oximes, semicarbazones, and 2,4-dinitrophenylhydrazones, respectively (Fig. 10.31). These compounds were frequently synthesized so as to identify a liquid aldehyde or ketone. The products are solid and crystalline, and by measuring their melting points, the original aldehyde or ketone could be identified by looking up melting point tables of these derivatives. Presently, it is easier to identify liquid aldehydes and ketones spectroscopically.

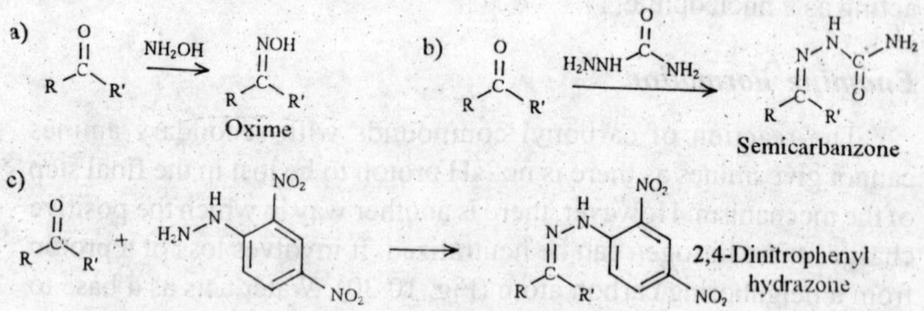

Fig. 10.31. Synthesis of Oximes, semicarbazones, and 2,4-dinitrophenylhydrazones.

10.7 Nucleophilic Addition—Oxygen and Sulfur Nucleophiles

Acetal and Ketal Formation

When an **aldehyde** or **ketone** is treated with an excess of **alcohol** in the presence of an acid catalyst, **two** molecules of alcohol are added to the carbonyl compound to give an **acetal** or a **ketal** respectively (Fig. 10.32). The final product is tetrahedral.

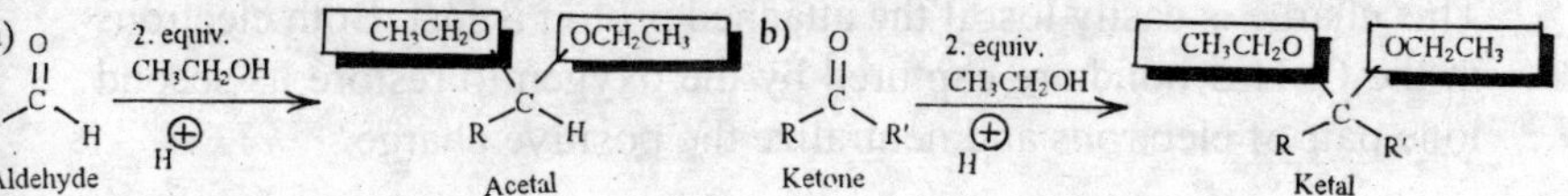

Fig. 10.32. Formation of an acetal and a ketal.

The reaction mechanism involves the nucleophilic addition of one molecule of alcohol to form a **hemiacetal** or **hemiketal.** Elimination of water occurs to form an **oxonium ion** and a second molecule of alcohol is then added (Fig. 10.33).

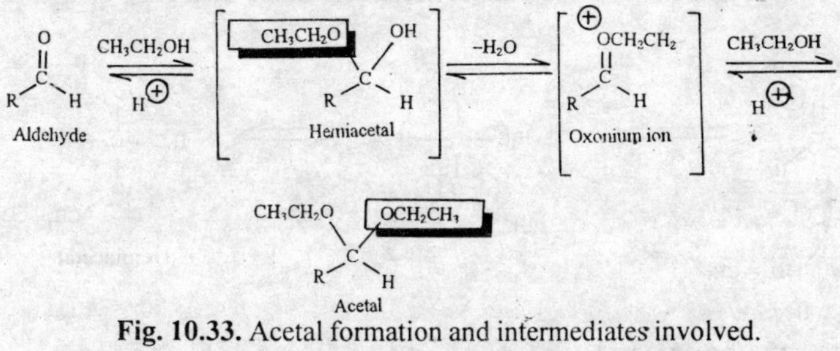

Fig. 10.33. Acetal formation and intermediates involved.

The mechanism is quite complex consider the reaction of methanol with acetaldehydes (ethanal; Fig 10.34). The aldehyde is the electrophile and the electrophilic center is the carbonyl carbon. Methanol is the nucleophile and the nucleophilic center is oxygen. However, methanol is a relatively weak nucleophile, so, the carbonyl group has to be activated by adding an acid catalyst if a reaction is to occur. The first step of the mechanism involves the oxygen of the carbonyl group using a lone pair of electrons to form a bond to a proton. This results in a charged intermediate where the positive charge is shared between the carbon and oxygen of the carbonyl group. Protonation increases the electrophilicity of the carbonyl group, making the carbonyl carbon even more electrophilic. Due to this it reacts better with the weakly nucleophilic alcohol. The alcoholic oxygen now uses one of its lone pairs of electrons to form a bond to the carbonyl carbon and carbonyl π bond breaks simultaneously with the π electrons moving onto the carbonyl oxygen and neutralizing the positive charge (Fig. 10.35). However, the alcoholic oxygen now has an unfavorable positive charge (which explain why methanol is a weak nucleophile in the first place). This charge is easily lost if the attached proton is lost. Both electrons in the O–H σ bond are captured by the oxygen to restore its second lone pair of electrons and neutralize the positive charge.

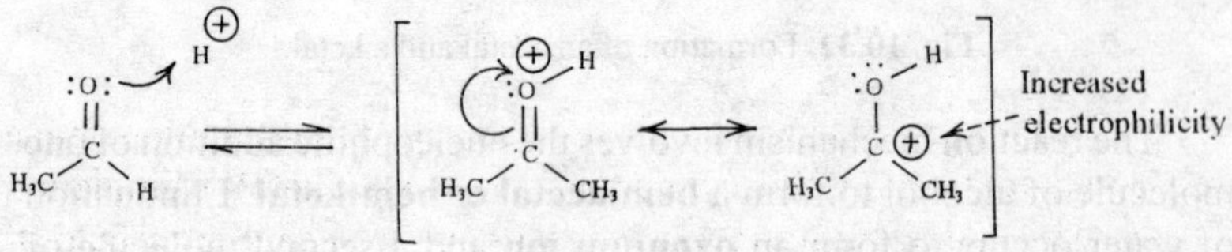

Increased electrophilicity

Fig. 10.34. Mechanism for acetal formation–Step 1.

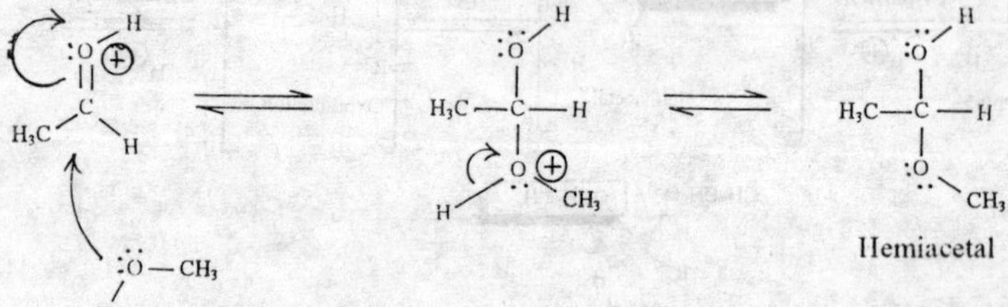

Hemiacetal

Fig. 10.35. Mechanism for acetal formation–Step 2 and 3.

The intermediate formed from this first nucleophilic addition is known as **hemiacetal**. When ketone is the starting material, the structure obtained is a **hemiketal**. Once the hemiacetal is formed, it is protonated and water is eliminated by the same mechanism described in the formation of imines with the only difference that oxygen donates a lone pair of electrons to force the removal of water rather than nitrogen (Fig. 10.36). The resulting oxonium ion is extremely electrophilic and a second nucleophilic addition of alcohol to forms the acetal.

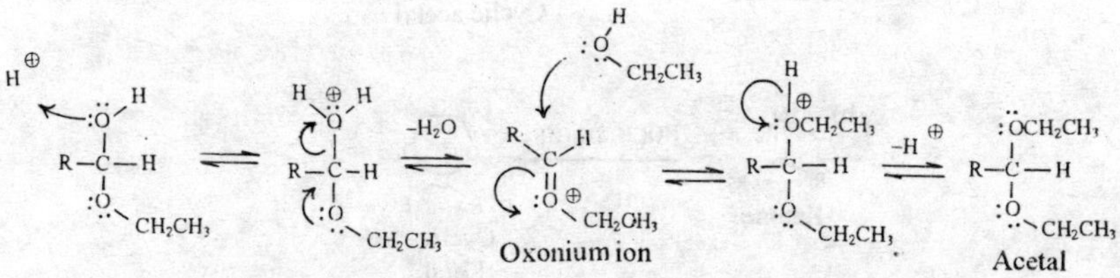

Oxonium ion

Acetal

Fig. 10.36. Mechanism of acetal formation from a hemiacetal.

All the stages in this mechanism are reversible and so it is possible to convert the acetal or ketal back to the original carbonyl compound using water and an aqueous acid as catalyst. Since water is added to the molecule in the reverse mechanism, this is a process known as **hydrolysis.**

Acid acts as a catalyst both for the formation and the hydrolysis of acetals and ketals, so to synthesize ketals and acetals in good yield, of the favorable conditions are the reaction is carried out in the absence of water using a small amount of concentrated sulfuric acid or an organic acid like *para*-toluenesulfonic acid. The yields are further boosted if the water formed during the reaction is removed from the reaction mixture.

To convert the acetal or ketal back to the original carbonyl compound, an aqueous acid is used such that there is a large excess of water present and the equilibrium is shifted towards the carbonyl compounds.

Both the synthesis and the hydrolysis of acetals and ketals can be done in high yield and so these functional groups are extremely good as protecting groups for aldehydes and ketones. Acetals and ketals are stable to nucleophiles and basic conditions and so the carbonyl group is 'disguised' and will not react with these reagents. Cyclic acetals and ketals are best used for the protection of aldehydes and ketones. These can be synthesized by using diols rather than alcohols (Fig. 10.37).

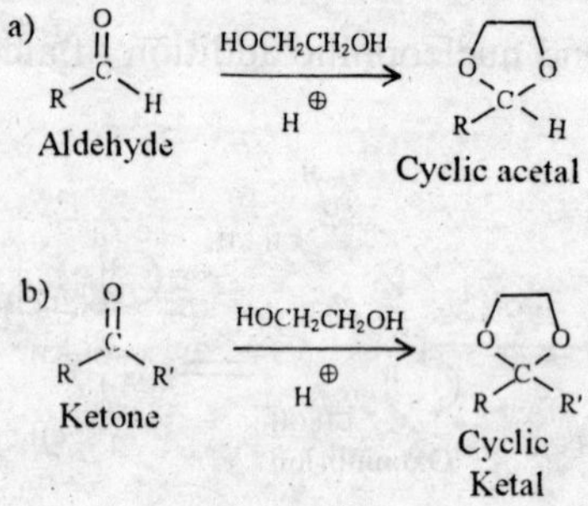

a) Aldehyde $\xrightarrow[H^{\oplus}]{HOCH_2CH_2OH}$ Cyclic acetal

b) Ketone $\xrightarrow[H^{\oplus}]{HOCH_2CH_2OH}$ Cyclic Ketal

Fig. 10.37. Synthesis of cyclic acetals and cyclic ketals.

Hemiacetals and Hemiketals

When aldehydes and ketones are dissolved in alcohol without an acid catalyst being present, only the first part of the above mechanism occurs with one alcohol molecule adding to the carbonyl group. An equilibrium is set up between the carbonyl group and the hemiacetal or hemiketal, with the equilibrium favoring the carbonyl compound (Fig. 10.38).

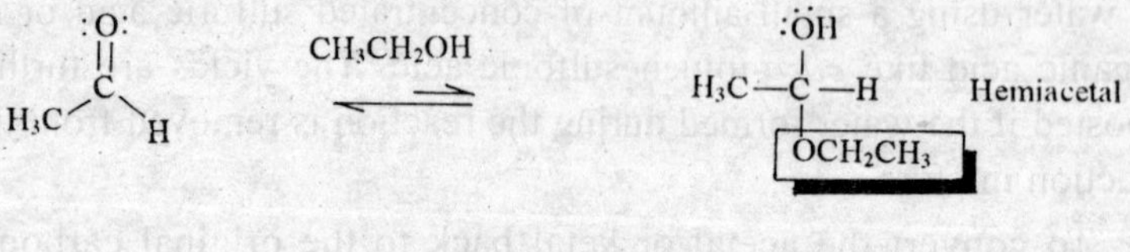

$H_3C-CHO \underset{}{\overset{CH_3CH_2OH}{\rightleftharpoons}} H_3C-CH(OH)(OCH_2CH_3)$ Hemiacetal

Fig. 10.38. Hemiacetal formation.

The reaction is not synthetically useful, because it is not generally possible to isolate the products. If the solvent is removed, the equilibrium is driven back to starting materials. However, cyclic hemiacetals are important in the chemistry of sugars.

Thioacetal and Thioketal Formation

Thioacetals and thioketals are the sulfur equivalents of acetals and ketals and are also prepared under acid conditions (Fig. 10.39). These can also be used to protect aldehydes and ketones, but the hydrolysis of these groups is more difficult. Moreover, the thioacetals and thioketals can be removed by reduction and this provides a method of reducing aldehydes and ketones.

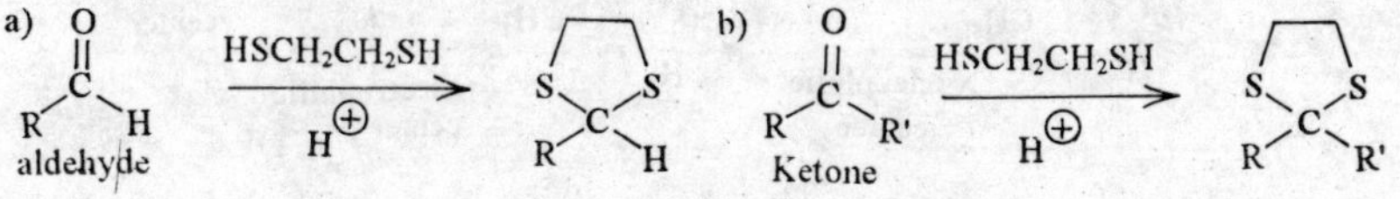

Fig. 10.39. Formation if (a) cyclic thioacetals and (b) cyclic thioketals.

10.8 Reactions of the Enolate Ions

Enolate Ions

Enolate ions are formed by treating aldehydes or ketones with a base. An α-proton must be present. Enolate ions can undergo a number of important reactions including alkylation and the **Aldol reactions.**

Alkylation

When an enolate ion is treated with an alkyl halide it results in a reaction called **alkylation** (Fig. 10.40). The overall reaction involves the replacement of an α-proton with an alkyl group. The nucleophilic and electrophilic centers of the enolate ion and methyl iodide are shown (Fig. 10.41). The enolate ion has its negative charge shared between the oxygen atom and the carbon atom because of resonance and so both of these atoms are nucleophilic centers. Iodomethane has a polar C–I bond where the iodine is a weak nucleophilic center and the carbon is a good electrophilic center.

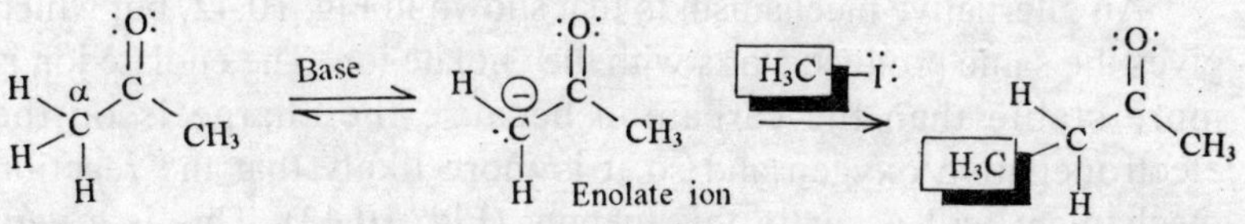

Fig. 10.40. Alkylation of a ketone.

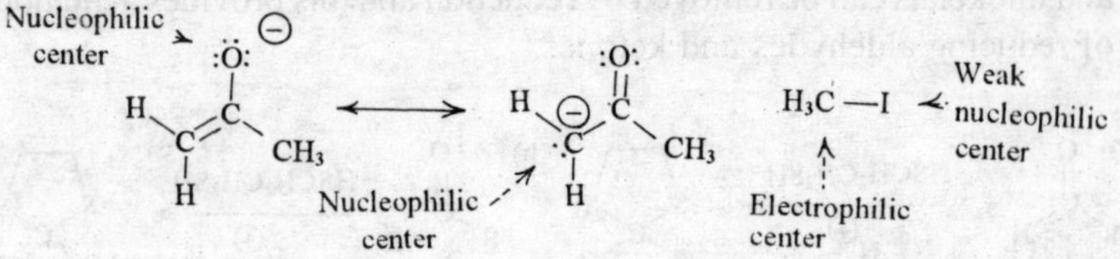

Fig. 10.41. Nucleophilic and electrophilic centers.

The possible reaction between these molecules involves the nucleophilic oxygen using one of its lone pairs of electrons to form a new bond to the electrophilic carbon on iodomethane (Fig. 10.42). At the same time, the C–I bond and both electrons move onto iodine to give it a fourth lone pair of electrons and a negative charge. This reaction is possible, but the actual product obtained is more like to arise from the reaction of the alternative carbanion structures reacting with methyl iodide (Fig. 10.42). This is a more useful reaction as it involves the formation of a carbon–carbon bond and allows the construction of more complex carbon skeletons.

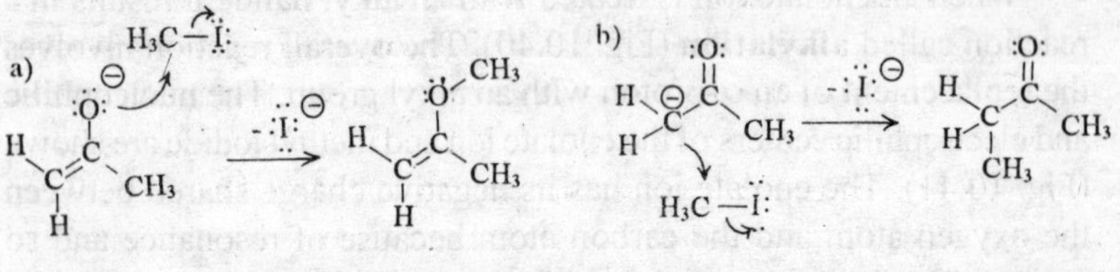

Fig. 10.42. (a) O-Alkylation; (b) C-Alkylation.

An alternative mechanism to that shown in Fig. 10.42, but which gives the same product, starts with the enolate ion. The enolate ion is more stable than the carbanion because the charge is on the electronegative oxygen and so it is more likely that the reaction mechanism will occur in this manner (Fig. 10.43). This is a very useful reaction in organic synthesis. However, there are limitations to the type of alkyl halide which can be used in the reaction. The reaction is S_{N2} with respect to the alkyl halide and so the reaction is suitable with primary alkyl, primary benzylic, and primary allylic halides. The enolate ion is a strong base and if it is reacted with secondary and tertiary halides, elimination of the alkyl halide occurs to give an alkene.

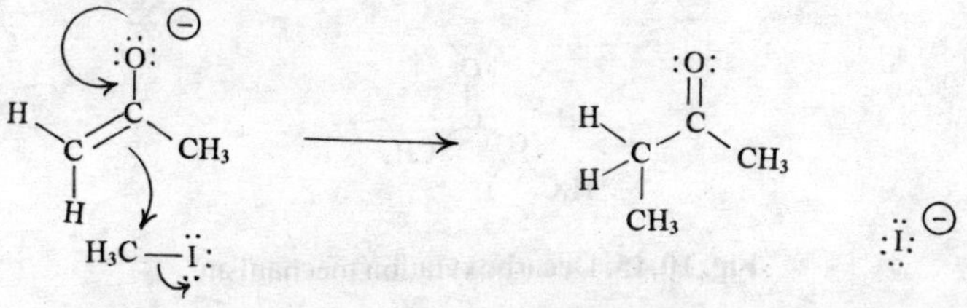

Fig. 10.43. Mechanism for C-alkylation of the enolate ion.

α-Alkylation occurs easily with ketones, but not so easily for aldehydes because the latter tend to undergo Aldol condensations instead.

The α-protons of a ketone like propanone are only weakly acidic and so a powerful base (e.g. lithium diisopropylamide) is required to generate the enolate ion needed for the alkylation. An alternative method of preparing the same product by using a milder base is to start with ethyl acetoacetate (a β-keto ester) (Fig. 10.44). The α-protons in this structure are more acidic because they are flanked by two carbonyl groups. Thus, the enolate can be formed using a weaker base like sodium ethoxide. Once the enolate has been alkylated, the ester group can be hydrolyzed and decarboxylated on heating with aqueous hydrochloric acid. The decarboxylation mechanism involves the β-keto group and would not take place if this group was absent (Fig.10.45). Carbon dioxide is lost and the enol tautomer is formed. This can then form the keto tautomers by the normal keto–enol tautomerism.

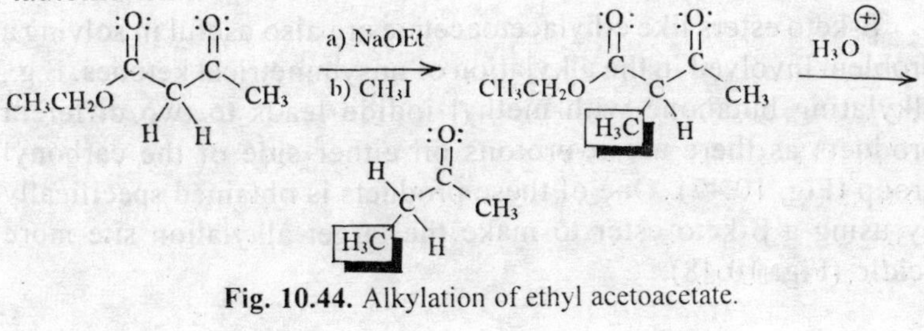

Fig. 10.44. Alkylation of ethyl acetoacetate.

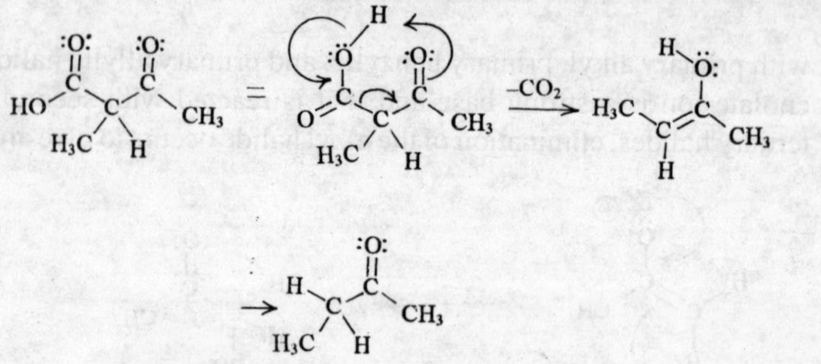

Fig. 10.45. Decarboxylation mechanism.

It is possible for two different alkylations to be done on ethyl acetoacetate as there is more than one α-proton present (Fig. 10.46).

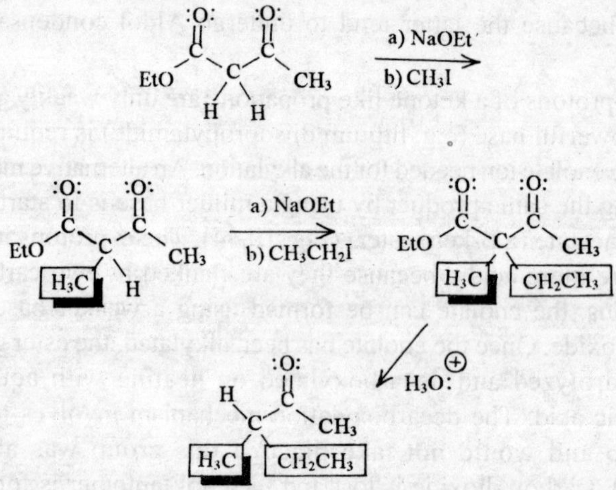

Fig. 10.46. Double alkylation of ethylacetoacetate.

β-keto esters like ethylacetoacetate are also useful in solving a problem involved in the alkylation of unsymmetrical ketones. E.g., alkylating butanone with methyl iodide leads to two different products as there are α-protons on either side of the carbonyl group (Fig. 10.47). One of these products is obtained specifically by using a β-keto ester to make the target alkylation site more acidic (Fig. 10.48).

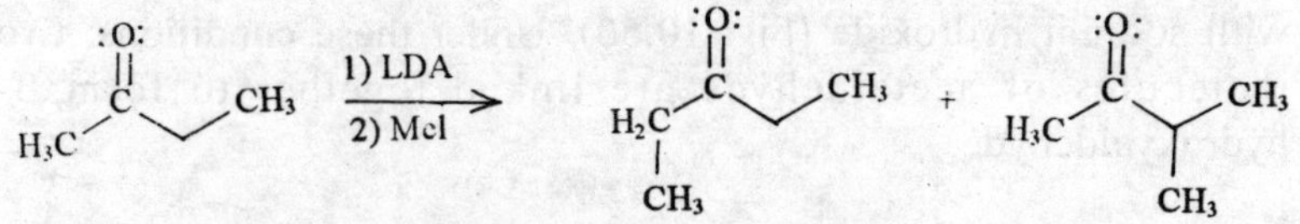

Fig. 10.47. Alkylation of butanone.

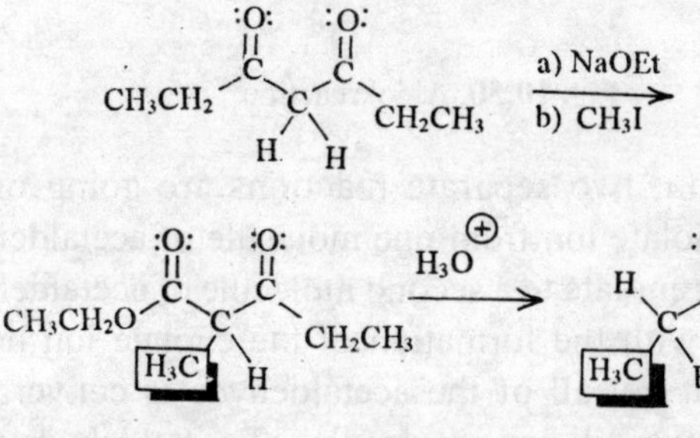

Fig. 10.48. Use of a β-keto ester to direct alkylation.

The alternative alkylation product could be obtained by using a different β-keto ester (Fig. 10.49).

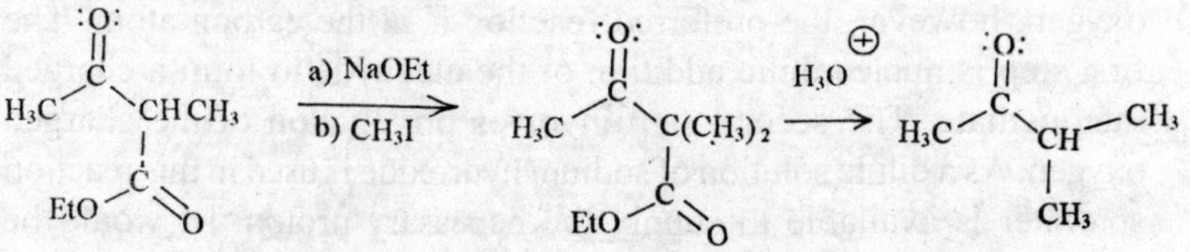

Fig. 10.49. Use of a β-keto ester to direct alkylation.

Aldol Reaction

Enolate ions can also react with aldehydes and ketones by nucleophilic addition. The enolate ion acts as the nucleophile while the aldehyde or ketone acts as an electrophile. As the enolate ion is formed from a carbonyl compound itself, and can then react with a carbonyl compound, it is possible for an aldehyde or ketone to react with itself. To illustrate this we look at the reaction of acetaldehyde with sodium hydroxide (Fig. 10.50). Under these conditions, two molecules of acetaldehyde are linked together to form β-hydroxyaldehyde.

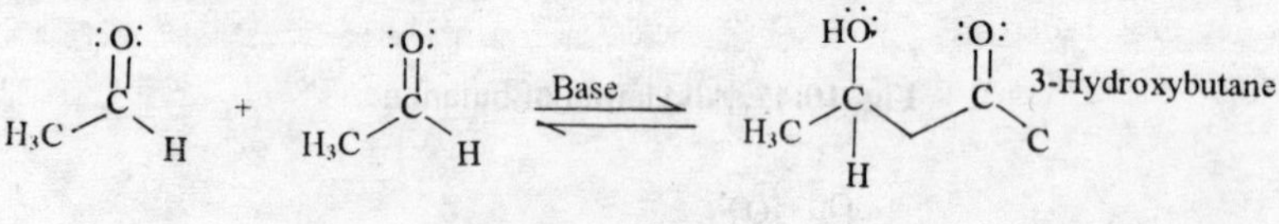

Fig. 10.50. Aldol reaction.

In this reaction, two separate reactions are going on i.e. the formation of an enolate ion from one molecule of acetaldehyde, and the addition of that enolate to a second molecule of acetaldehyde. The mechanism starts with the formation of the enolate ion however it must be noted that not all of the acetaldehyde is converted to the enolate ion and so we still have molecules of acetaldehyde present in the same solution as the enolate ions. Since acetaldehyde is susceptible to nucleophilic attack, the next stage in the mechanism is the nucleophilic attack of the enolate ion on acetaldehyde (Fig. 10.51). The enolate ion has two nucleophilic centers i.e. the carbon and the oxygen, however, the preferred reaction is at the carbon atom. The first step is nucleophilic addition of the aldehyde to form a charged intermediate. The second step involves protonation of the charged oxygen. As a dilute solution of sodium hydroxide is used in this reaction so water is available to supply the necessary proton. [It would be wrong to show a free proton (H^+) because the solution is alkaline.]

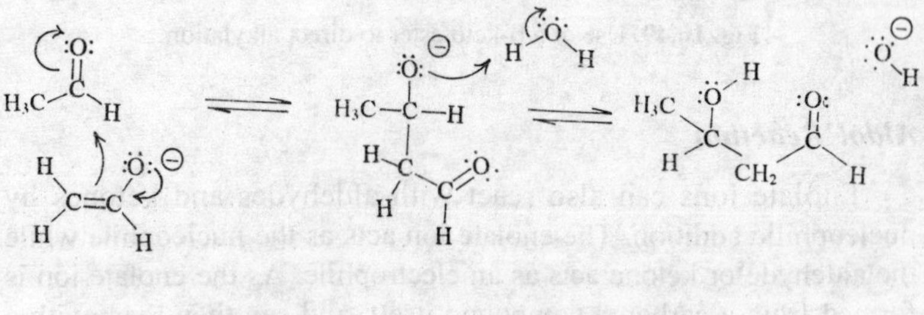

Fig. 10.51. Mechanism of the Aldol reaction.

If the above reaction is carried out with heating, then a different product is obtained (Fig. 10.52). This arises from elimination of a molecule of water from the *Aldol reaction product*. The two reasons for occurrence of this reaction are: First, the product still has an acidic proton (i.e. there is still a carbonyl group present and an α-hydrogen next to it). This proton is prone to attack from base. Second, the dehydration process results in a conjugated product that results in increased stability. The mechanism of dehydration is shown in Fig. 10.53. First of all, the acidic proton is removed and a new enolate ion is formed. The electrons in the enolate ion can then move in such a fashion that the hydroxyl group is expelled to give the final product i.e. an α,β-unsaturated aldehyde. In this example, it is possible to change the conditions such that one gets the Aldol reaction product or the α,β-unsaturated aldehyde, but in some cases only the α,β-unsaturated carbonyl product is obtained, particularly when extended conjugation is possible. The Aldol reaction is best carried out with aldehydes. Some ketones will undergo an Aldol reaction, but an equilibrium is set up between the products and starting materials and it is essential to remove the product as it is being formed so as take the reaction to completion.

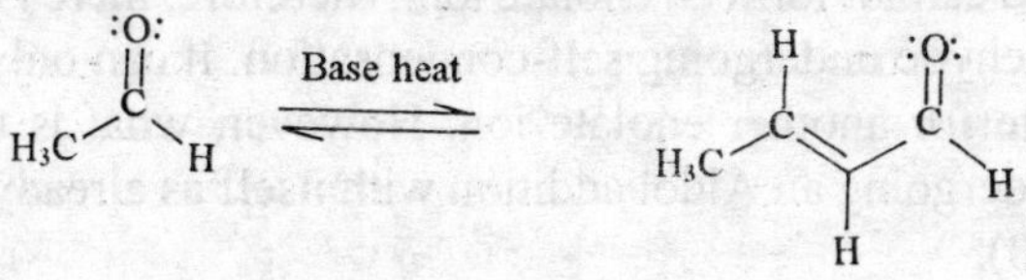

Base heat

Fig. 10.52. Formation of 2-butenal.

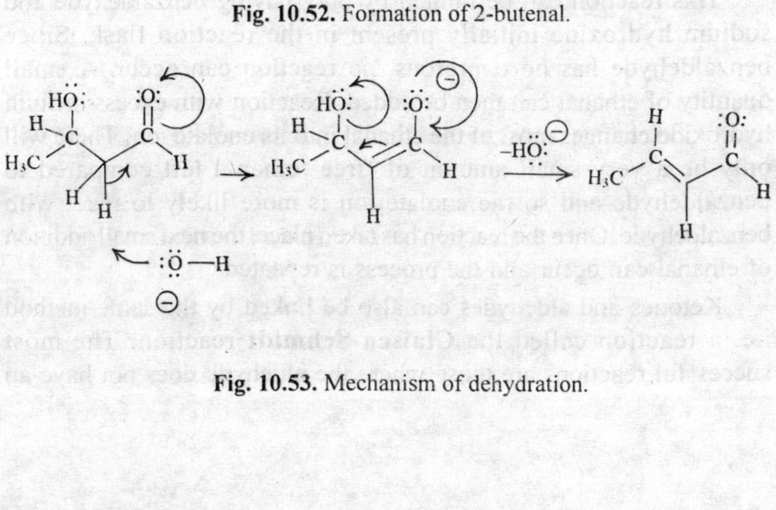

Fig. 10.53. Mechanism of dehydration.

Crossed Aldol Reaction

An Aldol reaction is used to link two molecules of the same aldehyde or ketone, however it is also possible to link two different carbonyl compounds. This is called a crossed Aldol reaction. E.g. benzaldehyde and ethanal can be linked in the presence of sodium hydroxide (Fig. 10.54). In this, ethanal reacts with sodium hydroxide to form the enolate ion which then reacts with benzaldehyde. Elimination of water takes place easily to give an extended conjugated system involving the aromatic ring, the double bond, and the carbonyl group.

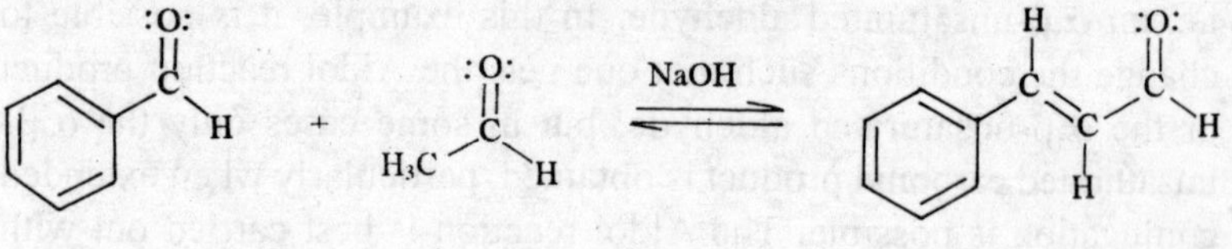

Fig. 10.54. Crossed Aldol reaction.

This reaction proceeds well because the benzaldehyde has no α-protons and cannot form an enolate ion. Therefore, there is no chance of benzaldehyde undergoing self-condensation. It can only act as the electrophile for another enolate ion. However, what is to stop the ethanal undergoing an Aldol addition with itself as already described (Fig. 10.50).

This reaction can be limited by only having benzaldehyde and sodium hydroxide initially present in the reaction flask. Since benzaldehyde has no α-protons, no reaction can occur. A small quantity of ethanal can then be added. Reaction with excess sodium hydroxide changes most of the ethanal into its enolate ion. There will only be a very small amount of 'free' ethanal left compared to benzaldehyde and so the enolate ion is more likely to react with benzaldehyde. Once the reaction has taken place, the next small addition of ethanal can occur and the process is repeated.

Ketones and aldehydes can also be linked by the same method i.e. a reaction called the **Claisen-Schmidt** reaction. The most successful reactions are those where the aldehyde does not have an α-proton (Fig.10.55).

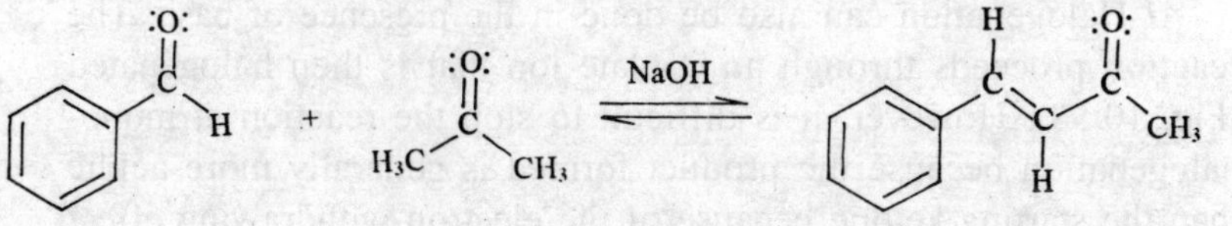

Fig. 10.55. Claisen-Schmidt reaction.

10.9 α–Halogenation

Definition

Aldehydes and ketones react with chlorine, bromine or iodine in acidic solution, this is called halogenation at the α-carbon (Fig. 10.56).

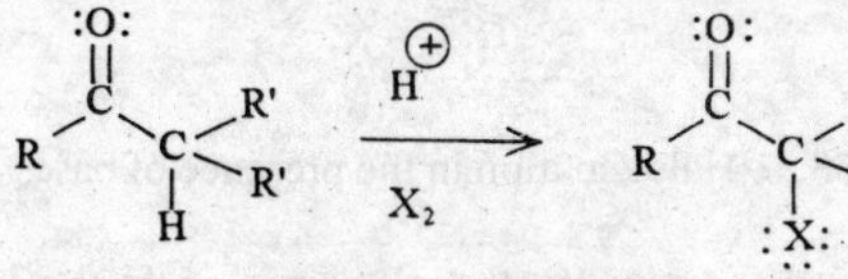

Fig. 10.56. α-Halogenation.

Mechanism

Because acid conditions are used so this process does not involve an enolate ion. Instead, the reaction occurs through the enol tautomers of the carbonyl compound. The enol tautomer acts as a nucleophile with a halogen by the mechanism shown (Fig. 10.57). In the final step, the solvent acts as a base to remove the proton.

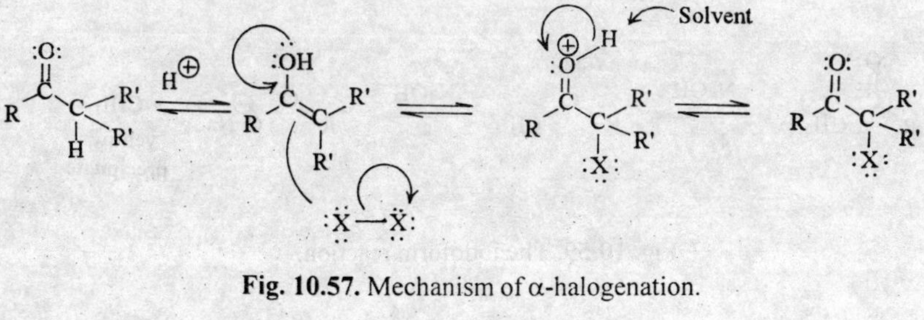

Fig. 10.57. Mechanism of α-halogenation.

Iodoform Reaction

α-Halogenation can also be done in the presence of base. The reaction proceeds through an enolate ion that is then halogenated (Fig. 10.58). However, it is difficult to stop the reaction at mono-halogenation because the product formed is generally more acidic than the starting ketone because of the electron-withdrawing effect of the halogen. Due to this, another enolate ion is quickly formed leading to further halogenation.

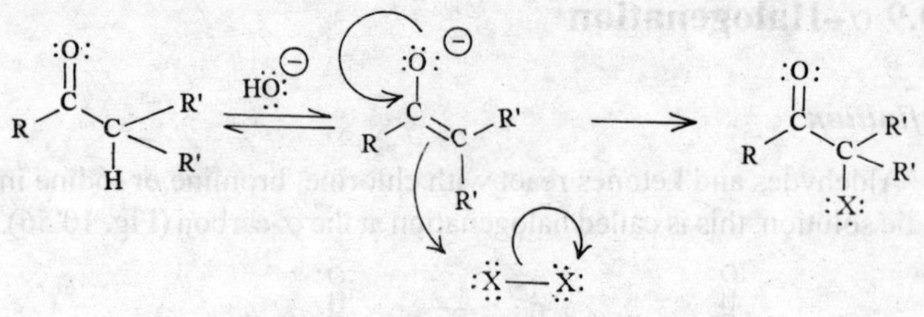

Fig. 10.58. α-Halogenation in the presence of base.

This tendency towards multiple halogenation is the basis for the ***iodoform test*** which is used to identify methyl ketones. The ketone to be tested is treated with excess iodine and base and if a yellow precipitate is formed, a methyl ketone is indicated. Under these conditions, methyl ketones undergo α-halogenation three times (Fig. 10.59). The product obtained is then susceptible to nucleophilic substitution whereby the hydroxide ion substitutes the tri-iodomethyl ($^{-}CI_3$) carbanion–a good leaving group because of the three electron-withdrawing iodine atoms. Tri-iodomethane is then formed as the yellow precipitate.

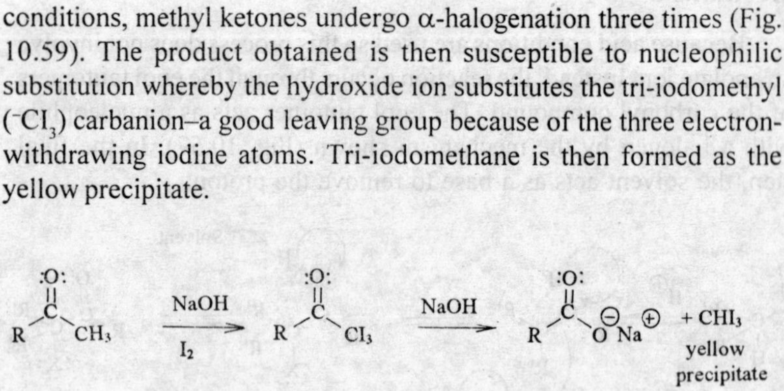

$$RCOCH_3 \xrightarrow[I_2]{NaOH} RCOCI_3 \xrightarrow{NaOH} RCOO^{\ominus}Na^{\oplus} + CHI_3$$

yellow precipitate

Fig. 10.59. The iodoform reaction.

10.10 Reduction and Oxidation

Reduction to Alcohols

Aldehydes and ketones when reduced yield alcohols with a hydride ion that is provided by reducing reagents like sodium borohydride or lithium borohydride. Primary alcohols are obtained from aldehydes and secondary alcohols from ketones.

Reduction to Alkanes

Aldehydes and ketones on complete reduction give alkanes by three different methods that are complementary to each other. The **Woff-Kishner** reduction is done under basic conditions and is suitable for compounds that might be sensitive to acid (Fig. 10.60). The reaction involves the nucleophilic addition by hydrazine followed by elimination of water to form a hydrazone. The mechanism is the same as that described for the synthesis of 2,4-dinitrophenylhydrazones.

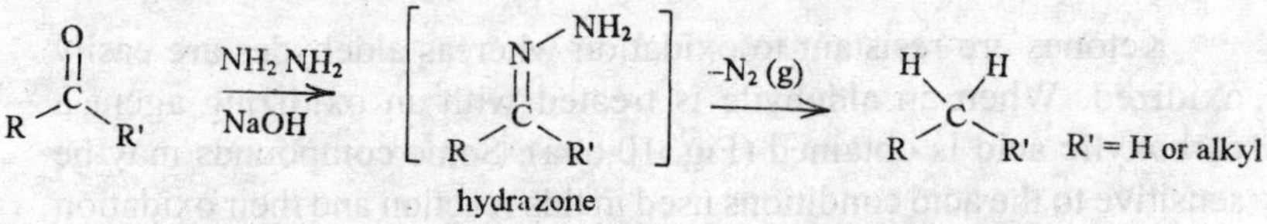

$$RC(=O)R' \xrightarrow[NaOH]{NH_2NH_2} \left[RC(=N{-}NH_2)R' \right]_{\text{hydrazone}} \xrightarrow{-N_2(g)} RCH_2R' \quad R' = H \text{ or alkyl}$$

Fig. 10.60. Wolf-Kishner reduction.

But the simple hydrazone formed under these reaction conditions spontaneously decomposes with the loss of nitrogen gas.

The **Clemmenson** reduction (Fig. 10.61) gives a similar product but is done under acid conditions and so this is a suitable method for compounds which are unstable to basic conditions.

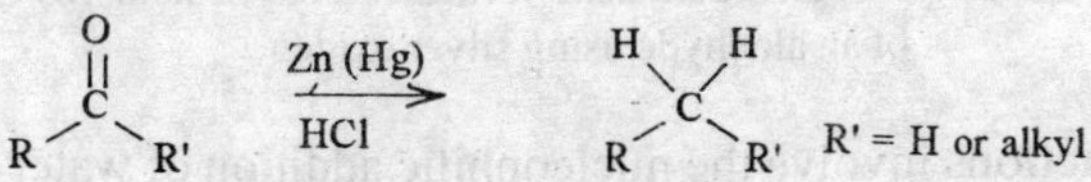

$$RC(=O)R' \xrightarrow[HCl]{Zn(Hg)} RCH_2R' \quad R' = H \text{ or alkyl}$$

Fig. 10.61. Clemmenson reduction.

Compounds that are sensitive to both acid and base can be reduced under neutral conditions by forming the thioacetal or thioketal, then reducing with Raney nickel (Fig. 10.62).

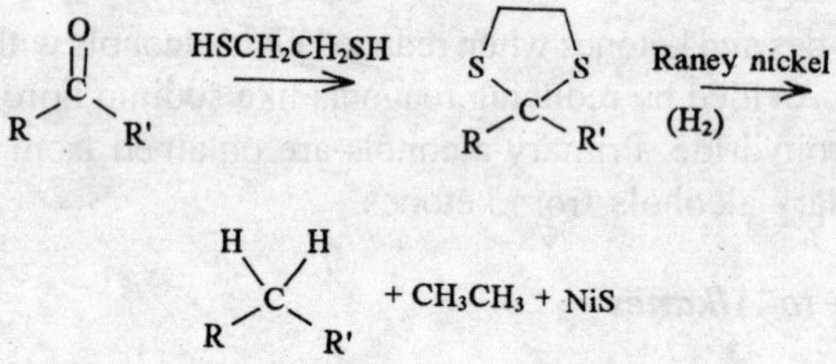

$$R-C(=O)-R' \xrightarrow{HSCH_2CH_2SH} \text{(cyclic thioketal, } R(R')C\text{ in S,S ring)} \xrightarrow[(H_2)]{\text{Raney nickel}} R-CH_2-R' + CH_3CH_3 + NiS$$

Fig. 10.62. Reduction via a cyclic thioketal.

Aromatic aldehydes and ketones can also be deoxygenated with hydrogen over a palladium charcoal catalyst. The reaction occurs because the aromatic ring activates the carbonyl group towards reduction. *Aliphatic aldehydes and ketones are not reduced* in this.

Oxidation

Ketones are resistant to oxidation whereas aldehydes are easily oxidized. When an aldehyde is treated with an oxidizing agent a carboxylic acid is obtained (Fig. 10.63a). Some compounds may be sensitive to the acid conditions used in this reaction and their oxidation can be carried out by using a basic solution of silver oxide (Fig. 10.63b).

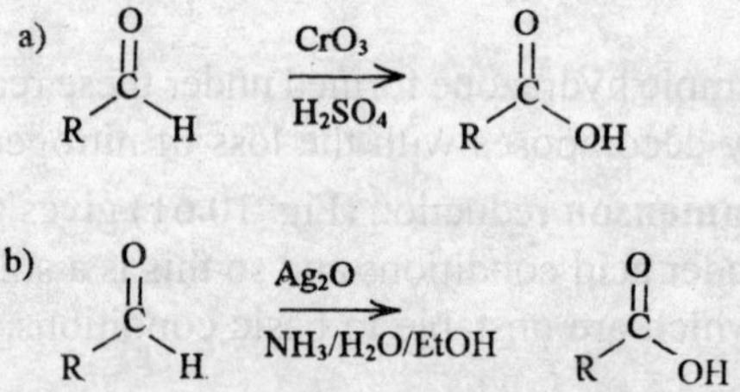

a) $$R-C(=O)-H \xrightarrow[H_2SO_4]{CrO_3} R-C(=O)-OH$$

b) $$R-C(=O)-H \xrightarrow[NH_3/H_2O/EtOH]{Ag_2O} R-C(=O)-OH$$

Fig. 10.63. (a) Oxidation of an aldehyde to form a carboxylic acid, (b) Oxidation of an aldehyde using silver oxide.

Both reactions involve the nucleophilic addition of water to form a 1,1-diol or hydrate which is then oxidized in the same way as an alcohol (Fig. 10.64).

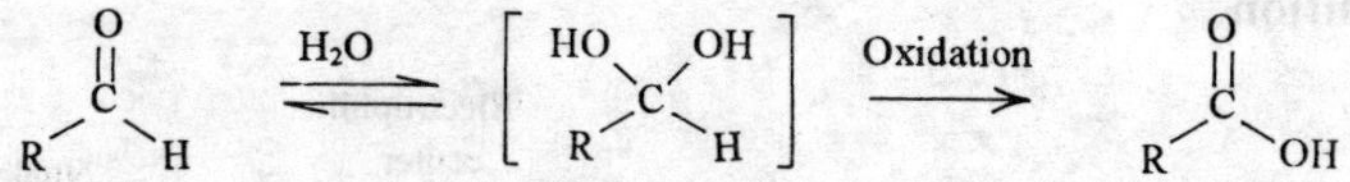

Fig. 10.64. 1,1-Diol intermediate.

10.11 α, β–Unsaturated Aldehydes and Ketones

Definition

α, β-Unsaturated aldehydes and ketones are aldehydes and ketones that are conjugated with a double bond. The α-position is defined as the carbon atom next to the carbonyl group, while the β-position is the carbon atom two bonds removed (Fig. 10.65).

Fig. 10.65. (a) α,β-unsaturated aldehyde; (b) α,β-unsaturated ketone.

Nucleophilic and Electrophilic Centers

The carbonyl group of α,β-unsaturated aldehydes and ketones are made up nucleophilic oxygen and an electrophilic carbon. However, α, β-unsaturated aldehydes and ketones also have another electrophilic carbon–the β-carbon. This is because of the influence of the electronegative oxygen that can result in the resonance shown (Fig. 10.66). Since two electrophilic centers are present, there are two places where a nucleophile can react. In both situations, an addition reaction occurs. If the nucleophile reacts with the carbonyl carbon, this is a normal nucleophilic addition to an aldehyde or ketone and is known as a **1,2-nucleophilic addition**. If the nucleophilc adds to the β-carbon, this is called a **1,4-nucleophilic addition** or a **conjugate addition**.

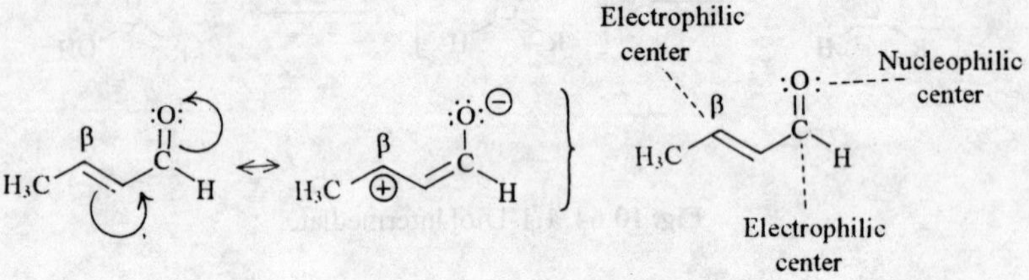

Electrophilic center

Nucleophilic center

Electrophilic center

Fig. 10.66. Nucleophilic and electrophilic centers.

1,2-Addition

The mechanism of 1,2-nucleophilic addition is the same as already described. It is found that Grignard reagents and organolithium reagents will react with α,β-unsaturated aldehydes and ketones in this way and do not attack the β-position (Fig. 10.67).

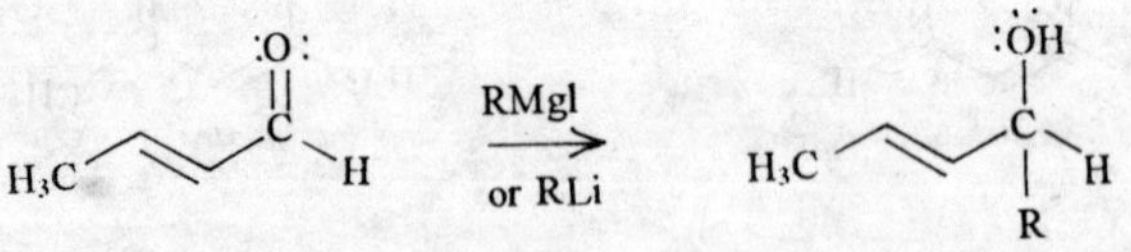

RMgl or RLi

Fig. 10.67. 1,2-Nucleophilic addition.

1,4-Addition

The mechanism for 1,4-addition involves two stages (Fig. 10.68), In the first stage, the nucleophile uses a lone pair of electrons to form a bond to the β-carbon. Simultaneously, the C=C π bond breaks and both electrons are used to form a new π bond to the carbonyl carbon. This in turn forces the carbonyl π bond to break with both of the electrons involved moving onto the oxygen as a third lone pair of electrons. The resulting intermediate is an enolate ion. Aqueous acid is now added to the reaction mixture. The carbonyl π bond is reformed, which forces open the C=C π bond. These electrons are now used to form a σ bond to a proton at the α carbon.

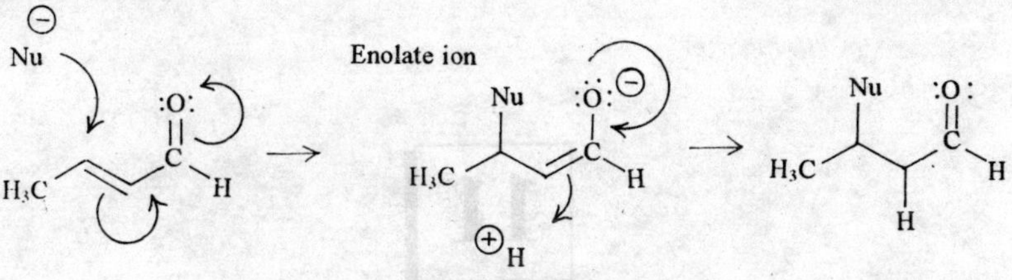

Fig. 10.68. Mechanism of 1,4-nucleophilic addition.

Conjugate addition reactions can be done with amines, or a cyanide ion. Alkyl groups can also be added to the β-position by using organocuprate reagents (Fig. 10.69). A large number of organocuprate reagents can be prepared allowing the addition of primary, secondary and tertiary alkyl groups, aryl groups, and alkenyl groups.

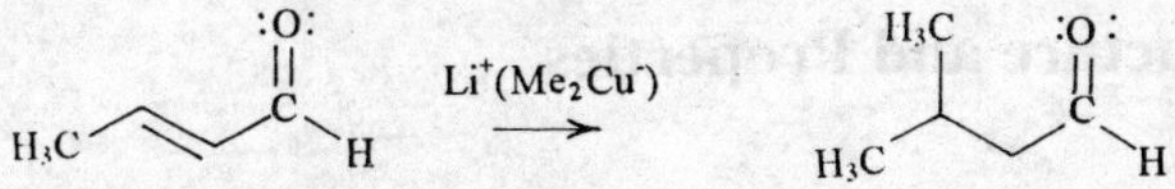

$Li^+(Me_2Cu^-)$

Fig. 10.69. Alkylation with organocuprate reagents.

Reduction

The reduction of α,β-unsaturated ketones to allylic alcohols can be easily done using lithium aluminum hydride under carefully controlled conditions (Fig. 10.70). With sodium borohydride, some reduction of the alkene also occurs.

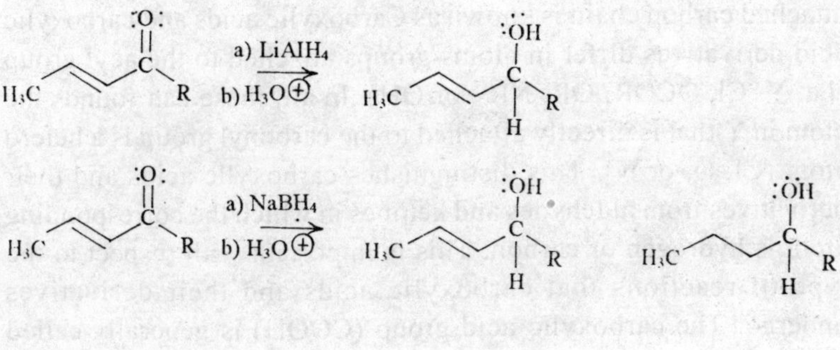

a) $LiAlH_4$
b) $H_3O^{\oplus}$

a) $NaBH_4$
b) $H_3O^{\oplus}$

11

Carboxylic Acids and their Derivatives

11.1 Structure and Properties

Structure

The structures of derivatives of carboxylic acids are derived from the parent carboxylic acid structure. The four common types of acid derivative are acid chlorides, acid anhydrides, esters, and amides (Fig. 11.1). These functional groups contain a carbonyl group (C=O) in which both atoms are sp^2 hybridized (Fig. 11.2). The carbonyl group along with the two neighboring atoms is planar with bond angles of 120°. The carbonyl group along with the attached carbon chain is known as Carboxylic acids and carboxylic acid derivatives differ in atoms/groups attached to the acyl group (i.e. Y=Cl, OCOR, OR, NR_2, or OH). In all, these can founds the atom in Y that is directly attached to the carbonyl group is a hetero atom (Cl, O, or N). This distinguishes carboxylic acids and their derivatives from aldehydes and ketones in which the corresponding atom is hydrogen or carbon. This is important with respect to the type of reactions that carboxylic acids and their derivatives undergo. The carboxylic acid group (COOH) is generally called **carboxyl group.**

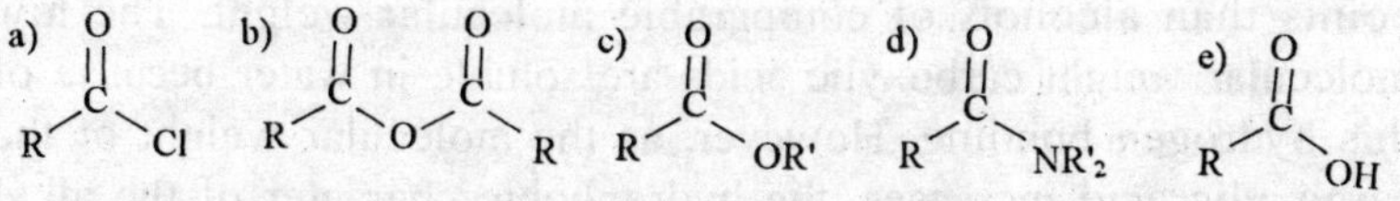

Fig. 11.1. (a) Acid chloride; (b) acid anhydride; (c) ester; (d) amide; (e) carboxylic acid.

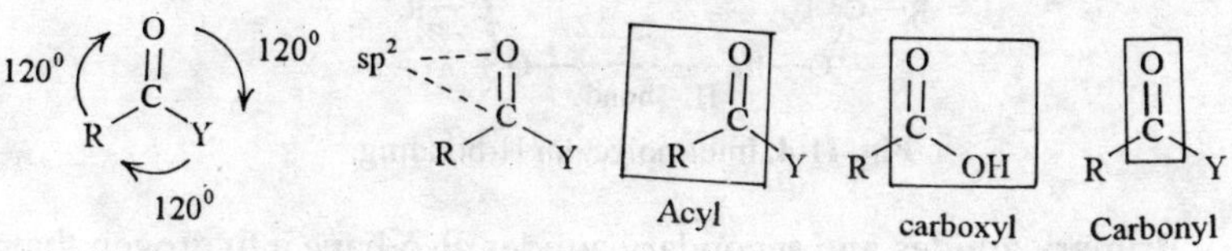

Fig. 11.2. Structure of the functional group.

Bonding

The bonds in the carbonyl C=O group consist of a strong σ bond and a weaker π bond (Fig. 11.3). As oxygen is more electronegative than carbon, the carbonyl group is polarized in such a way that the oxygen is slightly negative and the carbon is slightly positive, so oxygen can act as a nucleophilic center and carbon can act as an electrophilic center.

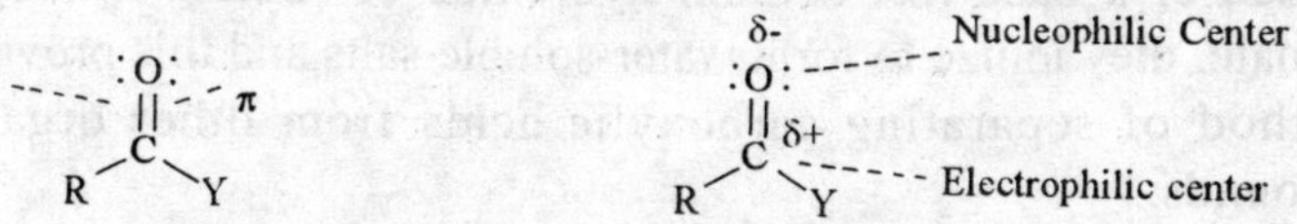

Fig. 11.3. Bonding and properties.

Properties

Carboxylic acids and their derivatives are polar molecules because of the polar nature of carbonyl group and the presence of a heteroatom in the group Y. Carboxylic acids can associate with each other as dimers (Fig. 11.4) through the formation of two intermolecular hydrogen bonds due to this the carboxylic acids have higher boiling points than alcohols of comparable molecular weight. The low molecular weight carboxylic acids are soluble in water because of this hydrogen bonding. However, as the molecular weight of the carboxylic acid increases, the hydrophobic character of the alkyl portion outweighs the polar character of the functional group and thus at higher molecular weight carboxylic acids are insoluble in water.

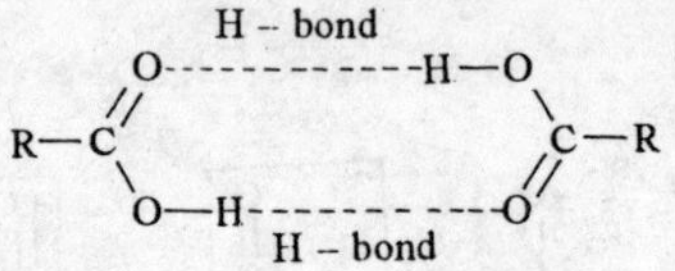

```
              H – bond
         O-----------H—O
        //              \
    R—C                  C—R
        \               //
         O—H-----------O
              H – bond
```

Fig. 11.4. Intermolecular H-bonding.

Primary amides and secondary amides also have a hydrogen that is capable of hydrogen bonding (i.e. $RCON\underline{H}R'$, $RCON\underline{H}_2$), due to this, these compounds also have higher boiling points for as compared to aldehydes and ketones of similar molecular weight. Acid chlorides, acid anhydrides, esters, and tertiary amides are polar, but due to lack a hydrogen atom that is capable of participating in hydrogen bonding, they have lower boiling points than carboxylic acids or alcohols of similar molecular weight, and similar boiling points to comparable aldehydes and ketones.

Carboxylic acids are weak acids in aqueous solution, forming an equilibrium between the free acids and the carboxylic ion. In the presence of a base like sodium hydroxide or sodium hydrogen carbonate, they ionize to form water-soluble salts and this provides a method of separating carboxylic acids from other organic compounds.

Reactions

Carboxylic acids and carboxylic acid derivatives commonly react with nucleophiles in a reaction called **nucleophilic substitution** (Fig. 11.5). These reaction involves replacement of one nucleophile with another. Nucleophilic substitution is possible because the displaced nucleophile contains an electronegative heteroatom (Cl, O, or N) that is capable of stabilizing a negative charge.

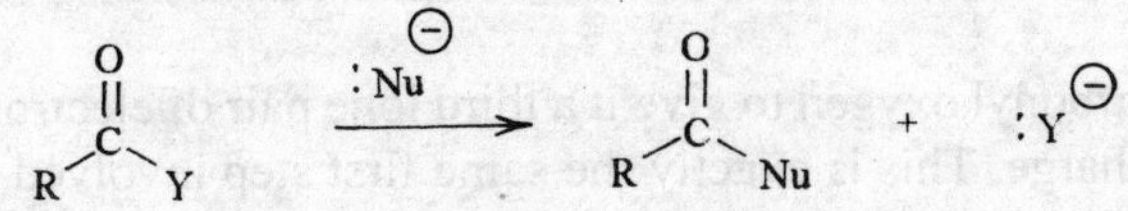

$$R-C(=O)-Y \xrightarrow{:Nu^{\ominus}} R-C(=O)-Nu + :Y^{\ominus}$$

Y = Cl, OCOR, OR, NR_2
Carboxylic acid derivatives

Fig. 11.5. Nucleophilic substitution.

11.2 Nucleophilic Substitution

Definition

Nucleophilic substitutions reactions are those reactions in which the substitution of one nucleophile for another is involved. Alkyl halides, carboxylic acids, and carboxylic acid derivatives undergo nucleophilic substitution. However, the mechanisms involved for alkyl halides are quite different from those involved for carboxylic acids and their derivatives. The reaction of a methoxide ion with ethanoyl chloride is a nucleophilic substitution reaction, (Fig. 11.6), In it one nucleophile (the methoxide ion) substitutes another nucleophile (Cl^-).

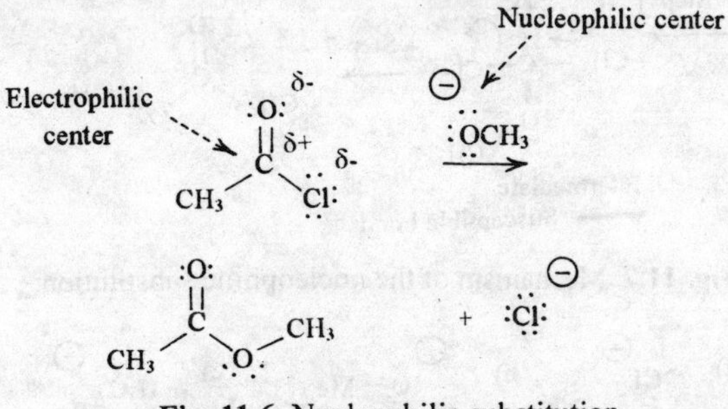

Nucleophilic center

Electrophilic center

$$CH_3-C(=O^{\delta-})^{\delta+}-Cl^{\delta-} \xrightarrow{:OCH_3^{\ominus}} CH_3-C(=O)-O-CH_3 + :Cl:^{\ominus}$$

Fig. 11.6. Nucleophilic substitution

Mechanism Charged Nucleophiles

Making use of the reaction in Fig. 11.6, we will illustrate the mechanism of nucleophilic substitution (Fig. 11.7). The methoxide ion uses of its lone pairs of electrons to form a bond to the electrophilic carbonyl carbon of the acid chloride. Simultaneously, the relatively weak π bond of the carbonyl group breaks and both of the π electrons move onto the carbonyl oxygen to give it a third lone pair of electrons and a negative charge. This is exactly the same first step involved in nucleophilic addition to aldehydes and ketones. However, with an aldehyde or a ketone, the tetrahedral structure is the final product. With carboxylic acid derivatives, the lone pair of electrons on oxygen return to reform the carbonyl π bond (Step 2). As this happens, the C–Cl bond breaks with both electrons moving onto the chlorine to form a chloride ion that departs the molecule. This explains how the products are formed, but why should the C–Cl σ bond break in preference to the C–OMe σ bond or the C–CH_3 σ bond can be explained by looking at the leaving groups which would be formed from these processes (Fig. 11.8). The leaving groups would be a chloride ion, a methoxide ion and a carbanion, respectively. The chloride ion is the best leaving group as it is the most stable. This is because chlorine is more electronegative than oxygen or carbon and can stabilize the negative charge. This same mechanism is involved in the nucleophilic substitutions of all other carboxylic acid derivatives and a general mechanism can be drawn as follow (Fig. 11.9).

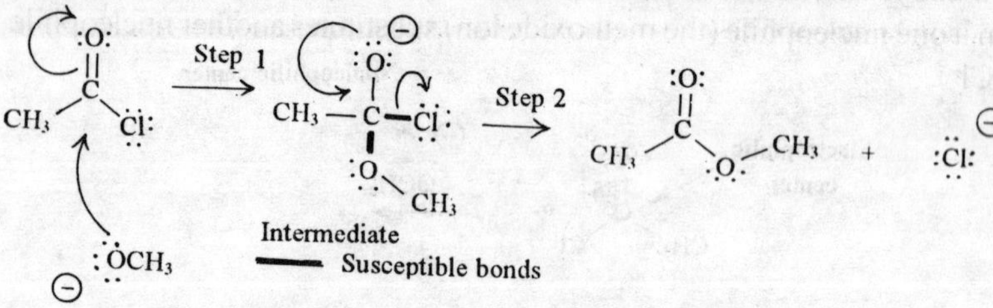

Fig. 11.7. Mechanism of the nucleophilic substitution.

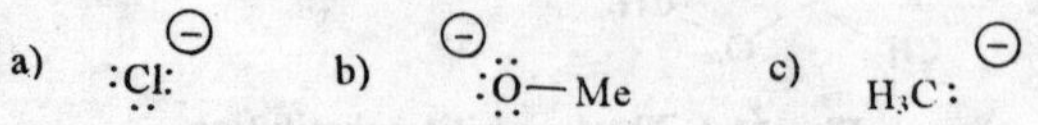

Fig. 11.8. Leaving groups; (a) chloride; (b) methoxide; (c) carbanion.

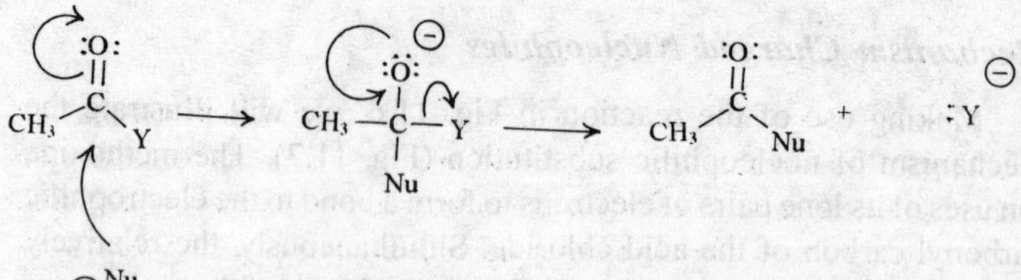

Fig. 11.9. General mechanism for nucleophilic substitution.

Mechanism Neutral Nucleophiles

Acid chlorides are quite reactive ana react with uncharged nucleophiles. For example, ethanoyl chloride react, with methanol to form an ester (Fig. 11.10). Oxygen is the nucleophilic center in methanol and uses one of its lone pairs of electrons to form a new bond to the electrophilic carbon of the acid chloride (Fig. 11.11). As this new bond forms, the carbonyl π bond breaks and both electrons move onto the carbonyl oxygen to give it a third pair of electrons and a negative charge (Step 1). The methanol oxygen gains a positive charge as it has effectively lost an electron by sharing its lone pair with carbon in the new bond. A positive charge on oxygen is not very stable and so the second stage in the mechanism is the loss of a proton. Both electrons in the O–H bond move onto the oxygen to restore a second lone pair of electrons and thus neutralize the charge. Methanol can help the process by acting as a base. The final stage in the mechanism is the same as before. The carbonyl π bond is reformed and as this happens, the C–Cl σ bond breaks with both electrons ending up in the departing chloride ion as a fourth lone pair of electrons. Finally, the chloride anion can remove a proton from $CH_3OH_2^+$ to form HCl and methanol.

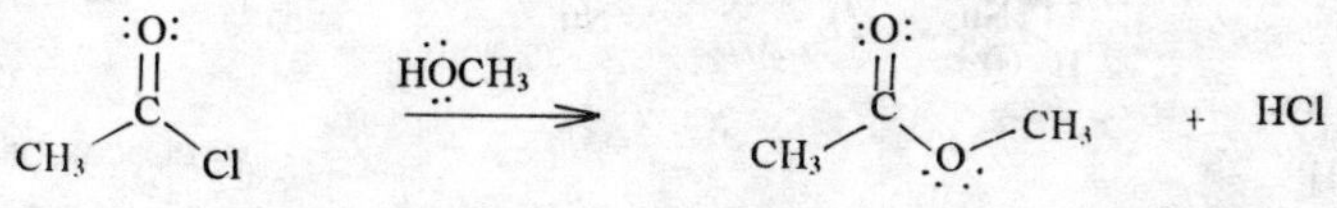

$$CH_3COCl \xrightarrow{HOCH_3} CH_3COOCH_3 + HCl$$

Fig. 11.10. Ethanoyl chloride reacting with methanol to form methyl ethanoate.

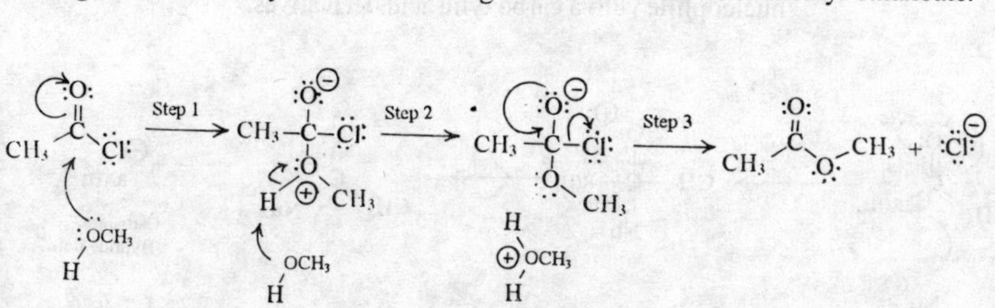

$$CH_3COCl + HOCH_3 \xrightarrow{\text{Step 1}} CH_3-C(O^-)(Cl)-O^+(H)CH_3 \xrightarrow{\text{Step 2}} CH_3-C(O^-)(Cl)-OCH_3 + CH_3OH_2^+ \xrightarrow{\text{Step 3}} CH_3COOCH_3 + Cl^-$$

Fig. 11.11. Mechanism for the reaction of an alcohol with an acid chloride.

The above mechanism is essentially the same mechanism involved in the reaction of ethanoyl chloride with sodium methoxide, the only difference being that we have to remove a proton during the reaction mechanism.

The same mechanism is true for nucleophilic substitutions of other carboxylic acid derivatives with neutral nucleophiles and we can write a general mechanism as follows (Fig. 11.12). In practice, acids or bases are generally added to improve yields.

Addition vs Substitution

Carboxylic acid derivatives undergo nucleophilic substitution whereas aldehydes and ketones undergo nucleophilic addition. This is because nucleophilic substitution of a ketone or an aldehyde would generate a carbanion or a hydride ion respectively (Fig. 11.13). These ions are unstable and highly reactive, so they are only formed with difficulty. Furthermore, C–C and C–H σ bonds are easily broken. Therefore, nucleophilic substitutions of aldehydes or ketones are not feasible.

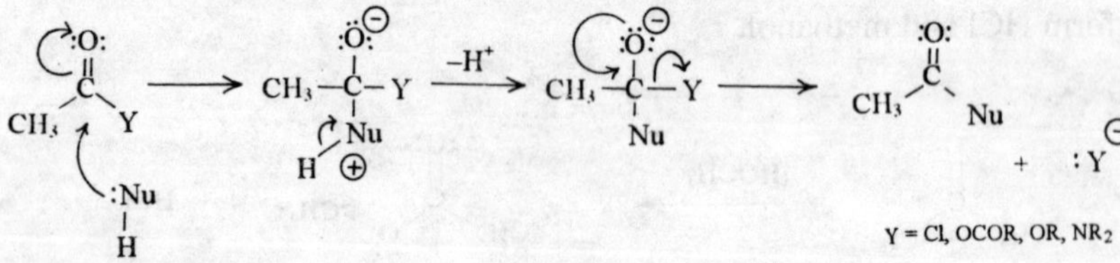

Y = Cl, OCOR, OR, NR_2

Fig. 11.12. General mechanism for the nucleophilic substitution of a neutral nucleophile with a carboxylic acid derivatives.

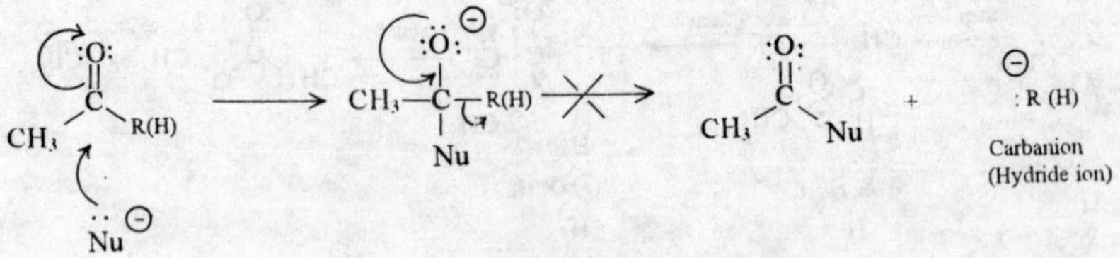

Fig. 11.13. Unfavorable formation of an unstable carbanion or hydride ion.

11.3 Reactivity

Reactivity Order

Acid chlorides can be converted to acid anhydrides, esters, or amides. These reactions are possible because acid chlorides are the most reactive of the four carboxylic acid derivatives. Nucleophilic substitutions of the other acid derivatives are more limited because they are less reactive. For example, acid anhydrides can be used to synthesize esters and amides, but cannot be used to synthesize acid chlorides. The possible nucleophilic reactions for each carboxylic acid derivative depends on its reactivity with respect to the other acid derivatives (Fig. 11.14). Reactive acid derivatives can be converted to less reactive (more stable) acid derivatives, but not the other way round. For example, an ester can be converted to an amide, but not to an acid anhydride.

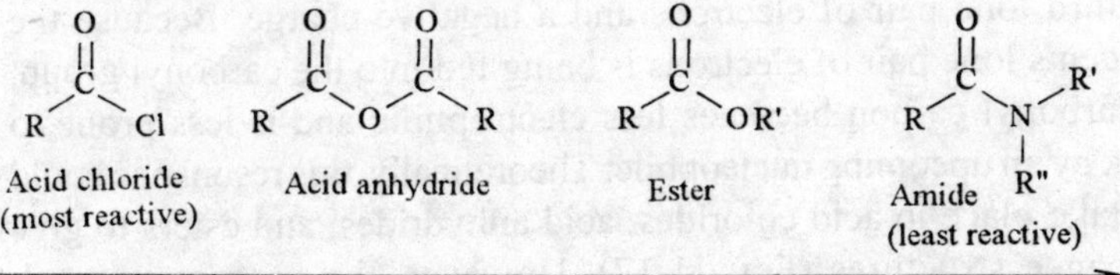

Fig. 11.14. Relative reactivity of carboxylic acid derivatives.

Electronic Factors

To understand this difference in reactivity of various acid derivatives look at the first step in the nucleophilic substitution mechanism (involving the addition of a nucleophile to the electrophilic carbonyl carbon) which is the rate-determining step. Therefore, the more electrophilic this carbon is, the more reactive it will be. The nature of Y has a significant effect in this respect (Fig. 11.15).

Y is linked to the acyl group by an electronegative heteroatom (Cl, O, or N) that makes the carbonyl carbon more electrophilic. The extent to which this happens depends on the electronegativity of Y. If Y is strongly electronegativity (e.g. chlorine), it has a strong electron-withdrawing effect on the carbonyl carbon making it more electrophilic and more reactive to nucleophiles. Because chlorine is more electronegative than oxygen, and oxygen is more electronegative than nitrogen, acid chlorides are more reactive than acid anhydrides and esters, while acid anhydrides and esters are more reactive than amides.

electrophilic center → $H_3C-C(=O)-Y$ with δ- on O, δ+ on C, δ- on Y

Y = Cl, OCOR, OR, NR_2

Fig. 11.15. The electrophilic center of a carboxylic acid derivative.

The electron-withdrawing effect of Y on the carbonyl carbon is an inductive effect. With amides, there is an important resonance contribution that **decreases** the electrophilicity of the carbonyl carbon (Fig. 11.16). The nitrogen has a lone pair of electrons that can form a bond to the neighboring carbonyl carbon. As this new bond is formed, the weak π bond breaks and both electrons move onto oxygen to give it a third lone pair of electrons and a negative charge. Because the nitrogen's lone pair of electrons is being fed into the carbonyl group, the carbonyl carbon becomes less electrophilic and is less prone to attack by an upcoming nucleophile. Theoretically, this resonance could also take place in acid chlorides, acid anhydrides, and esters to give resonance structures (Fig. 11.17). However, the process is much less important because oxygen and chlorine are less nucleophilic than nitrogen. In these structures, the positive charge ends up on an oxygen or a chlorine atom. These atoms are more electronegative than nitrogen and less able to stabilize a positive charge. These resonance structures might occur to a small extent with esters and acid anhydrides, but are far less likely in acid chlorides. This tend also matches the trend in reactivity.

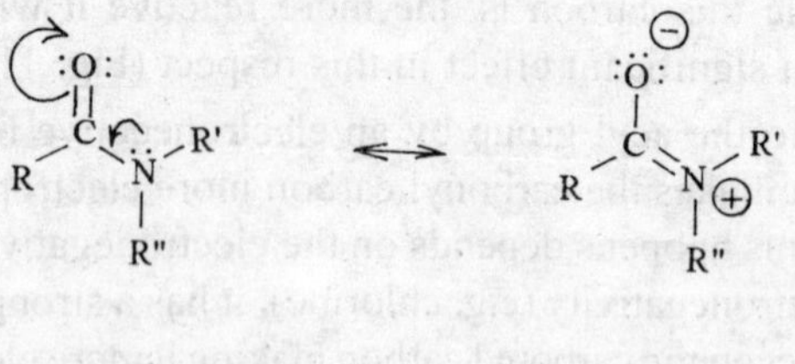

$R-C(=\ddot{O}:)-\ddot{N}R'R'' \longleftrightarrow R-C(-\ddot{O}:^{\ominus})=N^{\oplus}R'R''$

Fig. 11.16. Resonance contribution in an amide.

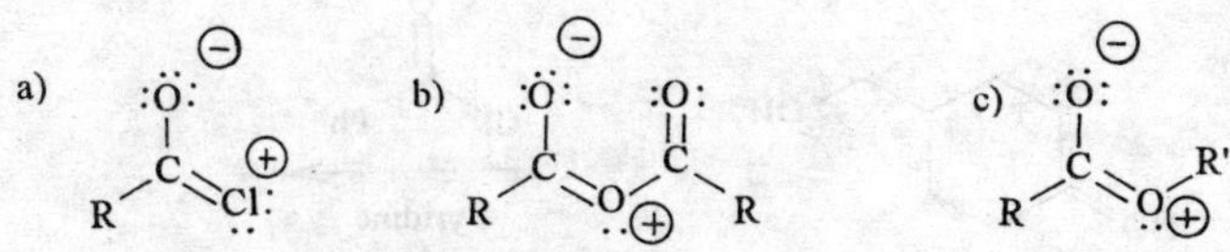

Fig. 11.17. Resonance structures for (a) an acid chloride; (b) an acid anhydride; (c) an ester

Although the resonance effect is weak in esters and acid anhydrides, it explain why acid anhydrides are more reactive than esters. Acid anhydrides have two carbonyl groups and so resonance can occur with either carbonyl group (Fig. 11.18). Due to this, the lone pair of the central oxygen is 'split' between both groups that means that the resonance effect is split between both carbonyl groups. This means that the effect of resonance at any one carbonyl group is diminished and it will remain strongly electrophilic. With an ester, there is only one carbonyl group and so it experiences the full impact of the resonance effect. Therefore, its electrophilic strength will be diminished relative to an acid anhydride.

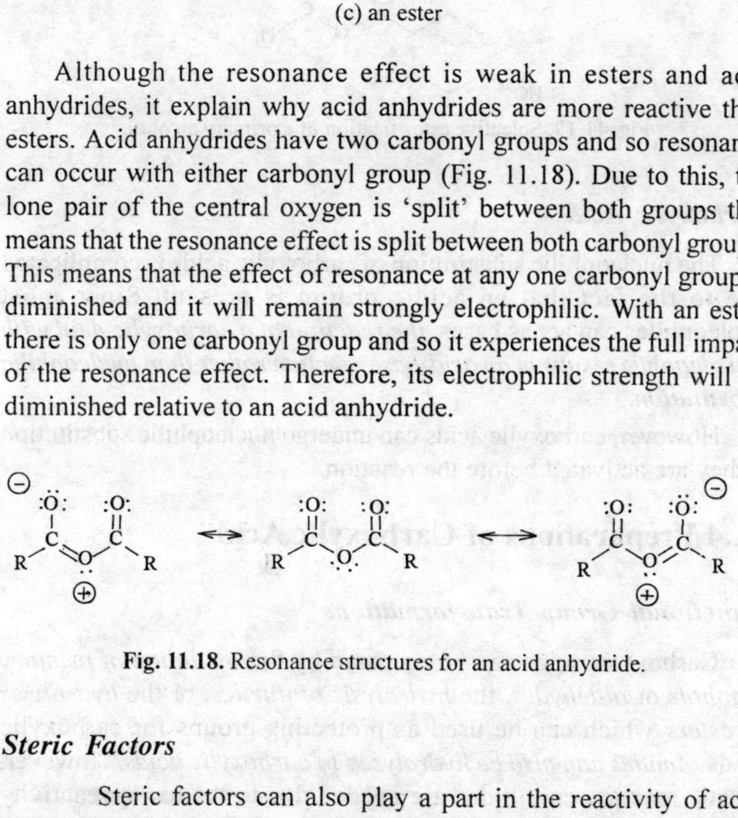

Fig. 11.18. Resonance structures for an acid anhydride.

Steric Factors

Steric factors can also play a part in the reactivity of acid derivatives. For example, a bulky group attached to the carbonyl group can hinder the approach nucleophiles and hence lower reactivity. The steric bulk of the nucleophile can also have an influence in slowing down the reaction. For example, acid chloride react faster with primary alcohols than they do with secondary or tertiary alcohols. This allows selective esterification if a molecule has more than one alcohol group present (Fig. 11.19).

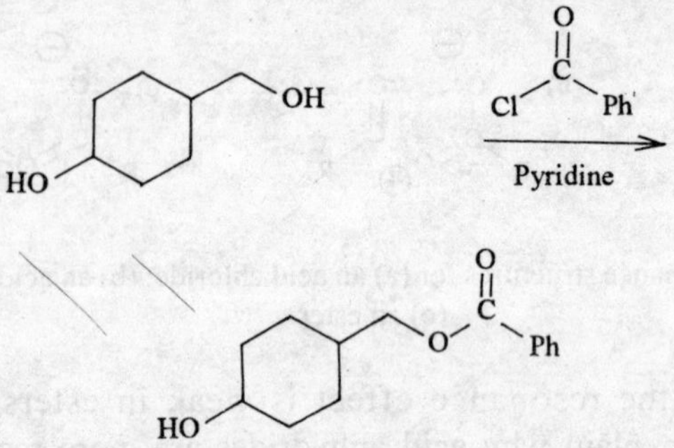

Fig. 11.19. Selective esterification of a primary alcohol.

Carboxylic Acids

The nucleophilic substitution of carboxylic acids is complicated due to the fact that an acidic proton is present. Since most nucleophiles can act as bases, *the reaction of a carboxylic acid with a nucleophile results in an acid-base reaction rather than nucleophilic substitution.*

However, carboxylic acids can undergo nucleophilic substitution if they are activated before the reaction.

11.4 Preparations of Carboxylic Acid

Functional Group Transformations

Carboxylic acids can be prepared by the *oxidation of primary alcohols* or *aldehydes*, the *hydrolysis of nitriles*, or the *hydrolysis of esters* which can be used as protecting groups for carboxylic acids. *Amides can also be hydrolyzed to carboxylic acids.* However, drastic reaction conditions are needed due to the lower reactivity of amides and so amides are less useful as carboxylic acid protecting groups.

Although acid chlorides and anhydrides can be easily hydrolyzed to carboxylic acids, the reaction serves no synthetic purpose because acid chlorides and acid anhydrides are synthesized from carboxylic acids in the first place and they are also very reactive to be used as protecting groups.

C–C Bond Formation

Aromatic carboxylic acid can be obtained *by oxidation alkyl benzenes*. It does not matter how large the alkyl group is, since they are all oxidized to a benzoic acid structure.

In both the methods by which alkyl halides can be converted to a carboxylic acid, the carbon chain is extended by one carbon. One method involves substituting the halogen with a cyanide ion, then hydrolyzing the cyanide group (Fig. 11.20a). This works best with primary alkyl halides. The other method involves the formation of a Grignard reagent which is then treated with carbon dioxide (Fig. 11.20b).

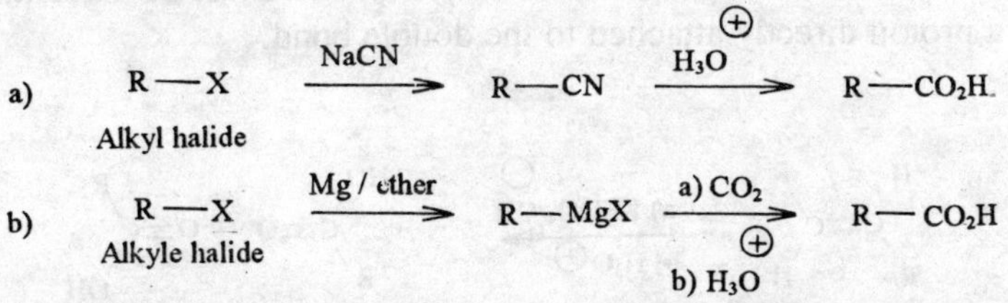

$$\text{a)}\quad \underset{\text{Alkyl halide}}{R{-}X} \xrightarrow{NaCN} R{-}CN \xrightarrow{H_3O^{\oplus}} R{-}CO_2H$$

$$\text{b)}\quad \underset{\text{Alkyle halide}}{R{-}X} \xrightarrow{Mg\,/\,ether} R{-}MgX \xrightarrow[\text{b) } H_3O^{\oplus}]{\text{a) } CO_2} R{-}CO_2H$$

Fig. 11.20. Synthetic routes from an alkyl halide to a carboxylic acid.

The mechanism for the Grignard reaction is similar to the nucleophilic addition of a Grignard reagent to an aldehyde or ketone (Fig. 11.21).

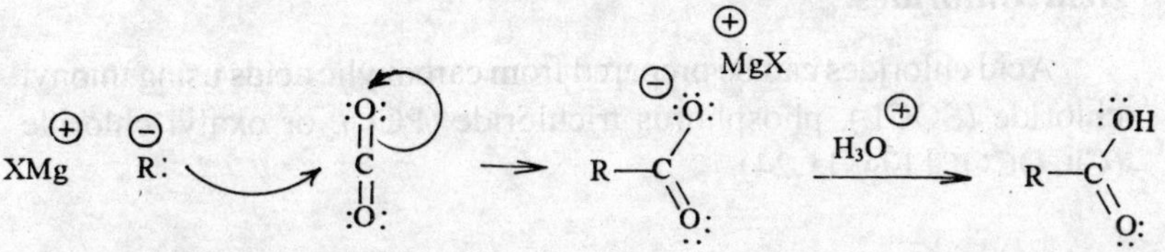

$$\overset{\oplus}{X}Mg \;\; \overset{\ominus}{R}{:} \;+\; O{=}C{=}O \longrightarrow R{-}C(=O){-}O^{\ominus}\;\overset{\oplus}{Mg}X \xrightarrow{H_3O^{\oplus}} R{-}C(=O){-}OH$$

Fig. 11.21. Mechanism for the Grignard reaction with carbon dioxide.

A range of carboxylic acids can be prepared by alkylating diethyl malonate, then hydrolyzing and decarboxylating the product (Fig. 11.22).

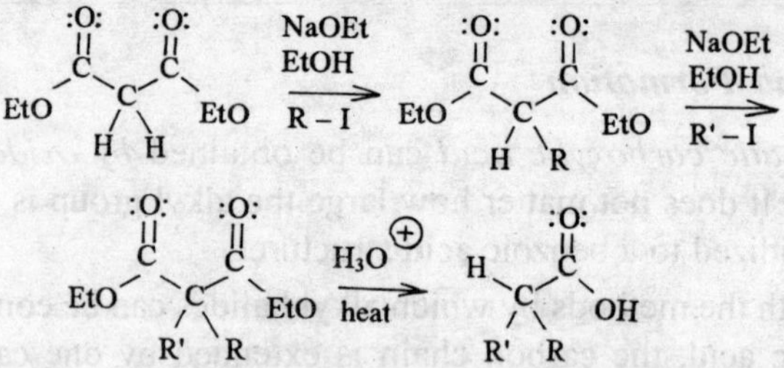

Fig. 11.22. Synthesis of carboxylic acids from diethyl malonate.

Bond Cleavage

Alkenes can be cleaved with potassium permanganate to produce carboxylic acids (Fig. 11.23). A vinylic proton must be present, that is a proton directly attached to the double bond.

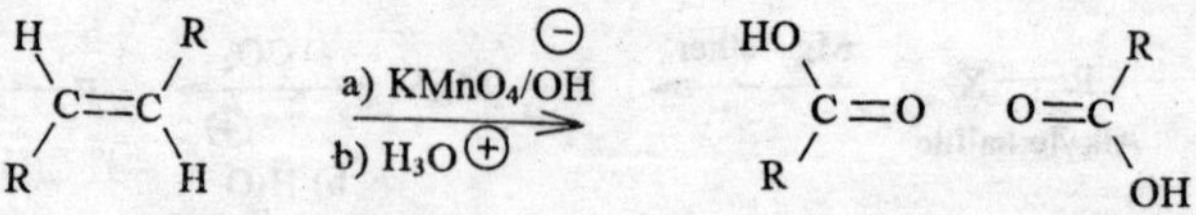

Fig. 11.23. Synthesis of carboxylic acids from alkenes.

11.5 Preparations of Carboxylic Acid Derivatives

Acid Chlorides

Acid chlorides can be prepared from carboxylic acids using thionyl chloride ($SOCl_2$), phosphorus trichloride (PCl_3), or oxalyl chloride (ClCOCOCl Fig. 11.24).

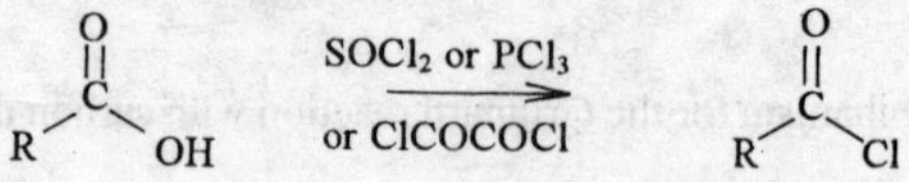

Fig. 11.24. Preparation of acid chlorides.

The mechanism for these reactions in general involves the OH group of the carboxylic acid acting as a nucleophile to form a bond to the reagent and displacing a chloride ion. This has three important consequences. First of all, the chloride ion can attack the carbonyl group to introduce the required chlorine atom. Secondly, the acidic proton is no longer present and so an acid-base reaction is prevented. Thirdly, the original OH group is converted into a good leaving group and is easily displaced once the chloride ion makes its attack. The reaction of a carboxylic acid with thionyl chloride follows the general mechanism shown in Fig. 11.25.

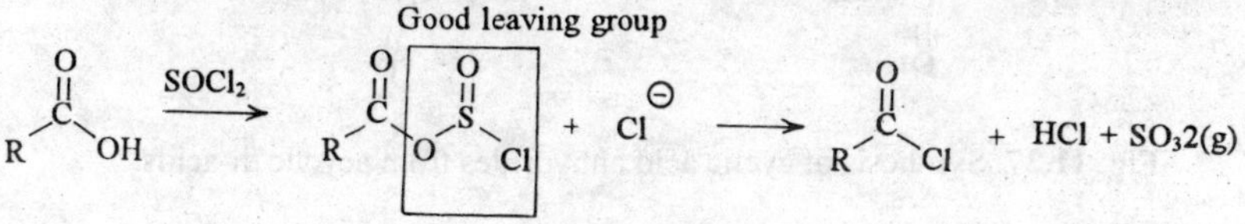

Good leaving group

$$RCOOH \xrightarrow{SOCl_2} R{-}C(=O){-}O{-}S(=O){-}Cl + Cl^{\ominus} \longrightarrow R{-}C(=O){-}Cl + HCl + SO_2(g)$$

Fig. 11.25. Intermediate in the thionyl chloride reaction to form an acid chloride.

The leaving group (SO_2Cl) spontaneously fragments to produce hydrochloric acid and suffer dioxide. The latter is lost as a gas which helps to take the reaction to completion. The detailed mechanism is shown in Fig. 11.26.

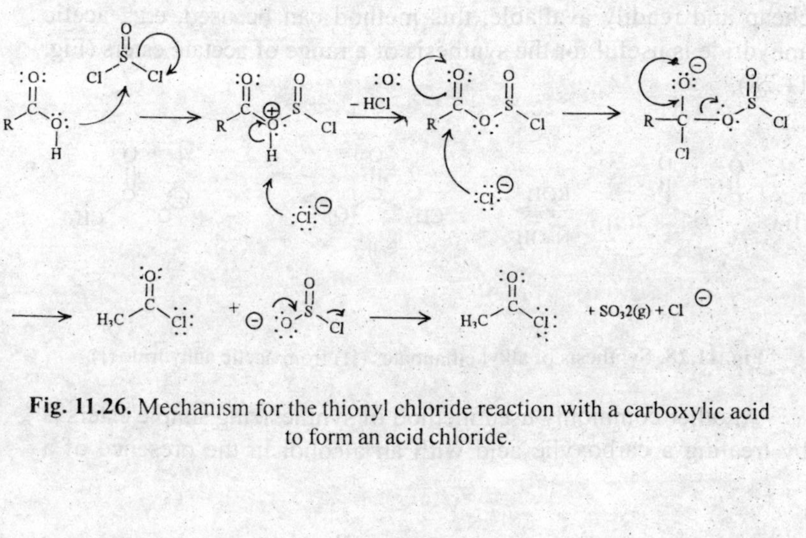

Fig. 11.26. Mechanism for the thionyl chloride reaction with a carboxylic acid to form an acid chloride.

Acid Anhydrides

Acid anhydrides can be prepared by treating acid chloride with a carboxylate salt. Carboxylic acids are not easily converted to acid anhydrides directly. However five-membered and six-membered cyclic anhydrides can be synthesized from diacids by heating structures to eliminate water (Fig. 11.27).

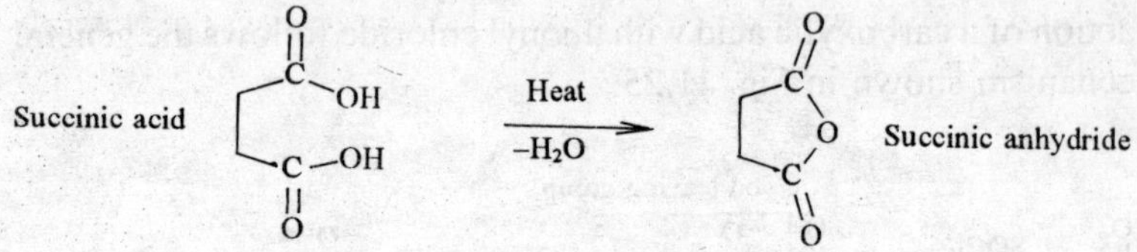

Succinic acid → (Heat, $-H_2O$) → Succinic anhydride

Fig. 11.27. Synthesis of cyclic acid anhydrides from acyclic di-acids.

Esters

There are various ways wherein esters can be synthesized. An effective method is to react an acid chloride with an alcohol in the presence of pyridine yield. Acid anhydrides also react with alcohols to esters, but are less reactive. Moreover, the reaction is wasteful because half of the acyl content on the acid anhydride is wasted as the leaving group (i.e. the carboxylate ion). If the acid anhydride is cheap and readily available, this method can be used, e.g., acetic anhydride is useful for the synthesis of a range of acetate esters (Fig. 11.28).

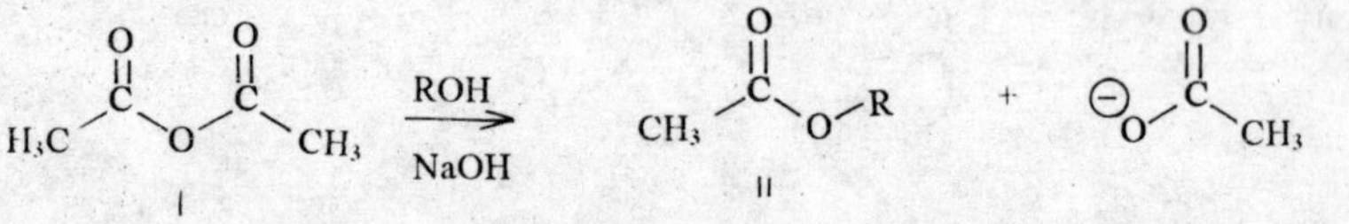

$$H_3C-CO-O-CO-CH_3 \; (I) \xrightarrow[NaOH]{ROH} CH_3-CO-O-R \; (II) + {}^{\ominus}O-CO-CH_3$$

Fig. 11.28. Synthesis of alkyl ethanoates (II) from acetic anhydride (I).

Another commonly used method of synthesizing simple esters is by treating a carboxylic acid with an alcohol in the presence of a catalytic amount of mineral acid (Fig. 11.29). The acid catalyst is needed because there are two difficult steps in the reaction mechanism. First, the alcohol molecule is not a good nucleophile and so the carbonyl group has to be activated. Secondly, the OH group of the carboxylic acid is not a good leaving group and this has to be converted into a better leaving group. The mechanism (Fig. 11.30) is another example of nucleophilic substitution. In it in the first step, the carbonyl oxygen forms a bond to the acidic proton which results in the carbonyl oxygen gaining a positive charge. Due to this the carbonyl carbon becomes more electrophilic and activates it to react with the weakly nucleophilic alcohol. In the second step, the alcohol its a lone pair of electrons to form a bond to the carbonyl carbon. Simultaneously, the carbonyl π bond breaks and both electrons move onto the carbonyl oxygen to form a lone pair of electrons and thereby neutralize the positive charge. Activation of the carbonyl group is important because the incoming alcohol gains an unfavorable positive charge during this step. In the third stage, a proton is transferred from the original alcohol portion to the OH group which we want to remove. By doing so, the latter moiety becomes a much better leaving group. Instead of a hydroxide ion, we can now remove a neutral water molecule. This is achieved in the fourth step where the carbonyl π bond is reformed and the water molecule is expelled.

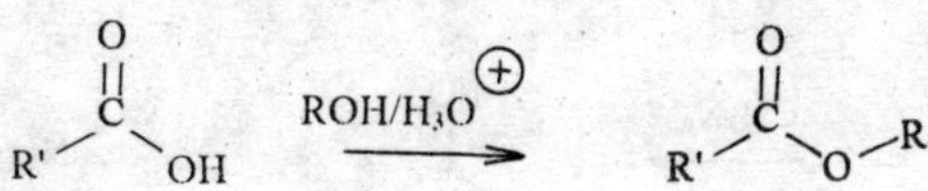

Fig. 11.29. Esterification of a carboxylic acid.

All the steps in the reaction mechanism are in equilibrium and so it is important to use the alcohol in large excess (i.e. as solvent) so as to drive the equilibrium to products. This is only practical with cheap and readily available alcohols such as methanol and ethanol. However, if the carboxylic acid is cheap and readily available it could be used in large excess instead.

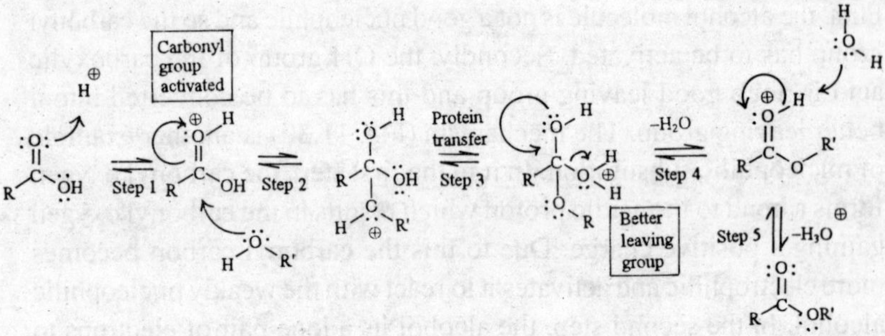

Fig. 11.30. Mechanism for the acid-catalyzed esterification of a carboxylic acid.

The method of preparing methyl esters is to treat carboxylic acids with diazomethane (Fig. 11.31). In this method good yields are obtained because nitrogen is formed as one of the products and because it is lost from the reaction mixture, the reaction is driven to completion. However, diazomethane is a very hazardous chemical that can explode, and strict precautions are essential when using it.

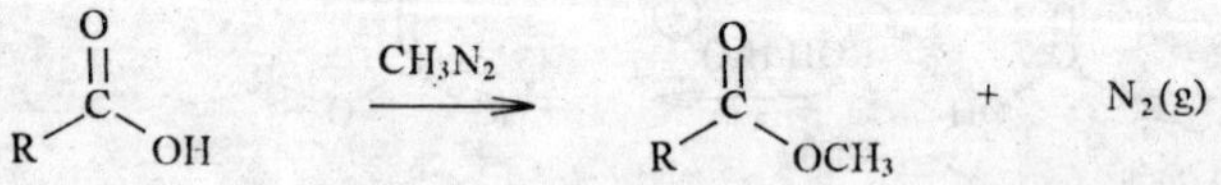

$$RCOOH \xrightarrow{CH_3N_2} RCOOCH_3 + N_2(g)$$

Fig. 11.31. Synthesis of methyl esters using diazomethane.

Lastly, the carboxylic acid can be converted to a carboxylate ion and then treated with an alkyl halide (Fig. 11.32). The reaction involves the S_N2 nucleophilic substitution of an alkyl halide and so the reaction works best with primary alkyl halides.

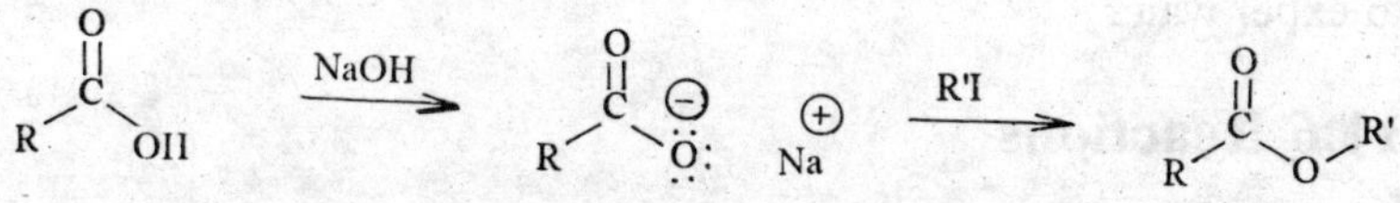

$$RCOOH \xrightarrow{NaOH} RCOO^{\ominus}\ Na^{\oplus} \xrightarrow{R'I} RCOOR'$$

Fig. 11.32. Synthesis of an ester by nucleophilic substitution of an alkyl halide.

Amides

Amides can be prepared from acid chloride by nucleophilic substitution when ammonia is treated with acid chlorides it gives a primary amide. Acid chloride on treatment with a primary amine gives a secondary amide, and on treatment with a secondary amine gives a tertiary amide. Tertiary amines cannot be used on this reaction because they do not form a stable product.

Two equivalent of amine are needed for the above reactions since one equivalent of the amine is used up in forming a salt with the hydrochloric acid that is produced in the reaction. To avoid wastage, one equivalent of sodium hydroxide can be added to the reaction so as to neutralize the HCl.

Amides can also be synthesized from acid anhydrides and esters but in general these reactions offer no advantage over acid chlorides because acid anhydrides and esters are less reactive. Moreover, with acid anhydrides, half of the parent carboxylic acid is lost as the leaving group. Thus, acid anhydrides are only used for the synthesis of amides if the acid anhydride is cheap and freely available (e.g. acetic anhydride).

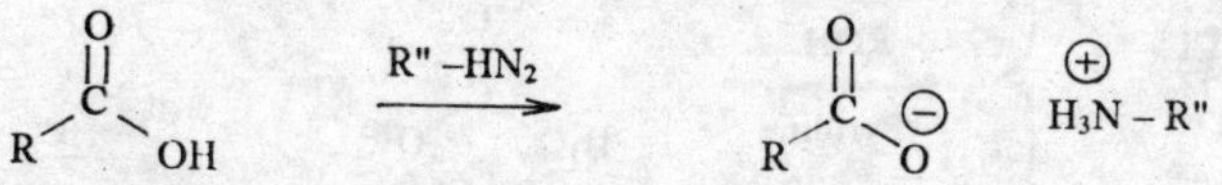

$$RCOOH \xrightarrow{R''-HN_2} RCOO^{\ominus}\quad \overset{\oplus}{H_3N}-R''$$

Fig. 11.33. Salt formation.

The synthesis of amides directly from carboxylic acids is not easy because the reaction of an amine with a carboxylic acid is a *typical acid-base reaction* resulting in the formation of a salt (Fig.

11.33). Some salts can be converted to an amide by heating strongly to expel water.

11.6 Reactions

Acid-base Reactions

Because carboxylic acids have an acidic proton ($CO_2\underline{H}$), they form water soluble carboxylate salts when treated with a base (e.g. sodium hydroxide or sodium bicarbonate Fig. 11.34).

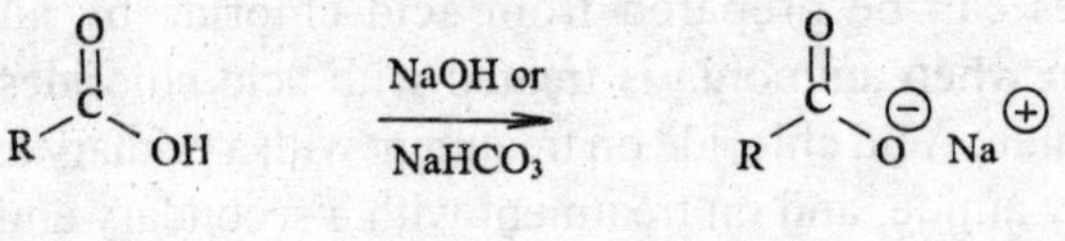

$$RCO_2H \xrightarrow[NaHCO_3]{NaOH\ or} RCO_2^{\ominus}\ Na^{\oplus}$$

Fig. 11.34. Salt formation

Interconversion of Acid Derivatives

Reactive acid derivatives can be converted to less reactive acid derivatives by nucleophilic substitution. Thus, acid chlorides can be converted to acid anhydride, esters, and amides (Fig. 11.35). Hydrochloric acid is released in all these reactions and this may lead to side reactions. Therefore, pyridine or sodium hydroxide may be added so as to mop up the hydrochloric acid (Fig. 11.36).

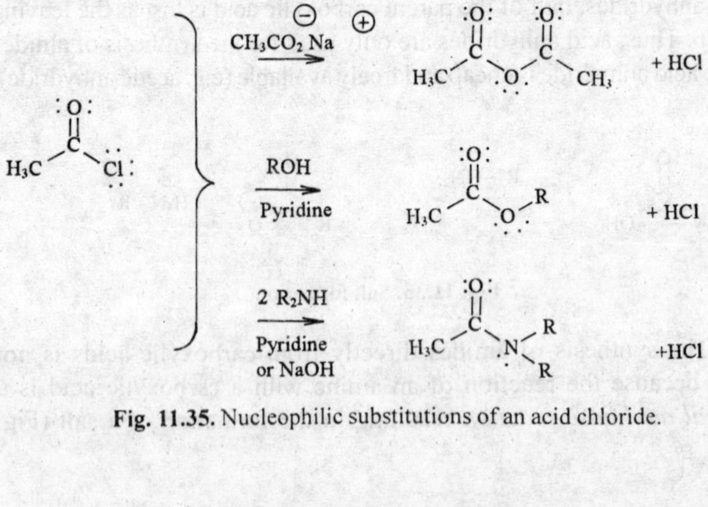

$$H_3C{-}COCl \xrightarrow{CH_3CO_2^{\ominus}\ Na^{\oplus}} H_3C{-}CO{-}O{-}CO{-}CH_3 + HCl$$

$$H_3C{-}COCl \xrightarrow[Pyridine]{ROH} H_3C{-}CO{-}OR + HCl$$

$$H_3C{-}COCl \xrightarrow[Pyridine\ or\ NaOH]{2\ R_2NH} H_3C{-}CO{-}NR_2 + HCl$$

Fig. 11.35. Nucleophilic substitutions of an acid chloride.

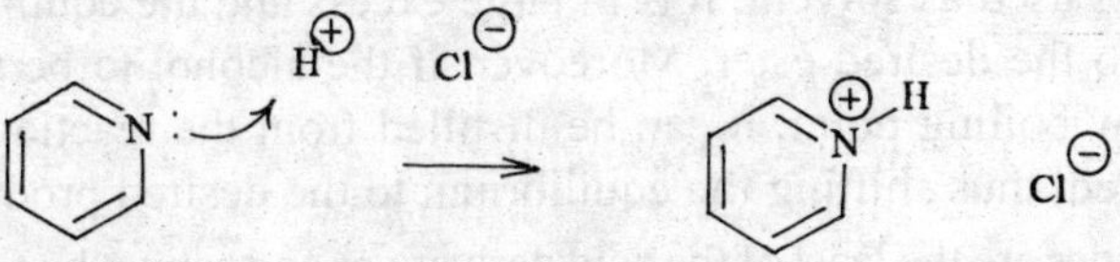

Fig. 11.36. Role of pyridine in 'mopping up' protons.

Acid anhydrides can be converted to esters and amides but not to acid chlorides (Fig. 11.37).

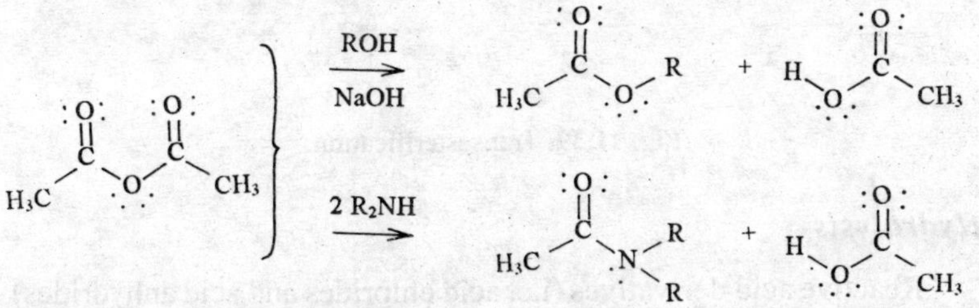

Fig. 11.37. Nucleophilic substitutions of acid anhydrides.

Esters can be converted to amides but not to acid chlorides or acid anhydrides (Fig. 11.38).

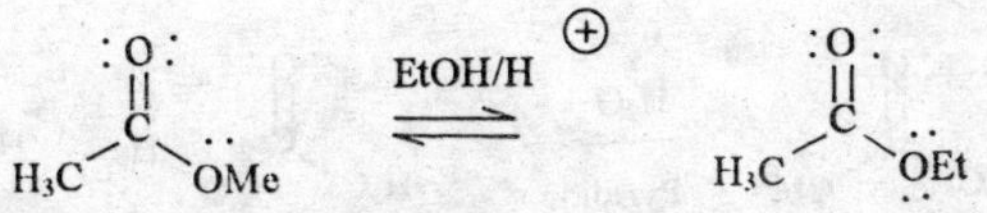

Fig. 11.38. Nucleophilic substitutions of an ester.

Esters can also be converted by nucleophilic substitution from one type of ester to another and this process is called **transesterification**. For example, a methyl ester can be dissolved in ethanol in the presence of an acid catalyst and converted to an ethyl ester (Fig. 11.39). The reaction is an equilibrium reaction, but if the alcohol is used as solvent, it is in large excess and the equilibrium is shifted to the desired ester. Moreover, if the alcohol to be replaced has a low boiling point, it can be distilled from the reaction as it is substituted, thus shifting the equilibrium to the desired product.

Amides are the least of the acid derivatives and cannot be converted to acid chlorides, acid anhydrides, or esters.

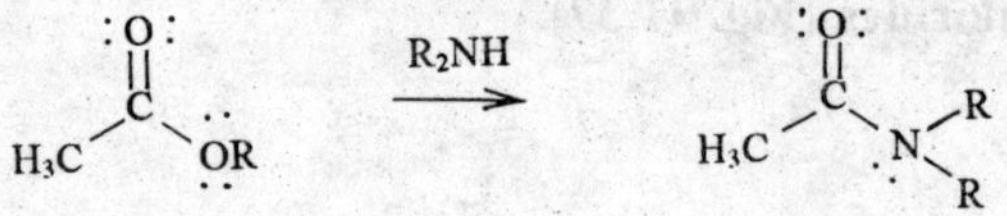

$$H_3C{-}C(=O){-}OR \xrightarrow{R_2NH} H_3C{-}C(=O){-}NR_2$$

Fig. 11.39. Transesterification.

Hydrolysis

Reactive acid derivatives (i.e. acid chlorides and acid anhydrides) get hydrolyzed by water to give the constituent carboxylic acids (Fig. 11.40). The reaction is an example of nucleophilic substitution where water acts as the nucleophile. Hydrochloric acid is a byproduct from the hydrolysis of an acid chloride, so pyridine is generally added to the reaction mixture to mop it up (Fig. 11.36).

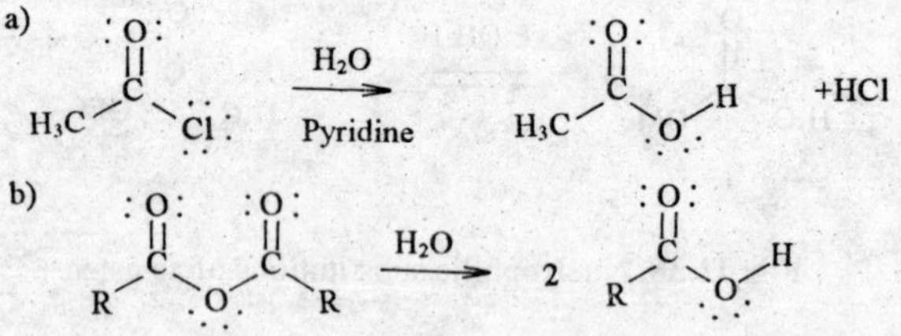

a) $$H_3C{-}C(=O){-}Cl \xrightarrow[\text{Pyridine}]{H_2O} H_3C{-}C(=O){-}O{-}H \quad +HCl$$

b) $$R{-}C(=O){-}O{-}C(=O){-}R \xrightarrow{H_2O} 2\ R{-}C(=O){-}O{-}H$$

Fig. 11.40. Hydrolysis of (a) an acid chloride; (b) an acid anhydride.

Esters and amides are less reactive and so the hydrolysis needs more drastic conditions using aqueous sodium hydroxide or aqueous acid with heating (Fig. 11.41).

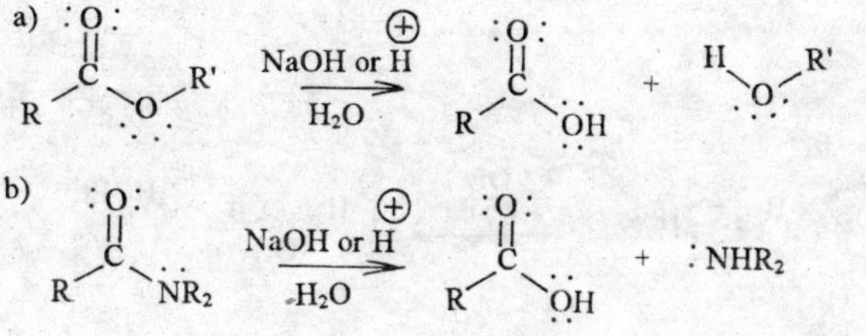

a) $RCOOR' \xrightarrow[H_2O]{NaOH \text{ or } H^{\oplus}} RCOOH + HOR'$

b) $RCONR_2 \xrightarrow[H_2O]{NaOH \text{ or } H^{\oplus}} RCOOH + :NHR_2$

Fig. 11.41. Hydrolysis of (a) esters; (b) amides.

Under basic conditions, the hydroxide ion acts as the nucleophile. For example, the mechanism of hydrolysis of ethyl acetate is as shown (Fig 11.42). However, the mechanism does not stop here. The carboxylic acid which is formed reacts with sodium hydroxide to form a water soluble carboxylate ion (Fig. 11.43a). Moreover, the ethoxide ion that is lost from the molecule is a stronger base than water and undergoes protonation (Fig. 11.43b). The basic hydrolysis of an ester is also called **saponification** and produces a water soluble carboxylate ion.

$CH_3COOEt + {}^{\ominus}OH \longrightarrow CH_3C(O^{\ominus})(OH)OEt \longrightarrow CH_3COOH + {}^{\ominus}OEt$

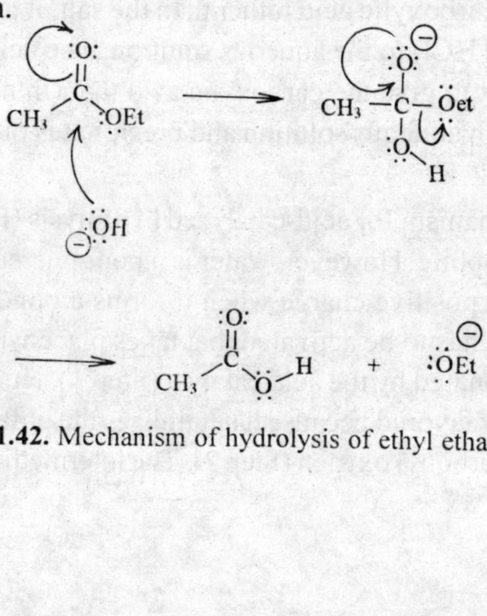

Fig. 11.42. Mechanism of hydrolysis of ethyl ethanoate.

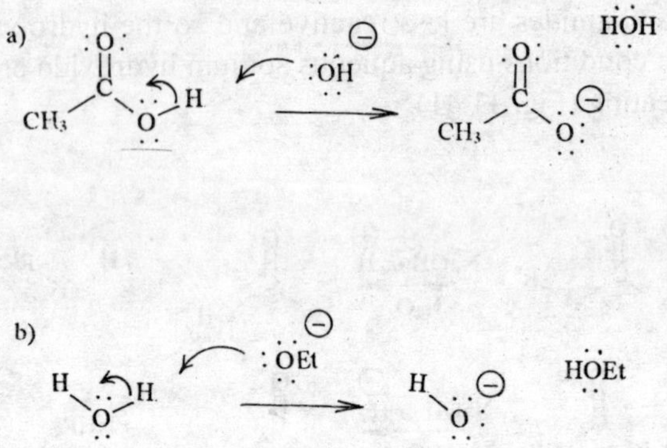

Fig. 11.43. (a) Ionization of a carboxylic acid; (b) neutralization of the ethoxide ion.

The same mechanism is involved in the basic hydrolysis of an amide and also results in the formation of a water soluble carboxylate ion. The leaving group from an amide is initially charged (i.e. $R_2N:^-$). However, this is a strong base and reacts with water to form a free amine and a hydroxide ion.

In the basic hydrolysis of esters and amides, the formation of carboxylate ion is irreversible and so serves to drive the reaction to completion.

To isolate carboxylic acid rather than the salt, it is essential to add acid (e.g. dilute HCl) to the aqueous solution. The acid protonates the carboxylate salt to give the carboxylic acid that (in most cases) is no longer soluble in aqueous solution and precipitates out as a solid or as an oil.

In the mechanism for acid-catalyzed hydrolysis (Fig. 11.44) water acts as a nucleophile. However, water is a poor nucleophile as it gains an unfavorable positive charge when it forms a bond. Therefore, the carbonyl group has to be activated that takes place when the carbonyl oxygen is protonated by the acid catalyst (Step 1). Nucleophilic attack by water in now favored because it neutralizes the unfavorable positive charge on the carbonyl oxygen (Step 2). The intermediate has a positive charge on the oxygen derived from water, but this is neutralized by losing the attached proton such that the oxygen gains the electrons in the O–H bond (Step 3). Another protonation now occurs (Step 4). This is necessary so as to convert a poor leaving group (the methoxide ion) into a good leaving group (methanol). The π bond can now be reformed (Step 5) with loss of methanol. Finally, water can act as a base to remove the proton from the carbonyl oxygen (Step 6).

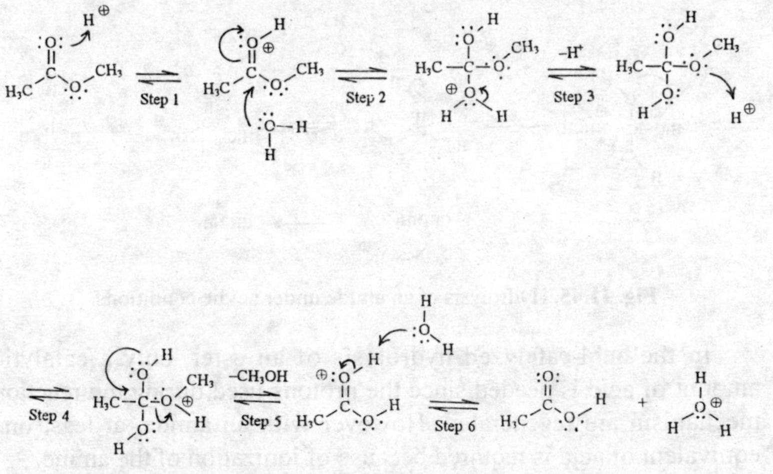

Fig. 11.44. Mechanism for the acid-catalyzed hydrolysis of an ester.

The acid-catalyzed hydrolysis of an ester is not as effective as basic hydrolysis because all the steps in the mechanism are reversible and there is no salt formation to pull the reaction through to products. Therefore, it is important to use an excess of water so as to shift the equilibria to the products. In contrast to esters, the hydrolysis of an amide in acid does result in the formation of an ion (Fig. 11.45). The leaving group here is an amine and since amines are basic, they will react with the acid to form a water soluble aminum ion. This is an irreversible step that pulls the equilibrium through to the products.

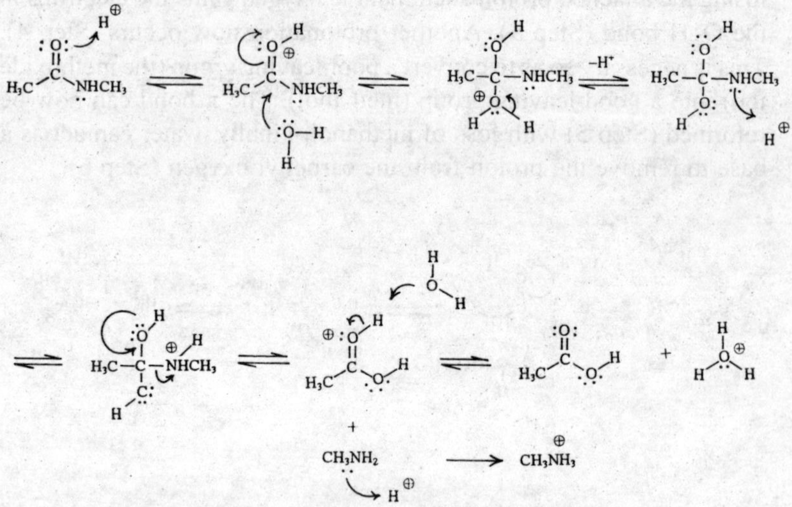

Fig. 11.45. Hydrolysis of an amide under acidic conditions.

In the acid-catalyzed hydrolysis of an ester, only a catalytic amount of acid is needed since the protons used during the reaction mechanism are regenerated. However with an amide, at least one equivalent of acid is required because of ionization of the amine.

Friedel-Crafts Acylation

Acid chlorides can be treated with aromatic rings in the presence of a Lewis acid to give aromatic ketones (Fig. 11.46). The reaction involves formation of an acylium ion from the acid chloride, followed by electrophilic substitution of the aromatic ring.

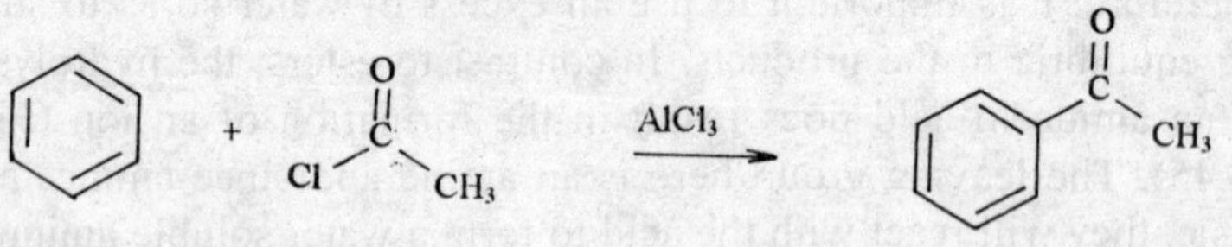

Fig. 11.46. Friedel-Crafts acylation.

Grignard Reaction

Acid chlorides and esters react with two equivalents of a Grignard reagent to produce a tertiary alcohol where two extra alkyl groups are provided by the Grignard reagent (Fig. 11.47).

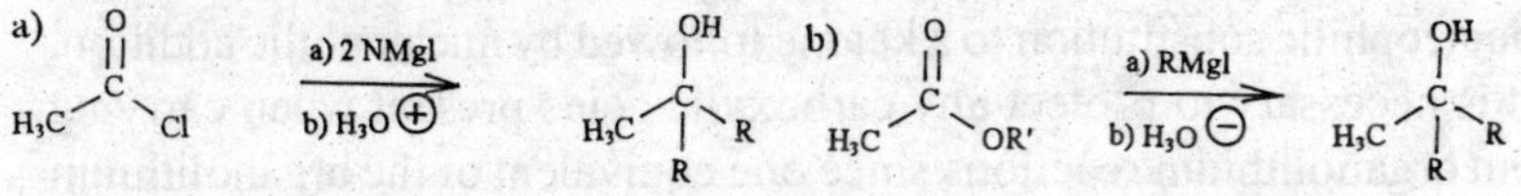

Fig. 11.47. Grignard reaction with (a) an acid chloride; and (b) an ester to produce a tertiary alcohol.

There are two reactions involved in this process (Fig. 11.48). The acid chloride reacts with one equivalent of Grignard reagent in a nucleophilic substitution to produce an intermediate ketone. This ketone is also reactive to Grignard reagents and immediately reacts with a second equivalent of Grignard reagent by the nucleophilic addition mechanism.

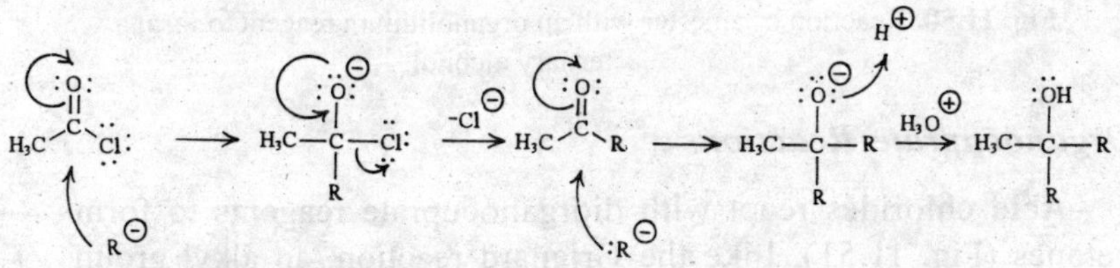

Fig. 11.48. Mechanism of the Grignard reaction with an acid chloride.

Carboxylic acids reacts with Grignard reagents in an acid-base reaction forming carboxylate ion and an alkane (Fig. 11.49). This has no synthetic use and it is important to protect carboxylic acids when carrying out Grignard reactions on another part of the molecule to avoid wastage of Grignard reagent.

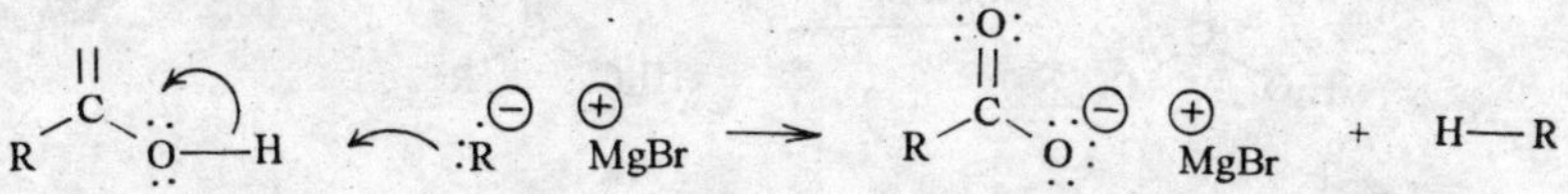

Fig. 11.49. Acid-base reaction of a Grignard reagent with a carboxylic acid.

Organolithium Reactions

Esters react with two equivalents of an organolithium reagent to yield a tertiary alcohol in which two of the alkyl groups are derived from the organolithium reagent (Fig. 11.50). The mechanism of the reaction is the same as that described in the Grignard reaction, i.e., nucleophilic substitution to a ketone followed by nucleophilic addition. It is necessary to protect any carboxylic acids present when carrying out organolithium reactions since one equivalent of the organolithium reagent would be wasted in an acid-base reaction with the carboxylic acid.

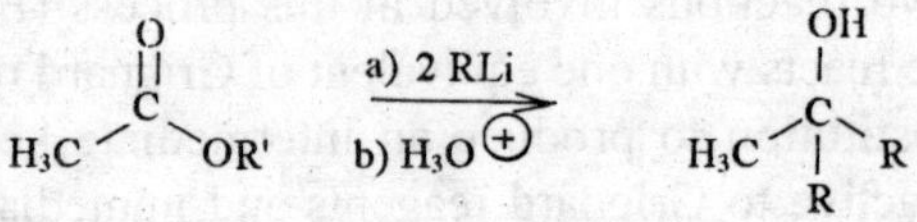

$$H_3C{-}C(=O){-}OR' \xrightarrow[\text{b) } H_3O^{\oplus}]{\text{a) 2 RLi}} H_3C{-}C(OH)(R){-}R$$

Fig. 11.50. Reaction of an ester with an organolithium reagent to form a tertiary alcohol.

Organocuprate Reactions

Acid chlorides react with diorganocuprate reagents to form ketones (Fig. 11.51). Like the Grignard reaction, an alkyl group displaces the chloride ion to form a ketone. However, unlike the Grignard reaction, the reaction stops at the ketone stage. The mechanism is believed to be radical based rather than nucleophilic substitution. The reaction does not occur with carboxylic acids, acid anhydrides, esters, or amides.

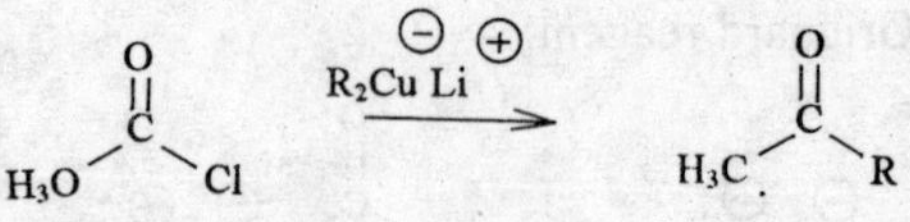

$$H_3O{-}C(=O){-}Cl \xrightarrow{R_2Cu^{\ominus}\,Li^{\oplus}} H_3C{-}C(=O){-}R$$

Fig. 11.51. Reaction of an acid chloride with a diorganocuprate reagent to produce a ketone.

Reduction

Carboxylic acids, acid chlorides, acid anhydrides and esters get reduced to primary alcohols when treated with lithium aluminum hydride ($LiAlH_4$; Fig. 11.52). The reaction involves nucleophilic substitution by a hydride ion to give an intermediate aldehyde. This cannot be isolated since the aldehyde immediately undergoes a nucleophilic addition reaction with another hydride ion (Fig. 11.53). The detailed mechanism is as shown in Fig. 11.54.

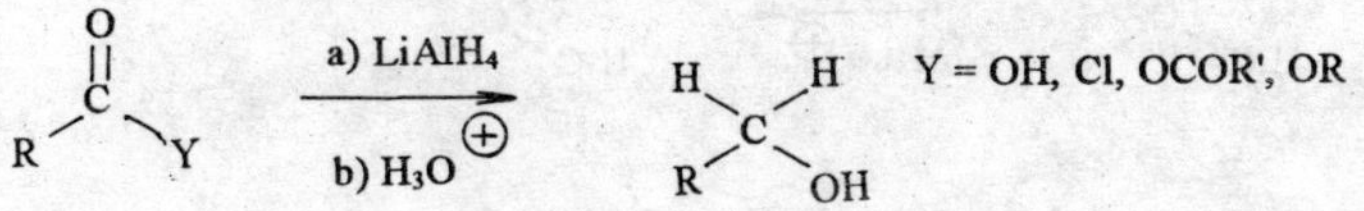

a) $LiAlH_4$ b) $H_3O^{\oplus}$ — Y = OH, Cl, OCOR', OR

Fig. 11.52. Reduction of acid chlorides, acid anhydrides, and esters with lithium aluminum hydride.

$LiAlH_4$ Nucleophilic substitution; $LiAlH_4$ Nucleophilic addition; 1^0 Alcohol

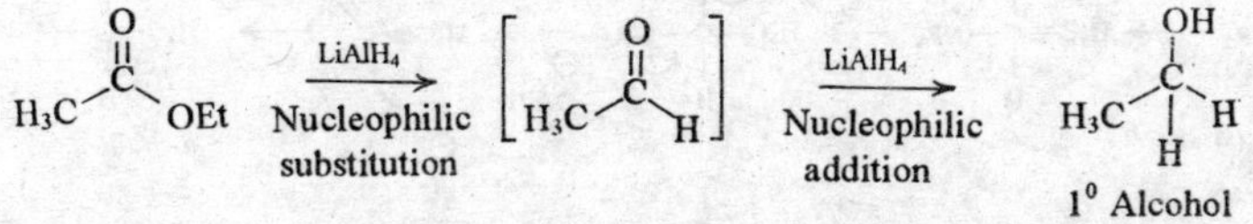

Fig. 11.53. Intermediate involved in the $LiAlH_4$ reduction of an ester.

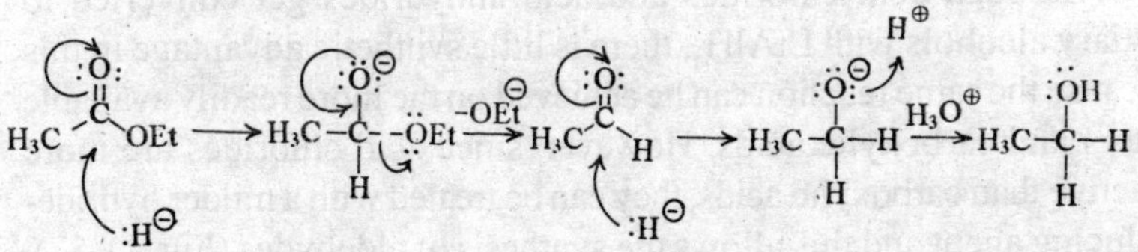

Fig. 11.54. Mechanism for the $LiAlH_4$ reduction of an ester to a primary alcohol.

Amides differ from carboxylic acids and other acid derivatives in their reaction with $LiAlH_4$. Instead of forming primary alcohols, amides are reduced to amines (Fig. 11.55). The mechanism (Fig.

11.56) involves addition of the hydride ion to form an intermediate that is converted to an organoaluminum intermediate. The difference in this mechanism is the intervention of the nitrogen's lone pair of electrons. These are fed into the electrophilic center to eliminate the oxygen that is then followed by the second hydride addition.

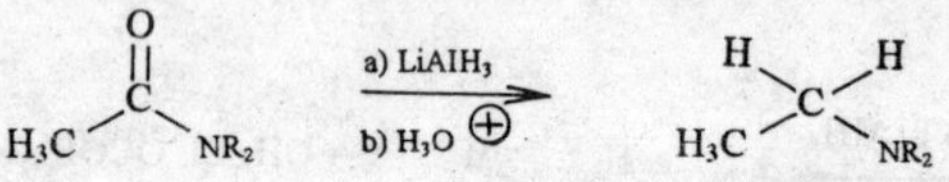

Fig. 11.55. Reduction of an amide to an amine.

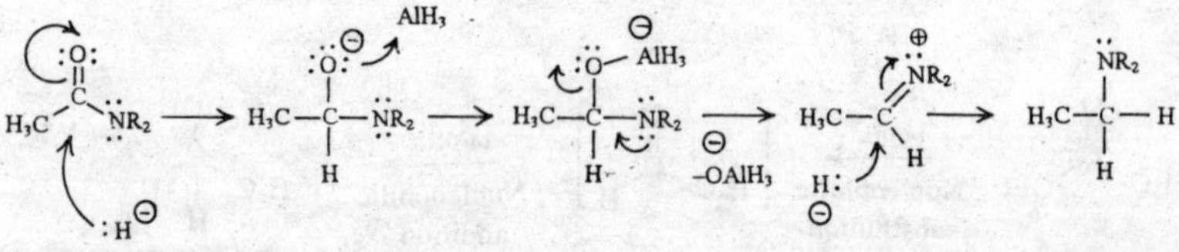

Fig. 11.56. Mechanism for the $LiAlH_4$ reduction of an amide to an amine.

Although acid chlorides and acid anhydrides get converted to tertiary alcohols with $LiAlH_4$, there is little synthetic advantage in this because the same reaction can be achieved on the more readily available esters and carboxylic acids. However, since acid chlorides are more reactive than carboxylic acids, they can be treated with a milder hydride-reducing agent and this allows the synthesis of aldehydes (Fig. 11.57). The hydride reagent used (*lithium of tri-tert-butoxyaluminum hydride*) contains three bulky alkoxy groups that lowers the reactivity of the remaining hydride ion. This means that the reaction stops after nucleophilic substitution with one hydride ion. Another sterically hindered hydride reagent–*diisobutylaluminum hydride (DIBAH)* can be used to reduce esters to aldehydes (Fig. 11.57). Normally low temperatures are required to avoid over-reduction.

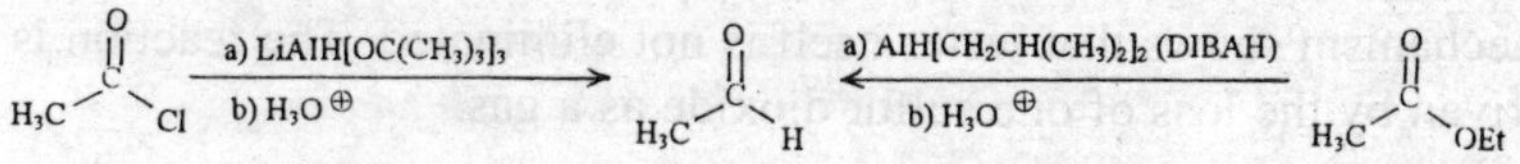

Fig. 11.57. Reduction of an acid chloride and an ester to an aldehyde.

Borane (B_2H_6) can be used as a reducing agent to convert carboxylic acids to primary alcohols. The advantage of using borane rather than $LiAlH_4$ is that the former does not reduce any nitrogroups that might be present. $LiAlH_4$ reduces a nitro group (NO_2) to an amino group (NH_2).

Carboxylic acids and acid derivatives are not reduced by the milder reducing agent such as sodium borohydride ($NaBH_4$). This reagent can therefore be used to reduce aldehydes and ketones without affecting any carboxylic acids or acid derivatives which might be present.

Dehydration of Primary Amides

Primary amides are dehydrated to nitriles using a dehydrating agent like thionyl chloride ($SOCl_2$), phosphorus pentoxide (P_2O_5), phosphoryl trichloride ($POCl_3$), or acetic anhydride (Fig. 11.58).

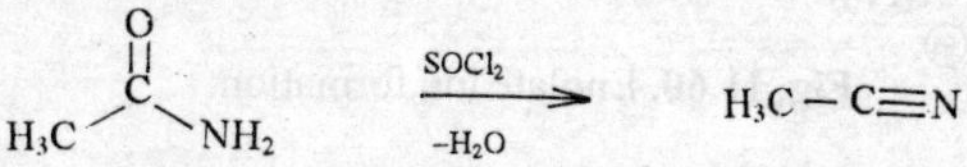

Fig. 11.58. Conversion of a primary amide to a nitrile.

The mechanism for the dehydration of an amide with thionyl chloride is shown below in Fig. 11.59.

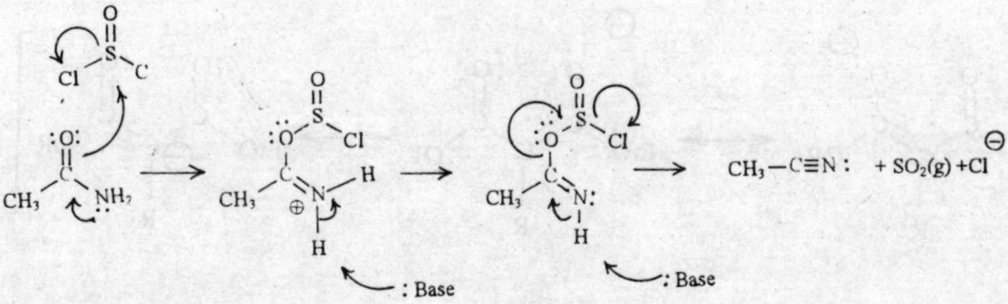

Fig. 11.59. Mechanism for the dehydration of a primary amide to a nitrile.

Although the reaction is the equivalent of a dehydration, the mechanism shows that water itself is not eliminated. The reaction is driven by the loss of one sulfur dioxide as a gas.

11.7 Enolate Reactions

Enolates

Enolate ions can be formed from aldehydes and ketones containing protons on an α-carbon. Enolate ions can also be formed from esters if they have protons on an α-carbon (Fig. 11.60). Such protons are slightly acidic and can be removed on treatment with a powerful base like lithium diisopropylamide (LDA). LDA acts as a base rather than as a nucleophile since it is a bulky molecule and this prevents it attacking the carbonyl group in a nucleophilic substitution reaction.

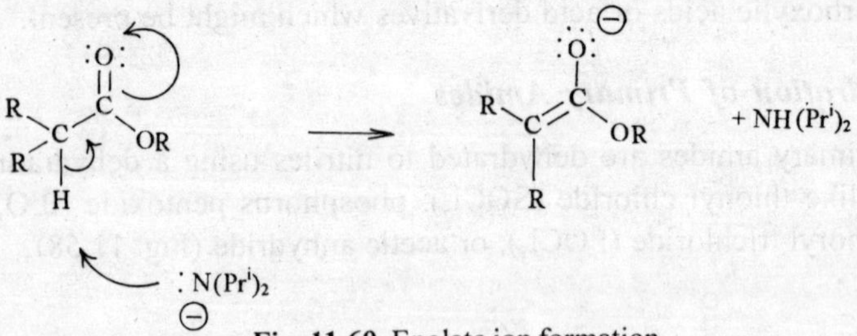

$+ NH(Pr^i)_2$

$:N(Pr^i)_2$

Fig. 11.60. Enolate ion formation.

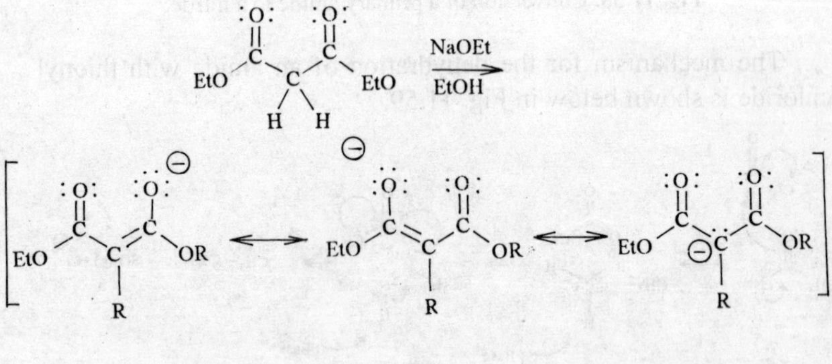

NaOEt

EtOH

Fig. 11.61. Formation of an enolate ion from diethyl malonate.

Formation of enolate is easier if there are two esters flanking the α-carbon since the α-proton will be more acidic. The acidic proton in diethyl malonate can be removed with a weaker base than LDA (e.g. sodium ethoxide; Fig. 11.61). The enolate ion is more stable since the charge can be delocalized over both carbonyl groups.

Alkylations

Enolate ions can be alkylated with alkyl halides through the S_N2 nucleophilic substitution of an alkyl halide (Fig. 11.62).

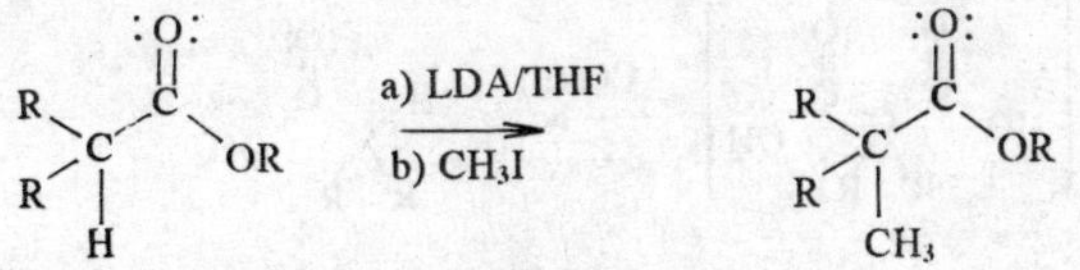

Fig. 11.62. α-alkylation of an ester.

Although simple ester can be converted to their enolate ions and alkylated, the use of a molecule like diethyl malonate is far more effective. This is because the α-protons of diethyl malonate (pK_a 10–12) are more acidic than the α-protons of a simple ester like ethyl acetate (pK_a 25) and can be removed by a milder base. It is possible to predict the base needed to carry out the deprotonation reaction by considering the pK_a value of the conjugate acid for that base. If this pK_a is higher than the pK_a value of the ester, then the deprotonation reaction is possible. For example, the conjugate acid of the ethoxide ion is ethanol (pK_a 16) and so any ester having a pK_a less than 16 will be deprotonated by the ethoxide ion. Therefore, diethyl malonate is deprotonated but not ethyl acetate. Moreover, the ethoxide ion is strong enough to deprotonate the diethyl malonate quantitatively such that all the diethyl malonate is converted to the enolate ion. This avoids the possibility of any competing **Claisen reaction** since that reaction needs the presence of unaltered ester. Diethyl malonate can be converted quantitatively to its enolate with ethoxide ion, alkylated with an alkyl halide, treated with another equivalent of base, then alkylated with a second different alkyl halide (Fig. 11.63). Subsequent hydrolysis and decarboxylation of the diethyl ester yields the carboxylic acid. The decarboxylation mechanism (Fig. 11.64) is dependent on the presence of the other carbonyl group at the β-position.

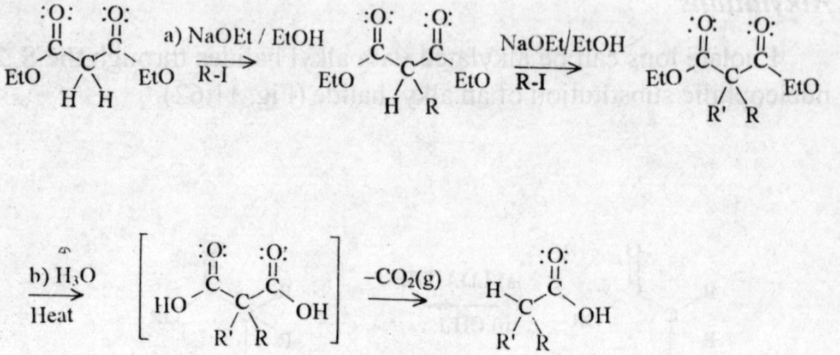

Fig. 11.63. Alkylations of diethyl malonate.

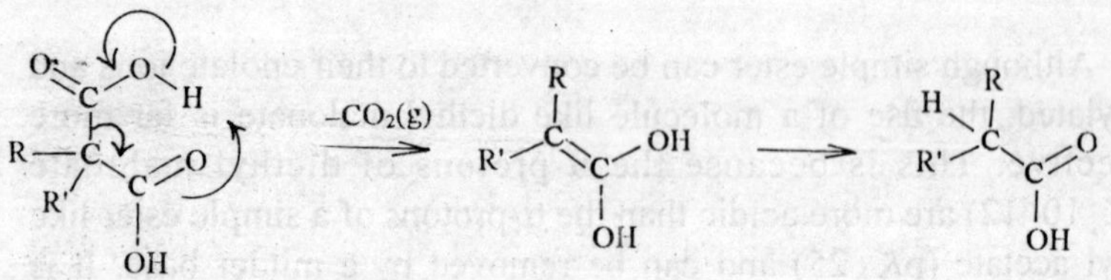

Fig. 11.64. Decarboxylation mechanism.

The final product can be considered as a disubstituted ethanoic acid. Theoretically, this product could also be synthesized from ethyl ethanoate. However, the use of diethyl malonate is better because the presence of two carbonyl groups permits easier formation of the intermediate enolate ions.

Claisen Condensation

The **Claisen reaction** involves the condensation or linking of two ester molecules to form a β-ketoester (Fig. 11.65). This reaction can be considered as the ester equivalent of the Aldol reaction. The reaction involves the formation of an enolate ion from one ester molecule which then undergoes nucleophilic substitution with a second ester molecule (Fig. 11.66, Step 1). The ethaoxide ion that is formed in step 2 removes an α-proton from the β-ketoester in step 3 to form a stable enolate ion and this drives the reaction to completion. The final product is isolated by protonating the enolate ion with acid.

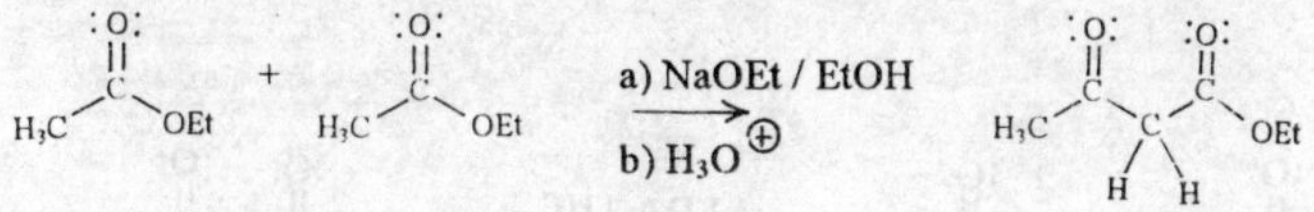

$H_3C-C(=O)-OEt$ + $H_3C-C(=O)-OEt$ —a) NaOEt / EtOH; b) $H_3O^{\oplus}$→ $H_3C-C(=O)-CH_2-C(=O)-OEt$

Fig. 11.65. Claisen condensation.

Two different esters can be used in the Claisen condensation as long as one of the esters has no α-protons and cannot form an enolate ion (Fig. 11.67). **β-Diketones** can be synthesized from the **mixed Claisen condensation of a ketone with an ester (Fig. 11.68).** It is better to use any ester that cannot form an enolate ion to avoid competing Claisen condensations.

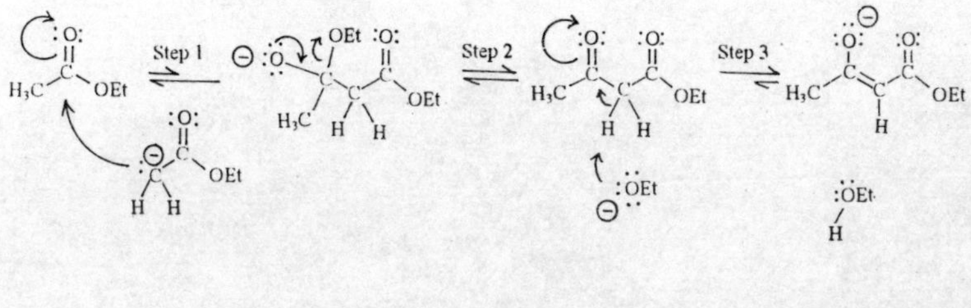

Fig. 11.66. Mechanism of the Claisen condensation.

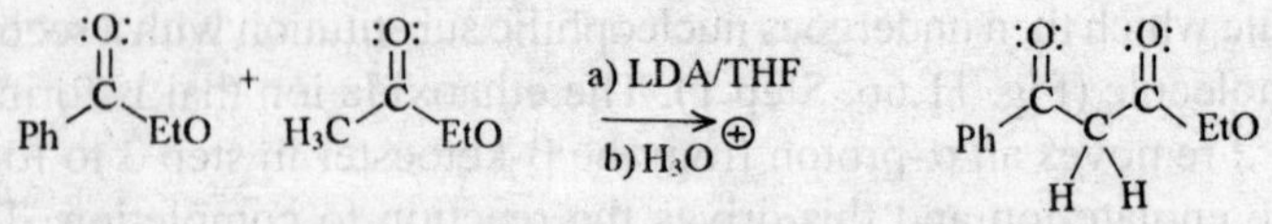

$PhCOOEt + H_3CCOOEt \xrightarrow[\text{b) } H_3O^{\oplus}]{\text{a) LDA/THF}} PhCOCH_2COOEt$

Fig. 11.67. Claisen condensation of two different esters.

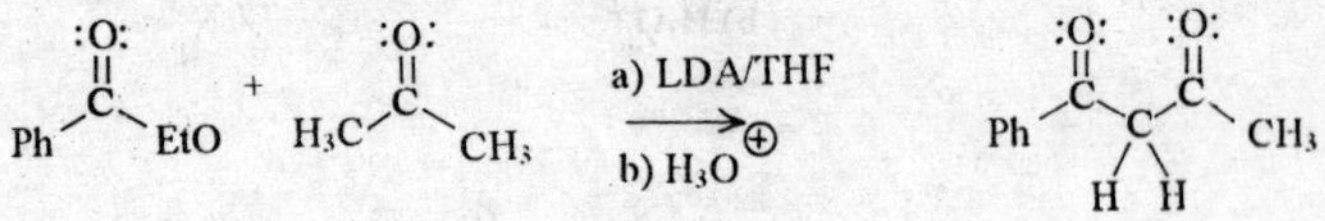

$PhCOOEt + H_3CCOCH_3 \xrightarrow[\text{b) } H_3O^{\oplus}]{\text{a) LDA/THF}} PhCOCH_2COCH_3$

Fig. 11.68. Claisen condensation of a ketone with an ester.

In both these last two examples, a very strong base is used in the form of LDA such that the enolate ion is formed quantitatively (from ethyl acetate and acetone respectively). This avoids the possibility of self-Claisen condensation and limits the reaction to the crossed Claisen condensation.

12

Alkyl Halides

12.1 Preparation and Physical Properties

Preparation

Alkenes when treated with hydrogen halides (HCl, HBr, and HI) or halogens (Cl_2 and Br_2) yield alkyl halides and dihaloalkanes respectively. Another useful method of preparing alkyl halides is by treating an alcohol with a hydrogen halide (HX = HCl, HBr, or HI). The reaction works best for tertiary alcohols. Primary and secondary alcohols can be converted to alky1 halides by treating them with thionyl chloride ($SOCl_2$) or phosphorus tribromide (PBr_3).

Structure

Alkyl halides are made up of an alkyl group linked to a halogen atom (F, Cl, Br, or I) by a single (σ) bond. The carbon atom linked to the halogen atom is sp^2 hybridized and it has a tetrahedral geometry with bond angles of approximately 109°. The carbon-hydrogen bond length increases with the size of the halogen atom and this is accompanied with a decrease in bond strength. For example, C–F bonds are shorter and stronger than C–Cl bonds.

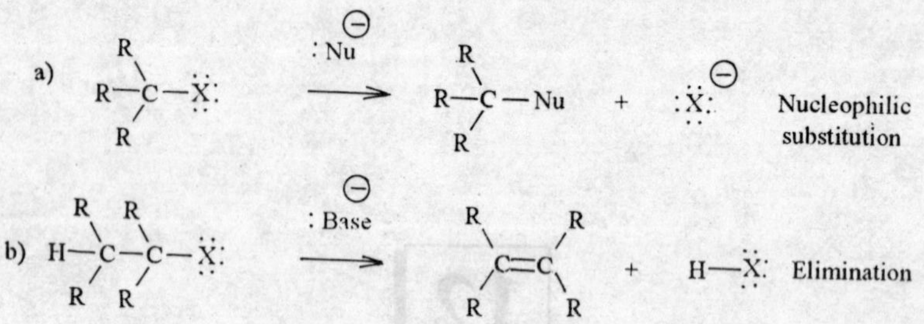

a) $$R_3C-\ddot{\underset{\cdot\cdot}{X}}: \xrightarrow{:Nu^{\ominus}} R_3C-Nu + :\ddot{\underset{\cdot\cdot}{X}}:^{\ominus}$$ Nucleophilic substitution

b) $$H-CR_2-CR_2-\ddot{\underset{\cdot\cdot}{X}}: \xrightarrow{:Base^{\ominus}} R_2C=CR_2 + H-\ddot{\underset{\cdot\cdot}{X}}:$$ Elimination

Fig. 12.1. Reactions of alkyl halides.

Bonding

The carbon-halogen bond is a σ bond. The bond is polar because the halogen atom is more electronegative than carbon. Due to this halogen is slightly negative and the carbon is slightly positive. Intermolecular hydrogen bonding or ionic bonding is not possible between alkyl halide molecules and the major intermolecular bonding force consists of weak van der Waals interactions.

Properties

The polar C–X bond present in alkyl halides has a substantial dipole moment. Alkyl halides are poorly soluble in water, but are soluble in organic solvents. They have boiling points that are similar to alkanes of comparable molecular weight. Due to polarity the carbon is an electrophilic center and the halogen is a nucleophilic center. Halogens are extremely weak nucleophilic centers and therefore, alkyl halides are more likely to react as electrophiles at the carbon center.

Reactions

The important reactions of alkyl halides are (a) nucleophilic substitution in which an attacking nucleophile replaces the halogen (Fig. 12.1a), and (b) elimination in which the alkyl halides loses HX and gets converted to an alkene (Fig. 12.1b).

12.2 Nucleophilic Substitution

Definition

Due to the presence of a strongly electrophilic carbon center alkyl halides are susceptible to nucleophilic attack a nucleophile displaces the halogen as a nucleophilic halide ion (Fig. 12.2). The reaction is called nucleophilic substitution and there are two types of mechanism i.e. the S_N1 and S_N2 mechanisms. Carboxylic acids and carboxylic acid derivatives also undergo nucleophilic substitutions, but the mechanisms are totally different.

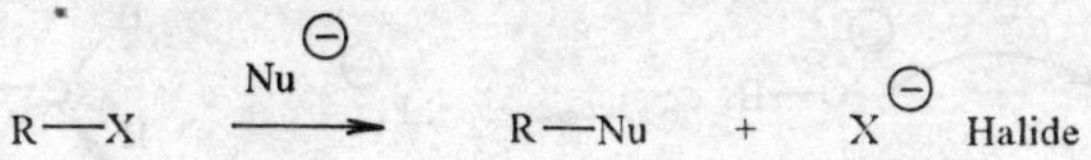

$$R{-}X \xrightarrow{Nu^{\ominus}} R{-}Nu + X^{\ominus} \text{ Halide}$$

Fig. 12.2. Nucleophilic substitution.

S_N2 Mechanism

The reaction between methyl iodide and a hydroxide ion is an example of the S_N2 mechanism (Fig. 12.3). The hydroxide ion is a nucleophile and uses one of its lone pair of electrons to form a new bond to the electrophilic carbon of the alkyl halide. Simultaneously, the C–I bond breaks. Both electrons in that bond move onto the iodine to give it a fourth lone pair of electrons and a negative charge. Since iodine is electronegative, it can stabilize this charge, so the overall process is favored.

In the *transition state* for this process (Fig. 12.4), the *new bond from the incoming nucleophile is partially formed and the C–X bond is partially broken.* The reaction center itself (CH_3) is planar. This transition state helps to explain various other features of the S_N2 mechanism. First, both the alkyl halide and the nucleophile are needed to form the transition state that means that the reaction rate is dependent on both components. Secondly, the hydroxide ion approaches iodomethane from one side while the iodide leaves from the opposite side. The hydroxide and the iodide ions are negatively charged and will repel each other, so they are as far apart as possible in the transition state. Moreover, the hydroxide ion has to gain access to the reaction center i.e. the electrophilic carbon. There is more room to attack from the 'rear' since the large iodine atom blocks approach from the other side. Lastly from an orbital point of view, it is proposed that the orbital from the incoming nucleophile starts to overlap with the empty antibonding orbital of the C–X bond (Fig. 12.5). As this interaction increases, the bonding interaction between the carbon and the halogen decreases until a transition state is reached where the incoming and outgoing nucleophiles are both partially bonded. The orbital geometry requires the nucleophiles to be on opposite sides of the molecule.

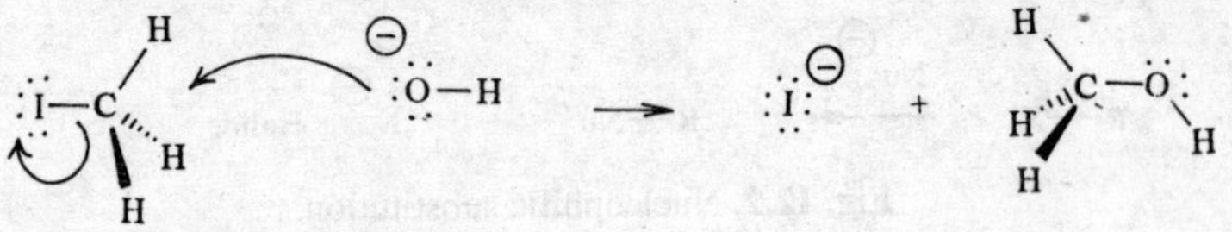

Fig. 12.3. S_N2 Mechanism for nucleophilic substitution.

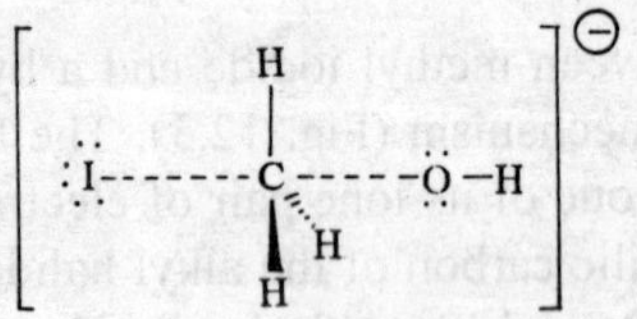

Fig. 12.4. Transition state for S_N2 nucleophilic substitution.

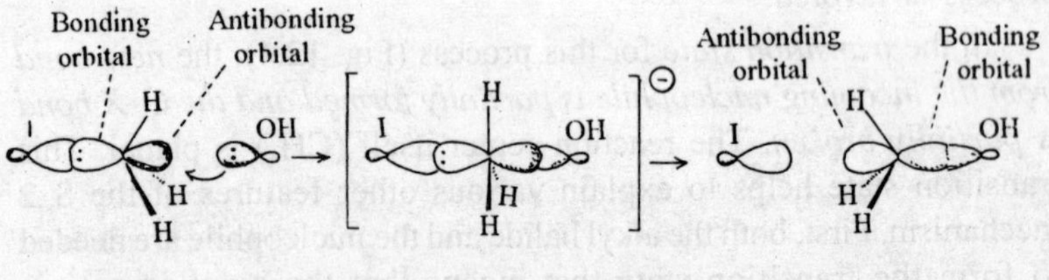

Fig. 12.5. Orbital interactions in the S_N2 mechanism.

A third interesting feature about this mechanism is about the three substituents on the carbon. Both the iodide and the alcohol product are tetrahedral compounds with the three hydrogens forming an 'umbrella' shape with the carbon (Fig. 12.6). However, the 'unmbrella' is pointing in a different direction in the alcohol product compared to the alkyl halide. *This means that the 'umbrella' has been turned inside out during the mechanism.* Hence, the carbon center has been 'inverted'. The transition state is the halfway house in this inversion.

There is no way of telling whether inversion has taken place in a molecule such as iodomethane, but proof of this inversion can be obtained by looking at the nucleophilic substitution of asymmetric alkyl halides with the hydroxide ion (Fig. 12.7). Measuring the optical activity of both alkyl halide and the alcohol permits the configuration of each enantiomer to be identified. This demonstrates that inversion of the asymmetric center occurs. This inversion is called the **'Walden Inversion'** and the mechanism called S_N2 mechanism. The S_N stands for 'substitution nucleophilic'. The 2 signifies that the rate of reaction is **second order** or **bimolecular** and depends on both the concentration of the nucleophile and the concentration of the alkyl halide. The S_N2 mechanism is possible for the nucleophilic substitutions of primary and secondary alkyl halides, but is difficult for tertiary alkyl halides. We can draw a general mechanism (Fig. 12.8) to account for a range of alkyl halides and charged nucleophiles. The mechanism is almost the same with nucleophiles like ammonia or amines with the only difference that a salt is formed and an extra step is needed to gain the free amine. For example, consider the reaction between ammonia and I-iodopropane (Fig. 12.9). Ammonia's nitrogen atom is the nucleophilic center for this reaction and uses its lone pair of electrons to form a bond to the alkyl halide. Due to this, the nitrogen will effectively lose an electron and will gain a positive charge. The C–I bond is broken and an iodide ion is formed as a leaving group, which then acts as a counterion to the alkylammonium salt. The free amine can be obtained by reaction with sodium hydroxide. This neutralize the amine to the free base that becomes insoluble in water and precipitates as a solid or as an oil.

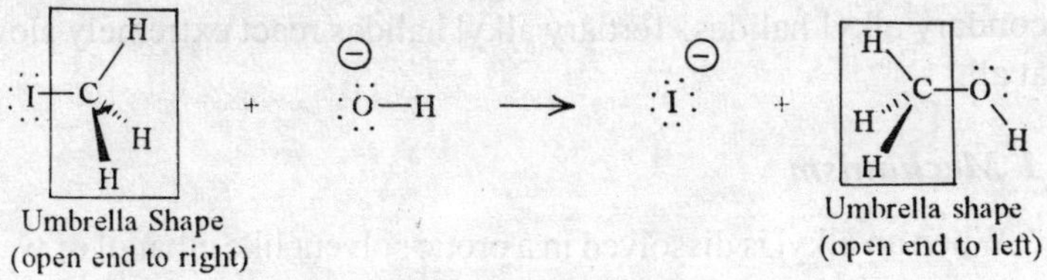

:I–C(H)(H)(H) + $\ominus$:O–H ⟶ $\ominus$:I: + H(H)(H)C–O:–H

Umbrella Shape (open end to right)

Umbrella shape (open end to left)

Fig 12.6. Walden inversion

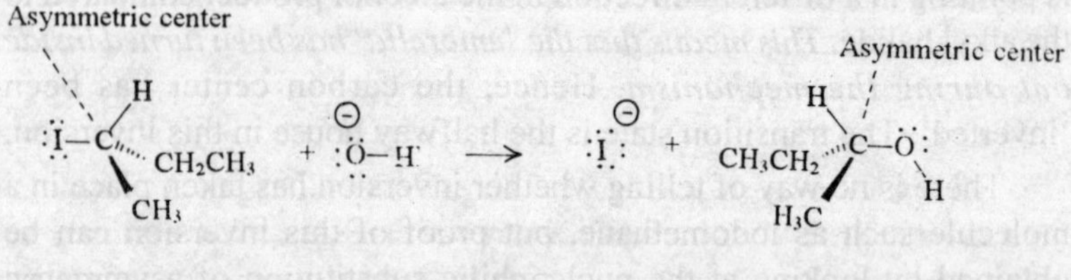

Asymmetric center

Asymmetric center

Fig. 12.7. Welden inversion of an asymmetric center.

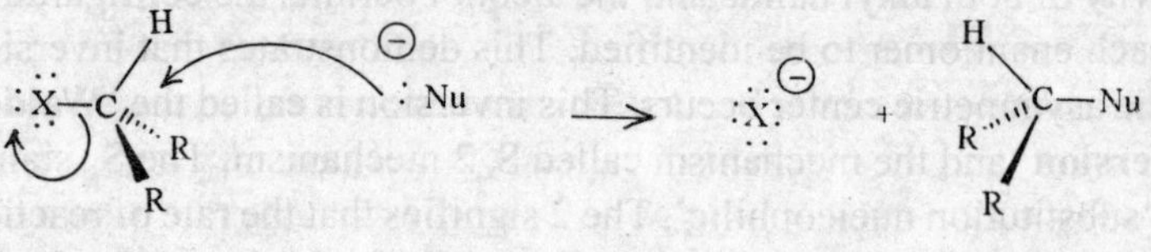

Fig. 12.8. General mechanism for the S_N2 nucleophilic substitution of alkyl halides.

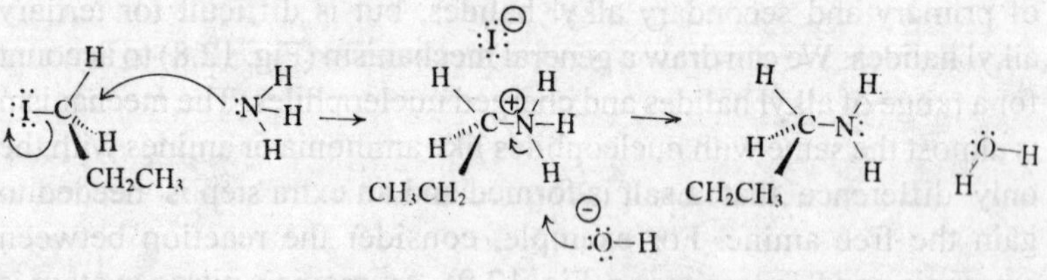

Fig. 12.9. S_N2 mechanism for the reaction of 1-iodopropane with ammonia.

The reaction of ammonia with an alkyl halide is a nucleophilic substitution as far as the alkyl halide is concerned. However, the same reaction can be considered as an alkylation from the ammonia's point of view. This is because the ammonia has gained an alkyl group from the reaction.

Primary alkyl halides undergo the S_N2 reaction faster than secondary alkyl halides. Tertiary alkyl halides react extremely slowly if at all.

S_N1 Mechanism

When an alkyl is dissolved in a protic solvent like ethanol or water, it gets exposed to a nonbasic nucleophile (i.e. the solvent molecule)

Under these conditions, the order of reactivity to nucleophilic substitution changes dramatically from that observed in the S_N2 reaction, such that tertiary alkyl halides are more reactive then secondary alkyl halides, with primary alkyl halides not reacting at all. Thus a different mechanism must be involved. For example, consider the reaction of 2-iodo-2-methylpropane with water (Fig. 12.10). In it, the rate of reaction depends on the concentration of the alkyl halide alone and the concentration of the attacking nucleophile has no effect. Thus, the nucleophile must present if the reaction is to occur, but it does not matter whether there is one equivalent of the nucleophile or an excess. Since the reaction rate depends only on the alkyl halide, the mechanism is called the S_N1 reaction, where S_N stands for substitution nucleophilic and the 1 shows that the reaction is **first order** or **unimolecular**, i.e. only one of the reactants affects the reaction rate.

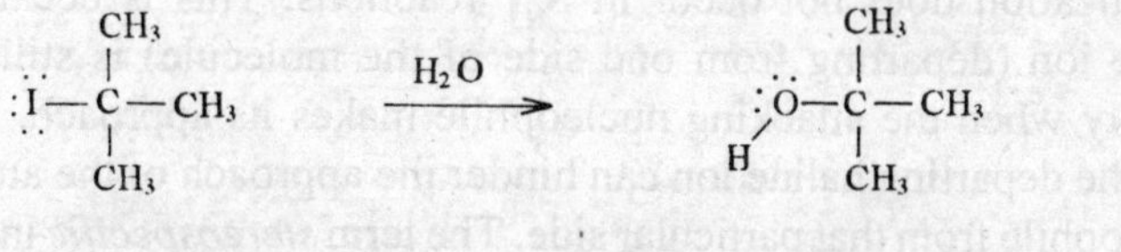

$$(CH_3)_3C{-}I \xrightarrow{H_2O} (CH_3)_3C{-}OH$$

Fig. 12.10. Reaction of 2-iodo-2-methylpropane with water.

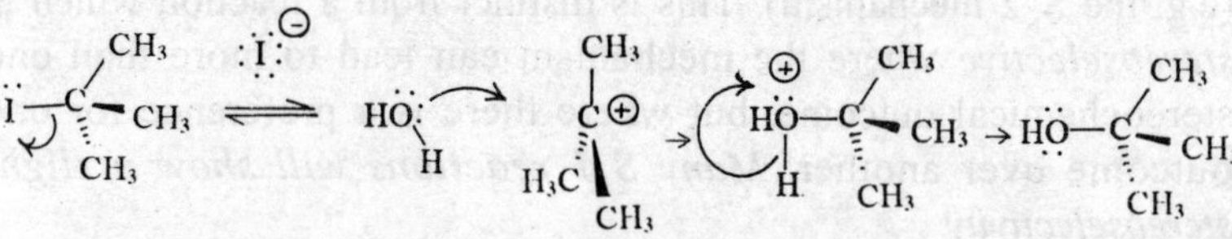

Fig. 12.11. S_N1 Mechanism.

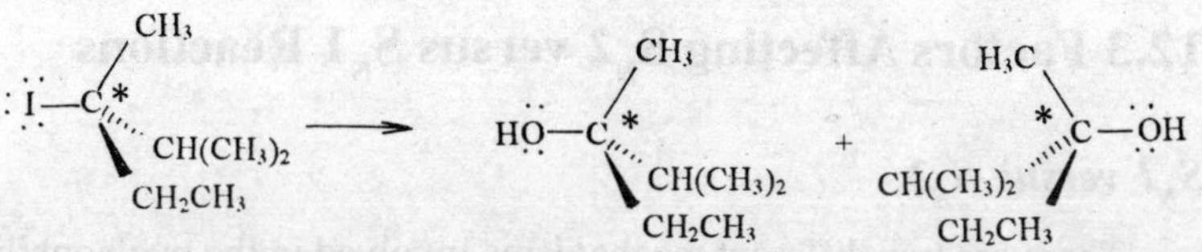

Fig. 12.12. Racemization of an asymmetric center during S_N1 nucleophilic substitution.

There are two steps in the S_N1 mechanism (Fig. 12.11). The first step is the rate-determining step and it involves loss of the halide ion.

The C–I bond breaks with both electrons on the bond moving onto the iodine atom to give it a fourth lone pair of electrons and a negative charge. The alkyl portion becomes a planar carbocation in which all three alkyl groups are as far apart from each other as possible. The central carbon atom is now sp^2 hybridized with an empty $2p_y$ orbital. In the second step, water acts as a nucleophile and reacts with the carbocation to form an alcohol.

In this mechanism the water molecule is coming in from the left hand side, but as the carbocation is planar, the water can attack equally well from the right hand side. Because the incoming nucleophile can attack from either side of the carbocation, so is no overall inversion of the carbon center. This is significant when the reaction is carried out on chiral molecules. For example, if a chiral alkyl halide reacts with water by the S_N1 mechanism, both enantiomeric alcohols would be formed resulting in a racemate (Fig. 12.12). However, **total** racemization does not occur in S_N1 reactions. This is because the halide ion (departing from one side of the molecule) is still in the vicinity when the attacking nucleophile makes its approach. Due to this, the departing halide ion can hinder the approach of the attacking nucleophile from that particular side. The term *stereospecific* indicates that the mechanism results in one specific stereochemical outcome (e.g. the S_N2 mechanism). This is distinct from a reaction which is *stereoselective* where the mechanism can lead to more than one stereochemical outcome, but where there is a preference for one outcome over another. *Many S_N1 reactions will show a slight stereoselectivity.*

12.3 Factors Affecting S_N2 versus S_N1 Reactions

S_N1 versus S_N2

There are two different mechanisms involved in the nucleophilic substitution of alkyl halides. When polar aprotic solvents are used, the S_N2 mechanism is preferred. Primary alkyl halides react more quickly than secondary alkyl halides, with tertiary alkyl halides hardly reacting at all. Under protic solvent conditions with nonbasic nucleophiles (e.g. dissolving the alkyl halide in water or alcohol), the

S_N1 mechanism is preferred and the order of reactivity is reversed. Tertiary alkyl halides are more reactive than secondary alkyl halides, and primary alkyl halides do not react at all.

There are various factors that determine if substitution will be S_N1 or S_N2 and they also control the rate at which these reactions occur. These include the nature of the nucleophile and the type of solvent is used. *The reactivity of primary, secondary, and tertiary alkyl halides is controlled by electronic and steric factors.*

Solvent

The S_N2 reaction is suitable in polar aprotic solvents (i.e. solvents with a high dipole moment, but with no O–H or N–H groups). These include solvents like acetonitrile (CH_3CN) or dimethylformamide (DMF). These solvents are polar enough to dissolve the ionic reagents needed for nucleophilic substitution, but they do so by solvating the metal cation rather than the anion. Anions are solvated by hydrogen bonding and because the solvent is incapable of hydrogen bonding, the anions remain unsolvated. Such 'naked' anions retain their nucleophilicity and react more strongly with electrophiles.

Polar, protic solvents like water or alcohols can also dissolve ionic reagents but they solvate both the metal cation and the anion. Thus, the anion is 'caged' in by solvent molecules. Thus stabilizes the anion, makes it less nucleophilic and makes it less likely to react by the S_N2 mechanism. Due to this, the S_N1 mechanism becomes more important.

The S_N1 mechanism is specially favored when the polar protic solvent is also a nonbasic nucleophile. Therefore, it is most likely to take place when an alkyl halide is dissolved in water or alcohol. Protic solvents are bad for the S_N2 mechanism because they solvate the nucleophile, but they are good for the S_N1 mechanism. This is because polar protic solvents can solvate and stabilize the carbocation intermediate. If the carbocation is stabilized, the transition state leading to it will also be stabilized and this determines whether the S_N1 reaction is favored or not. Protic solvents will also solvate the nucleophile by hydrogen bonding, but unlike the S_N2 reaction, this does not affect the reaction rate since the rate of reaction is independent of the nucleophile.

Nonpolar solvents are of no use in either the S_N1 or the S_N2 reaction as they cannot dissolve the ionic reagents needed for nucleophilic substitution.

Nucleophilicity

The relative nucleophilic strengths of incoming nucleophiles will affect the rate of the S_N2 reaction with stronger nucleophiles reacting faster. A charged nucleophile is stronger than the corresponding uncharged nucleophile (e.g. alkoxide ions are stronger nucleophiles than alcohols). Nucleophilicity is also related to base strength when the nucleophilic atom is the same (e.g. $RO^- > HO^- > RCO_2^- > ROH > H_2O$) (topic G3). In polar aprotic solvents, the order of nucleophilic strength for the halides is $F^- > Cl^- > Br^- > I^-$.

Because the rate of the S_N1 reaction is independent of the incoming nucleophile, the nucleophilicity of the incoming nucleophile is not so important.

Leaving Group

The nature of the leaving group is important to both the S_N1 and S_N2 reactions the better the leaving group, the faster the reaction. In the transition states of both reactions, the leaving group has gained a partial negative charge and the better that can be stabilized, the more stable the transition state and the faster the reaction. Therefore, the best leaving groups are those that form the most stable anions. This is also related to basicity in that the more stable the anion, the weaker the base. Iodide and bromide ions are stable ions and weak bases, and prove to be good leaving groups. The chloride ion is less stable, more basic and a poorer leaving group. The fluoride ion is a very poor leaving group and thus alkyl fluorides do not undergo nucleophilic substitution. The need for a stable leaving group explains why alcohols, ethers, and amines do not undergo nucleophilic substitutions since they would involve the loss of a strong base (e.g. RO^- or R_2N^-).

Alkyl Halides — S_N2

There are two factors that affect the rate at which alkyl halides undergo the S_N2 reaction. These are *electronic* and *steric*. To illustrate why different alkyl halides react at different rates in the S_N2 reaction, let us compare a primary, secondary, and tertiary alkyl halide (Fig. 12.13).

Alkyl groups have an inductive, electron-donating effect that tends to lower the electrophilicity of the neighboring carbon center. Lowering the electrophilic strength means that the reaction center will be less reactive to nucleophiles. Therefore, tertiary alkyl halides will be less likely to react with nucleophiles than primary alkyl halides, since the inductive effect of three alkyl groups is greater than one alkyl group.

Steric factors also play a role in making the S_N2 mechanism difficult for tertiary halides. An alkyl group is a bulky group compared to a hydrogen atom, and can therefore act like a shield against any incoming nucleophile (Fig. 12.14). A tertiary alky halide has three alkyl shields compared to the one alkyl shield of a primary alkyl halide. Therefore, a nucleophile is more likely to be deflected when it approaches a tertiary alkyl halide and fails to reach the electrophilic center.

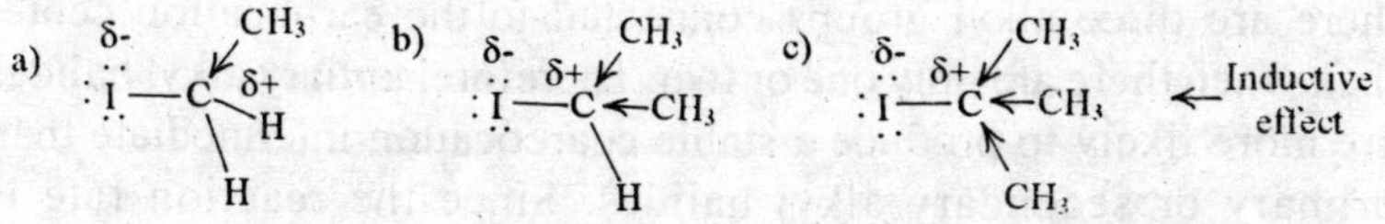

a) δ- :I—C δ+ CH₃ H H b) δ- δ+ :I—C←CH₃ CH₃ H c) δ- δ+ :I—C←CH₃ CH₃ CH₃ ← Inductive effect

Fig. 12.13. (a) Iodoethane; (b) 2-iodopropane; (c) 2-iodo-2-methylpropane.

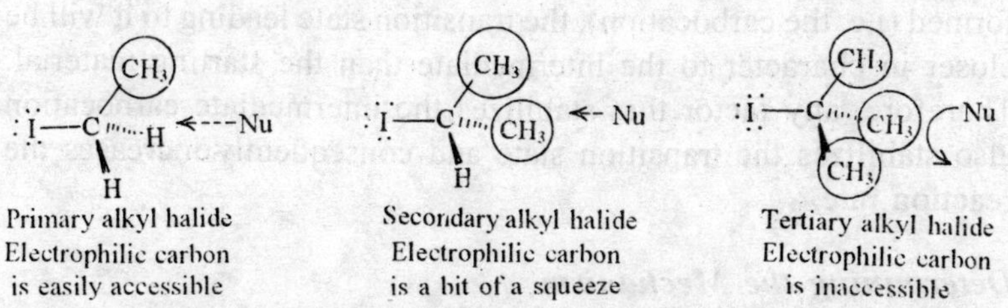

:I—C H CH₃ H ←---Nu
Primary alkyl halide
Electrophilic carbon
is easily accessible

:I—C CH₃ CH₃ H ←- Nu
Secondary alkyl halide
Electrophilic carbon
is a bit of a squeeze

:I—C CH₃ CH₃ CH₃ Nu
Tertiary alkyl halide
Electrophilic carbon
is inaccessible

Fig. 12.14. Steric factors affecting nucleophilic substitution.

Alkyl Halides — S_N1

Steric and electronic factors also play role in the rate of the S_N1 reaction because the steric bulk of three alkyl substituents makes it very difficult for a nucleophile to reach the electrophilic carbon center of tertiary alky halides, these structures undergo nucleophilic substitution by the S_N1 mechanism. In this mechanism, the steric problem is relieved because loss of the halide ion creates a planar carbocation where the alkyl groups are much further apart and where the carbon center is more accessible. Formation of the carbocation also relieves steric strain between the substituents.

Electronic factors also help in the formation of the carbocation because the positive charge can be stabilized by the inductive and hyperconjugative effects of the three alkyl groups (Fig. 12.15).

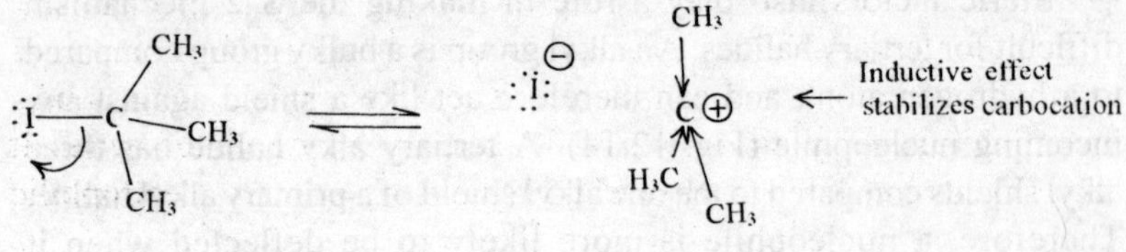

$$:\ddot{I}-C(CH_3)_3 \rightleftharpoons :\ddot{I}:^{\ominus} + (CH_3)_3C^{\oplus}$$

Inductive effect stabilizes carbocation

Fig. 12.15. Inductive effects stabilizing a carbocation.

Both the inductive and hyperconjugation effects are greater when there are three alkyl groups connected to the carbocation center than when there are only one or two. Therefore, tertiary alkyl halides are more likely to produce a stable coarbocation intermediate than primary or secondary alkyl halides. Since the reaction rate is determined by how well the **transition state** of the rate determining step stabilized. In a situation in which a high energy intermediate is formed (i.e. the carbocation), the transition state leading to it will be closer in character to the intermediate than the starting material. Therefore, any factor that stabilizes the intermediate carbocation also stabilizes the transition state and consequently increases the reaction rate.

Determining the Mechanism

Generally the nucleophilic substitution of primary alkyl halides will occur via the S_N2 mechanism, whereas nucleophilic substitution of tertiary alkyl halides will occur by the S_N1 mechanism. Generally secondary alkyl halides are more likely to react by the S_N2 mechanism, but it is not possible to predict this with certainty. The only way to find out for certain is to try out the reaction and see whether the reaction rate depends on the concentration of both reactants (S_N2) or whether it depends on the concentration of the alkyl halide alone (S_N1).

If the alkyl halide a chiral the optical rotation of the product could be measured to see whether it is a pure enantiomer or not. If it is, the mechanism is S_N2. If not, it is S_N1.

12.4 Elimination

Definition

Alkyl halides having a proton attached to a neighboring β-carbon atom can undergo an elimination reaction to produce an alkene and a hydrogen halide (Fig. 12.16). This reaction is the reverse of the electrophilic addition of a hydrogen halide to an alkene. There are two mechanisms by which this elimination can occur. These are E2 mechanism and the E1 mechanism.

The E2 reaction is the most effective for the synthesis of alkenes from alkyl halides and can be used on primary, secondary, and tertiary alkyl halides. The E1 reaction is not so useful from a synthetic point of view and occurs in competition with the S_N1 reaction of tertiary alkyl halides. Primary and secondary alkyl halides do not generally react by this mechanism.

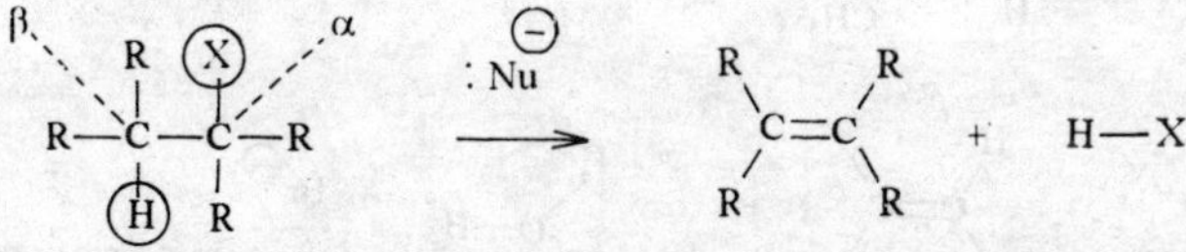

Fig. 12.16. Elimination of an alkyl halide.

Susceptible B-protons

An alkyl halide can undergo an elimination reaction if it has a susceptible proton situated on a β-carbon i.e. the carbon next to the C–X group. This proton is lost during the elimination reaction along with the halide ion. In some respects, there is similarity here between alkyl halides and carbonyl compounds (Fig. 12.17). Alkyl halides can have susceptible protons at the β-position whilst carbonyl compounds can have acidic protons at their α-position. By comparing both structures, it can be seen that the acidic/susceptible proton is attached to a carbon neighboring an electrophilic carbon.

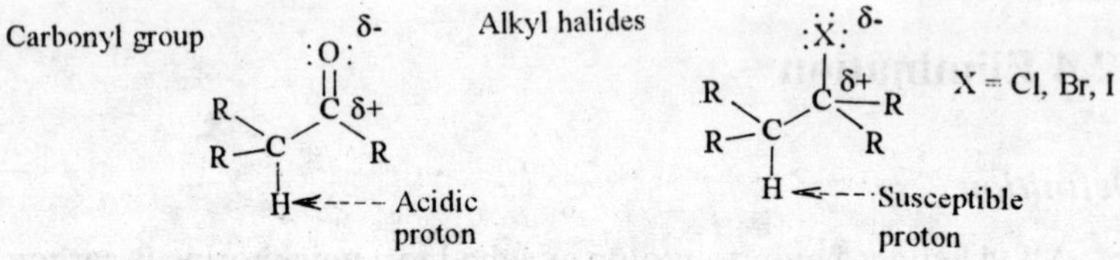

Carbonyl group δ- :O: R C δ+ R—C R H←---- Acidic proton

Alkyl halides :X: δ- δ+ R C—R R—C R H←--- Susceptible proton X = Cl, Br, I

Fig. 12.17. Comparison of carbonyl compound and an alkyl halide.

E2 Mechanism

The E2 mechanism is a *concerted mechanism* and involves both the alkyl halide and the nucleophile. Due to this, the reaction rate depends on the concentration of both reagents and is called second order (E2 = Elimination second order). To illustrate the mechanism, we shall look at the reaction of 2-bromopropane with a hydroxide ion (Fig. 12.18).

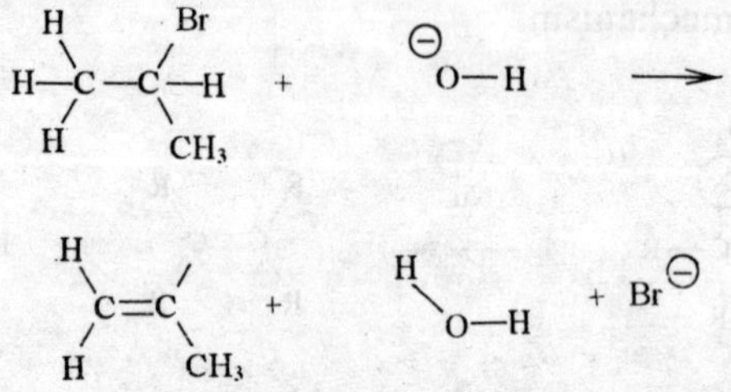

$$H_3C{-}CH(Br){-}CH_3 + {}^{\ominus}O{-}H \longrightarrow H_2C{=}CH{-}CH_3 + H{-}O{-}H + Br^{\ominus}$$

Fig. 12.18. Reaction of 2-bromopropane with the hydroxide ion.

The mechanism (Fig. 12.19) involves the hydroxide ion forming a bond to the susceptible proton. As the hydroxide ion forms its bond, the C–H bond breaks. Both electrons in that bond could move onto the carbon, but there is a neighboring electrophilic carbon that attracts the electrons and so the electrons move in to form a π bond between the two carbons. Simultaneously as this π bond is formed, the C–Br bond breaks and both electrons end up on the bromine atom that is lost as a bromide ion.

The E2 elimination is stereospecific, with elimination taking place in an antiperiplanar geometry. The diagrams (Fig. 12.20) show that the four atoms involved in the reaction are in plane with the H and Br on opposite sides of the molecule.

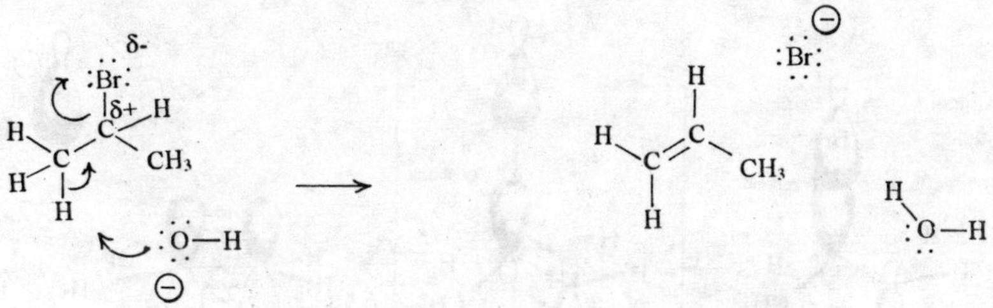

Fig. 12.19. E2 Elimination mechanism.

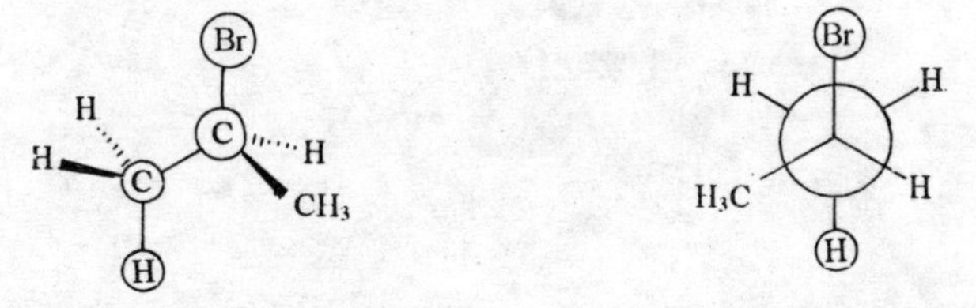

Circled atoms are in one plane

Antiperiplanar arrangement

Fig. 12.20. Relative geometry of the atoms involved in the E2 elimination mechanism.

The reason for this stereospecificity can be explained using orbital diagrams (Fig.12.21). In the transition state of this reaction, the C–H and C–Br σ bonds are in the process of breaking. As they do so, the sp^3 hybridized orbitals which were used for these σ bonds are changing into *p* orbitals that begins to interact with each other to form the eventual π bond. For all this to happen in the one transition state, an antiperiplanar arrangement is essential.

E1 Mechanism

The E1 mechanism generally occurs when an alkyl halide is dissolved in protic solvent where the solvent can act as a nonbasic nucleophile. These are the same conditions for the S_N1 reaction and so both these reactions generally take place simultaneously forming a mixture of products. The example the E1 mechanism, is the reaction of 2-iodo-2-methyl-butane with methanol (Fig. 12.22).

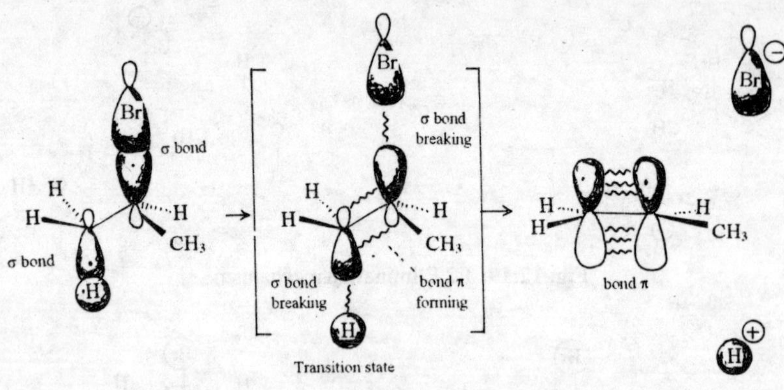

Fig. 12.21. Orbital diagram of the E2 elimination process.

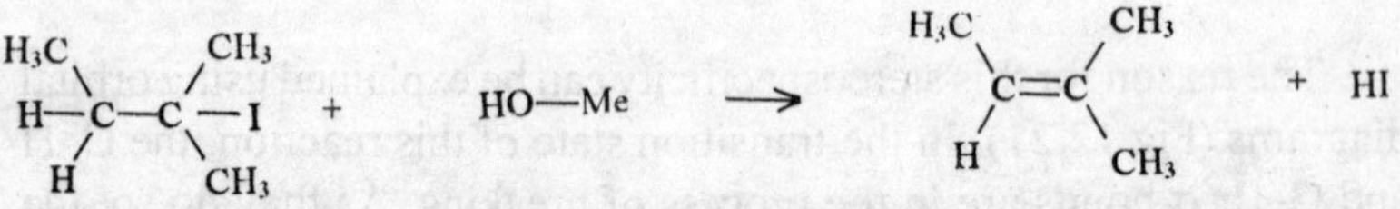

$$H_3C\text{–}CH(H)\text{–}C(CH_3)_2\text{–}I \;+\; HO\text{—}Me \;\longrightarrow\; (H_3C)(H)C{=}C(CH_3)_2 \;+\; HI$$

Fig. 12.22. Elimination reaction of 2-iodo-2-methyl-butane.

There are two stages to the mechanism (Fig. 12.23). The first stage is exactly the same as described for the S_N1 mechanism and that is cleavage of the C–X bond to form a planar carbocation intermediate in which the positive charge is stabilized by the three alkyl groups surrounding it. In the second stage, the methanol forms a bond to the susceptible proton on the β-carbon. The C–H bond breaks and both electrons are used to form a π bond to neighboring carbocation. The first step of the reaction mechanism is the rate-determining step and as this is dependent only on the concentration of the alkyl halides, the reaction is first order (E1 = elimination first order). There is no stereospecificity involved in this reaction and a mixture of isomers can be obtained with the more stable (more substituted) alkene being favored.

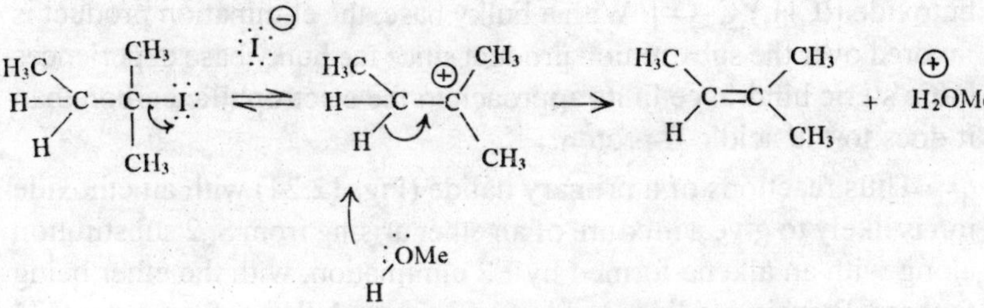

Fig. 12.23. The E1 mechanism.

E2 versus E1

The E2 elimination occurs with a strong base (like a hydroxide or ethoxide ion) in a protic solvent (like ethanol or water). The E2 reaction is more common than the E1 elimination and more useful. All types of alkyl halide can undergo the E2 elimination and the method is useful for preparing alkenes.

The conditions that favour E1 are the same which that favour the S_N1 reaction (i.e. a protic solvent and a nonbasic nucleophile). Therefore, the E1 reaction normally only takes place with tertiary alkyl halides and will be in competition with the S_N1 reaction.

12.5 Elimination versus Substitution

Introduction

Alkyl halides can undergo both elimination and substitution reactions and so generally both substitution and elimination products are present. The ratio of the products will depend on the reaction conditions, the nature of the nucleophile and the nature of the alkyl halide.

Primary alkyl halides undergo the S_N2 reaction with a large range of nucleophiles (e.g. RS^-, I^-, CN^-, NH_3, or Br^-) in polar aprotic solvents like hexamethyl phosphoramide (HMPA; $[(CH_3)_2N]_3PO$). However, there is always the possibility of some E2 elimination taking place as well. Neverthless, substitution is usually favored over elimination, even when using strong bases like HO^- or EtO^-. If E2 elimination of a primary halide is desired, it is best to use a strong bulky base like *tert*-butoxide $[(CH_3)_3C–O^-]$. With a bulky base, the elimination product is favored over the substitution product since the bulky base experiences more steric hindrance in its approach to the electrophilic carbon than it does to the acidic β-proton.

Thus reactions of a primary halide (Fig. 12.24) with an ethoxide ion is likely to give a mixture of an ether arising from S_N2 substitution along with an alkene formed by E2 elimination, with the ether being favored. By using sodium *tert*-butoxide instead, the preferences would be reversed.

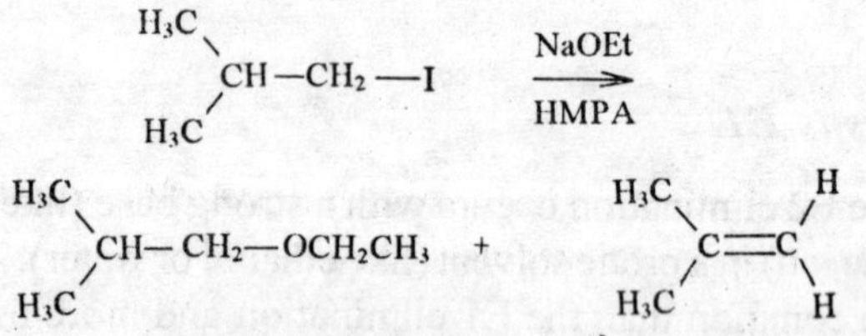

$$(H_3C)_2CH—CH_2—I \xrightarrow[\text{HMPA}]{\text{NaOEt}} (H_3C)_2CH—CH_2—OCH_2CH_3 + (H_3C)_2C{=}CH_2$$

Fig. 12.24. Reaction of 1-iodo-2-methylpropane with sodium ethoxide.

Increasing the temperature of the reaction shifts the balance from the S_N2 reactions to the elimination reaction. This is because the elimination reaction has a higher activation energy because of more bonds being broken. The S_N1 and E1 reactions do not occur for primary alkyl halides.

Secondary Alkyl Halides

Secondary alkyl halides can undergo both S_N2 and E2 reactions to give a mixture of products. However, the substitution product predominates if a polar aprotic solvent is used and the nucleophile is a weak base. Elimination will predominate if a strong base is used as the nucleophile in a polar, protic solvent. In this case, bulky bases are not so crucial and the use of ethoxide in ethanol will give more elimination product than substitution product. Increasing the temperature of the reaction favors E2 elimination over S_N2 substitution as explained above.

If weakly basic or nonbasic nucleophiles are used in protic solvents, elimination and substitution may occur by the S_N1 and E1 mechanisms to give mixtures.

Tertiary Alkyl Halides

Tertiary alkyl halides are essentially unreactive to strong nucleophiles in polar, aprotic solvents i.e. the conditions for the S_N2 reaction. Tertiary alkyl halides can undergo E2 reactions when treated with a strong base in protic solvent and will do so in good yield since the S_N2 reaction is so highly disfavored. Under nonbasic conditions in a protic solvent, E1 elimination and S_N1 substitution both occur.

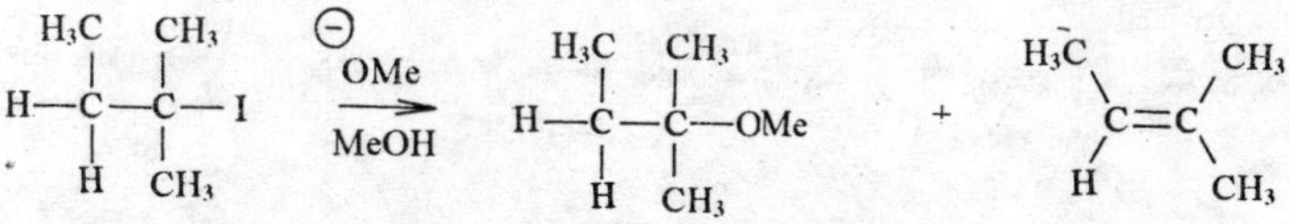

Fig. 12.25. Reaction of 1-iodo-2-methylbutane with methoxide ion.

A tertiary alkyl halide when treated with sodium methoxide forms an ether or an alkene (Fig. 12.25). A protic solvent is used here and this favors both the S_N1 and E1 mechanisms. However, a strong base is also being used and this favors the E2 mechanism. Therefore, the alkene would be expected to be the major product with only a very small amount of substitution product arising from the S_N2 reaction. Heating the same alkyl halide in methanol alone means that the reaction is being done in a protic solvent with a nonbasic nucleophile (MeOH). These conditions would yield a mixture of substitution and elimination products arising from the S_N1 and E1 mechanism. The substitution product would be favored over the elimination product.

12.6 Reactions of Alkyl Halides

Nucleophilic Substitution

The nucleophilic substitution of alkyl halides is an important method of obtaining a wide variety of different functional groups.

Therefore, it is possible to convert a variety of primary and secondary alkyl halides to alcohol, ethers, thiols, thioethers, esters, amines, and azides (Fig. 12.26). Alkyl iodides and alkyl chlorides can also be synthesized from other alkyl halides.

It is also possible to construct larger carbon skeletons using alkyl halides. A simple example is the reaction of an alkyl halide with a cyanide ion (Fig. 12.27). This is an important reaction because the nitrile product can be hydrolyzed to yield a carboxylic acid.

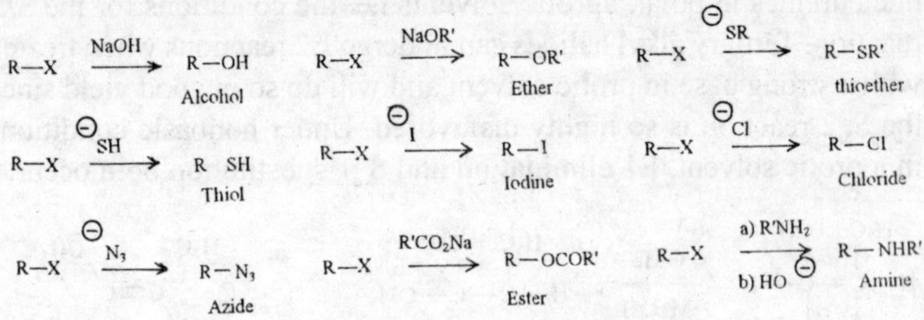

$R{-}X \xrightarrow{NaOH} R{-}OH$ (Alcohol)　　$R{-}X \xrightarrow{NaOR'} R{-}OR'$ (Ether)　　$R{-}X \xrightarrow{^{\ominus}SR} R{-}SR'$ (thioether)

$R{-}X \xrightarrow{^{\ominus}SH} R\ SH$ (Thiol)　　$R{-}X \xrightarrow{^{\ominus}I} R{-}I$ (Iodine)　　$R{-}X \xrightarrow{^{\ominus}Cl} R{-}Cl$ (Chloride)

$R{-}X \xrightarrow{^{\ominus}N_3} R{-}N_3$ (Azide)　　$R{-}X \xrightarrow{R'CO_2Na} R{-}OCOR'$ (Ester)　　$R{-}X \xrightarrow[b)\ HO^{\ominus}]{a)\ R'NH_2} R{-}NHR'$ (Amine)

Fig. 12.26. Nucleophilic substitutions of alkyl halides.

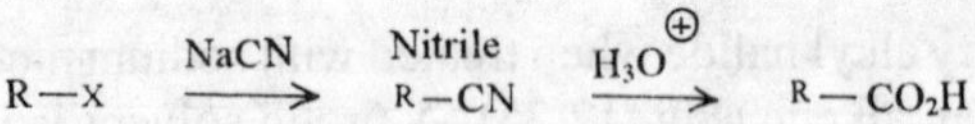

$$R{-}X \xrightarrow{NaCN} \underset{}{\overset{Nitrile}{R{-}CN}} \xrightarrow{H_3O^{\oplus}} R{-}CO_2H$$

Fig. 12.27. Constructing larger carbon skeletons from an alkyl halide.

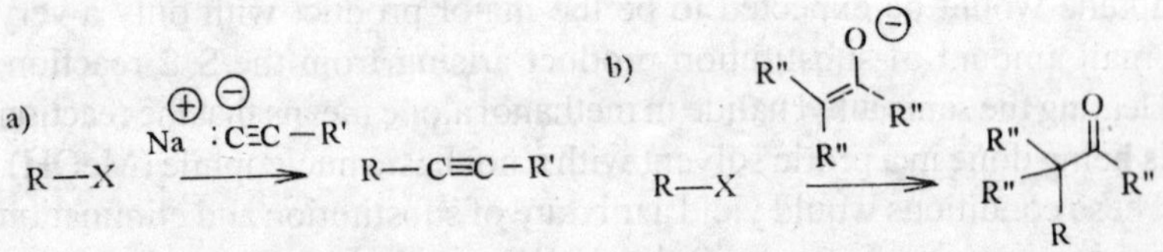

a) $R{-}X \xrightarrow{Na^{\oplus}\ {}^{\ominus}:C{\equiv}C{-}R'} R{-}C{\equiv}C{-}R'$

b) $R{-}X \xrightarrow{R''_2C{=}C(O^{\ominus})R''} R''_2C(R){-}C(=O){-}R''$

Fig. 12.28. (a) Synthesis of a disubstituted alkyne; (b) alkylation of an enolate ion.

The reaction of an acetylide ion with a primary alkyl halide allows the synthesis of disubstituted alkynes (Fig. 12.28a). The enolate ions of esters or ketones can also be alkylated with alkyl halides to create larger carbon skeletons (Fig 12.28b). The most successful nucleophilic substitutions are with primary alkyl halides. With secondary and tertiary alkyl halides, the elimination reaction may compete, particularly when the nucleophile is a strong base. The substitution of tertiary alkyl halides is best done in a protic solvent with weakly basic nucleophiles. However, yields may be poor.

Elimination

Elimination reactions of alkyl halides (**dehydrohalogenations**) are a useful method of synthesizing alkenes. For good yield, a strong base (e.g. NaOEt) should be used on a protic solvent (EtOH) with a secondary or tertiary alkyl halide. The reaction proceeds by an E2 mechanism. Heating increases the chances of elimination over substitution.

For primary alkyl halides, a strong, bulky base (e.g. NaOBu') should be used. The bulk hinders the possibility of the S_N2 substitution and encourages elimination by the E2 mechanism. The advantage of the E2 mechanism is that it is higher yielding than the E1 mechanism and is also stereospecific. The geometry of the product obtained is determined by the antiperiplanar geometry of the transition state. For example, the elimination in Fig 12.29 gives the (*E*)-isomer and none of the (Z)-isomer.

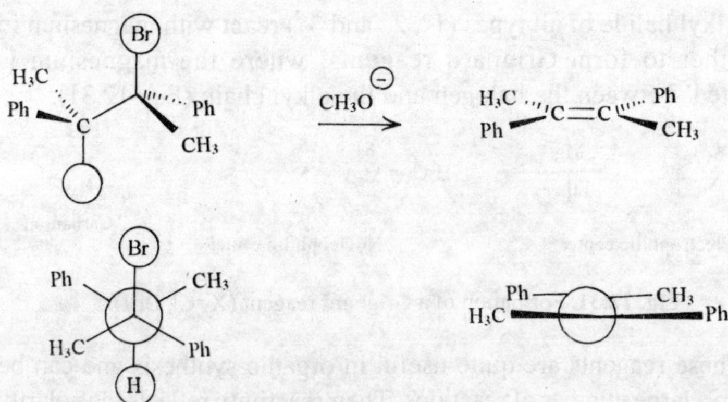

Fig. 12.29. Stereochemistry of the E2 elimination reaction.

If the elimination takes place by the E1 mechanism, the reaction is more likely to compete with S_N1 reaction and we get a mixture of substitution and elimination products.

The E2 elimination needs the presence of a β-proton. If there are various options available, a mixture of alkenes will be obtained, but the favored alkene will be the most substituted (and most stable) one (**Zaitsev's rule;** Fig. 12.30).

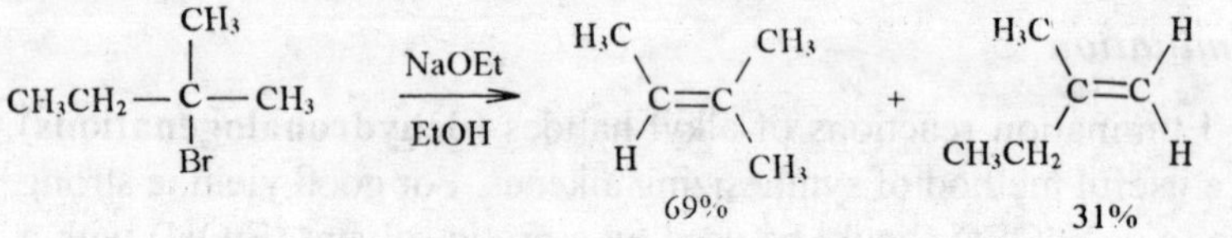

$$CH_3CH_2-C(CH_3)(Br)-CH_3 \xrightarrow[EtOH]{NaOEt} (H_3C)(H)C{=}C(CH_3)(CH_3) + (H_3C)(CH_3CH_2)C{=}CH_2$$

69% 31%

Fig. 12.30. Example of Zaitsev's rule.

The transition state for the reaction resembles the product more than the reactant and so the factors that stabilize the product also stabilize the transition state and make that particular route more likely. However, the opposite preference is found when potassium *tert*-butoxide is used as base, and the less substituted alkene is favored.

12.7 Organometallic Reactions

Grignard Reagents

Alkyl halide of all types (1°, 2° and 3°) react with magnesium in dry ether to form Grignard reagents, where the magnesium is 'inserted' between the halogen and the alkyl chain (Fig. 12.31).

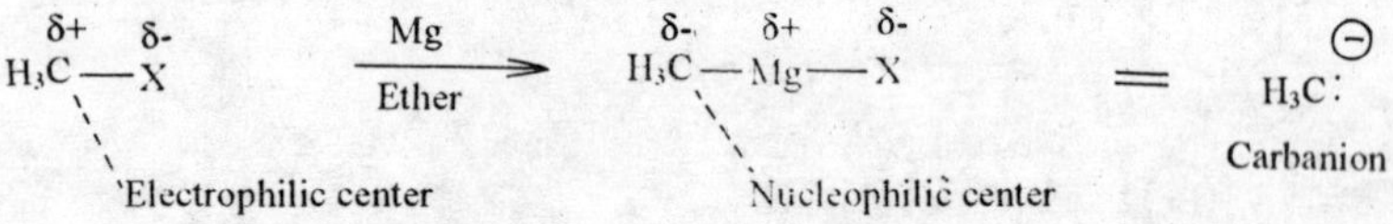

$$\overset{\delta+}{H_3C}-\overset{\delta-}{X} \xrightarrow[Ether]{Mg} \overset{\delta-}{H_3C}-\overset{\delta+}{Mg}-\overset{\delta-}{X} \equiv H_3C:^{\ominus}$$

Electrophilic center; Nucleophilic center; Carbanion

Fig. 12.31. Formation of a Grignard reagent (X=Cl, Br,I).

These reagents are quite useful in organic synthesis and can be used in a large number of reactions. Their reactivity reflects the polarity of the atoms present. Since magnesium is a metal it is electropositive, it means that the electrons in the C–Mg bond spend more of their time closer to the carbon making it slightly negative and a nucleophilic center. This reverses the character of this carbon since it is an electrophilic center in the original alkyl halide. In essence, a Grignard reagent can be viewed as providing the equivalent of a carbanion. The carbanion is not a distinct species, but the reactions which occur can be explained as if carbanion was present.

A Grignard reagent can react as a base with water to form an alkane. This is one way of converting an alkyl halide to an alkane. The same acid-base reaction can occur with a number of proton donors (Bronsted acids) including functional groups like alcohols, carboxylic acids, and amines (Fig. 12.32). This can prove a disadvantage if the Grignard reagent is intended to react at some other site on the target molecule. In such cases, functional groups containing an X–H bond (where X = a heteroatom) would have to be protected before the Grignard reaction is carried out.

$$R{-}X \xrightarrow{Mg} R{-}MgX \xrightarrow{H_2O \text{ or } ROH \text{ or } RCO_2H \text{ or } RNH_2} R{-}H$$

Fig. 12.32. Conversion of a Grignard reagent to an alkane.

The reactions of Grignard reagents with aldehydes and ketones give alcohols, reaction with acid chlorides and esters give tertiary alcohols, reaction with carbon dioxide to give carboxylic acids, reaction with nitriles give ketones, and reaction with epoxides give alcohols.

Organolithium Reagents

Alkyl halides can be converted to organolithium reagents using lithium metal in an alkane solvent (Fig. 12.33).

$$R{-}X \xrightarrow[\text{Pentane}]{Li} R{-}Li$$

Fig. 12.33. Formation of organolithium reagents.

Organocuprate Reagents

Organolithium reagents reacts just like Grignard reagents. For example, reaction with aldehydes and ketones proceeds by nucleophilic addition to yield secondary and tertiary alcohols respectively.

Organocuprate reagents (another source of carbanion equivalents) are obtained by the reaction of one equivalent or cuprous iodide with two equivalents of an organolithium reagent (Fig. 12.34).

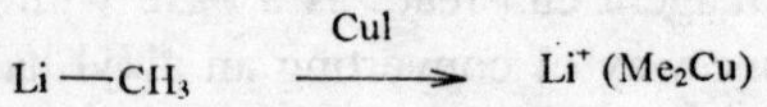

$$Li—CH_3 \xrightarrow{CuI} Li^+ (Me_2Cu)$$

Fig. 12.34. Formation of organocuprate reagents.

$$R—X \xrightarrow{R''_2\overset{\ominus}{Cu}\ \overset{\oplus}{Li}} R—R'' \qquad X = Br, I$$

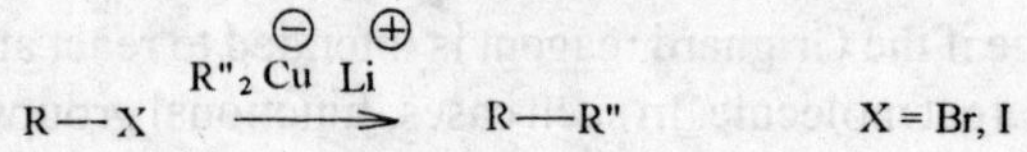

Fig. 12.35. Reaction of an organocuprate reagent with an alkyl halide.

These reagents are useful in the 1,4-addition of alkyl groups to α,β-unsaturated carbonyl systems and can also be reacted with alkyl halides to produce larger alkanes (Fig. 12.35). The mechanism is considered to be radical based.

13

Alcohols, Phenols and Thiols

13.1 Preparation of Alcohols

Functional Group Transformation

Alcohols can be prepared by nucleophilic substitution of alkyl halides, hydrolysis of esters, reduction of carboxylic acids or esters, reduction of aldehydes or ketones, electrophilic addition of alkenes, hydroboration of alkenes, or substitution of ethers.

C–C Bond Formation

Alcohols can also be obtained from epoxides, aldehydes, ketones, esters, and acid chloride as a consequence of C–C bond formation. These reactions involve the addition of carbanion equivalents through the use of Grignard or organolithium reagents.

13.2 Preparation of Phenols

Incorporation

Phenol groups can be introduced into an aromatic ring by sulfonation of the aromatic ring followed by reacting the product with sodium hydroxide to convert the sulfonic acid group to a phenol (Fig. 13.1). The reaction conditions are drastic and only alkyl-substituted phenols can be prepared by this method.

Another general method of preparing phenols is to hydrolyze a diazonium salt, prepared from an aniline group (NH_2) (Fig. 13.2).

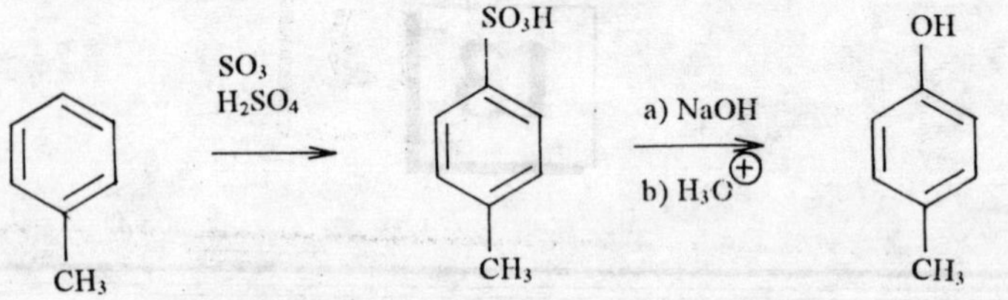

Fig. 13.1. Synthesis of a phenol via sulfonation.

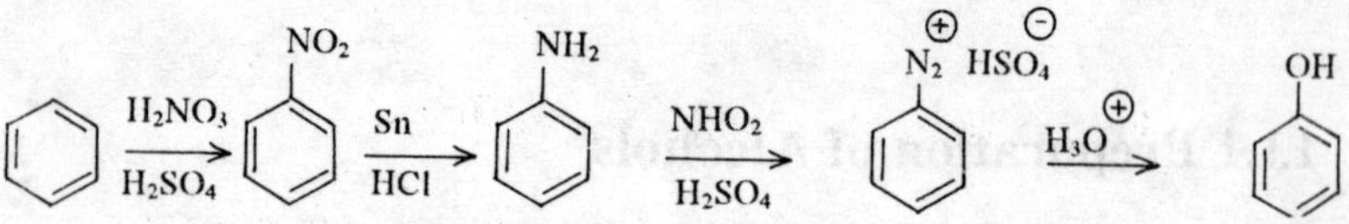

Fig. 13.2. Synthesis of a phenol via diazonium salt.

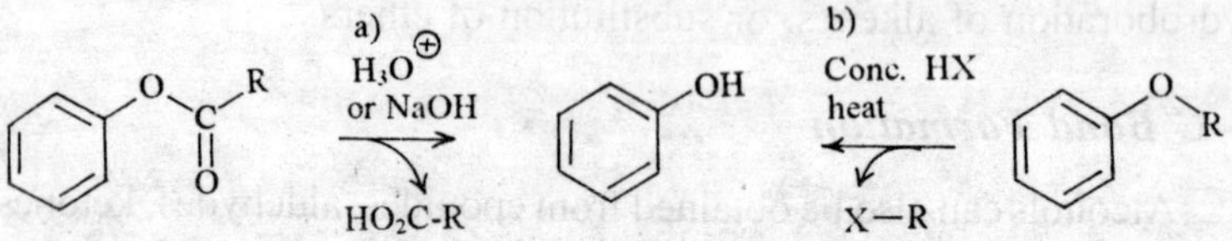

Fig. 13.3. Functional group transformations to a phenol.

Functional Group Transformation

A number of functional group can be converted to phenols e.g. Sulfonic acids and amino groups which have already been mentioned. Phenyl esters can be hydrolyzed (Fig. 13.3a). Aryl ethers can be cleaved (Fig. 13.3b). The bond between the alkyl group and oxygen is specifically cleaved because the Ar–OH bond is too strong to be cleaved.

13.3 Properties of Alcohols and Phenols

Alcohols

The alcohol functional group (R_3C–OH) has the same geometry as water, with a C–O–H bond angle of approximately 109°. Both the carbon and the oxygen are *sp*^{3} hybridized. Due to the presence of the O–H group intermolecular hydrogen bonding is possible that accounts for the higher boiling points of alcohols compared with alkanes of similar molecular weight. Due to hydrogen bonding alcohols are more soluble in protic solvents than alkenes of similar molecular weight. Actually, the smaller alcohols (methanol, ethanol, propanol, and *tert*-butanol) are completely miscible in water. With larger alcohols, the hydrophobic character of the bigger alkyl chain takes precedence over the polar alcohol group and so larger alcohols are insoluble in water.

The O–H and C–O bonds are both polarized because of the electronegative oxygen, in such a way that oxygen is slightly negative and the carbon and hydrogen atoms are slightly positive. Due to this the oxygen serves as a nucleophilic center while the hydrogen and the carbon atoms serve as weak electrophilic centers (Fig. 13.4).

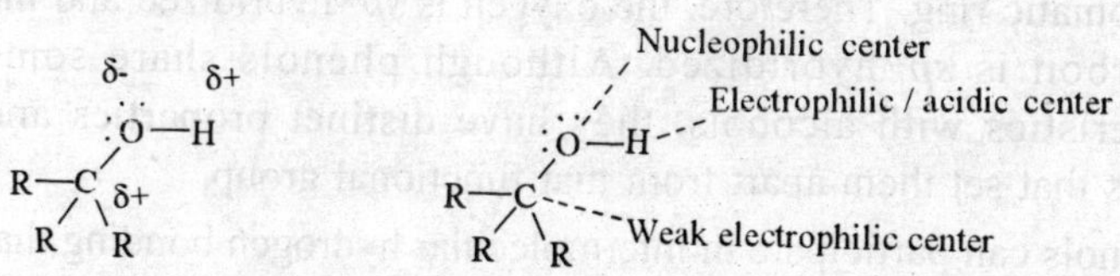

Nucleophilic center

Electrophilic / acidic center

Weak electrophilic center

Fig. 13.4. Bond polarization and nucleophilic and electrophilic centers.

Because of the presence of the nucleophilic oxygen and electrophilic proton, alcohols can act both as weak acids and as weak bases when dissolved in water (Fig. 13.5). However, the equilibrium in both cases is virtually completely weighted to the unionized form.

Alcohols generally react with stronger electrophiles than water. However, they are less likely to react with nucleophiles unless the latter are also strong bases, in that case the acidic proton is abstracted to form an **alkoxide** ion (RO^-; Fig. 13.6). alkoxide ions are quite

the oxygen atom acting as the nucleophilic center. The intermediate formed can then react more readily as an electrophile at the carbon center.

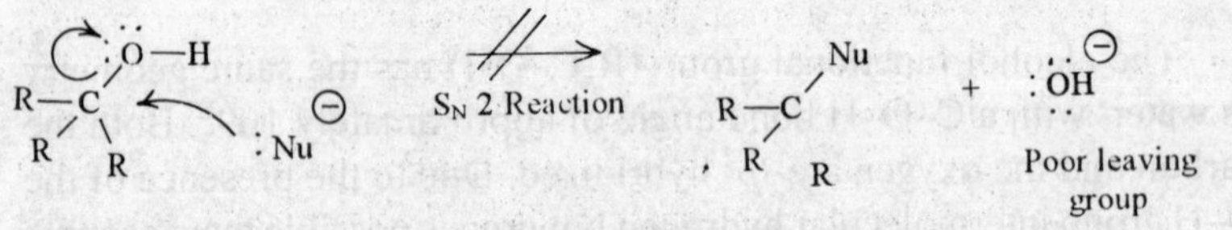

Fig. 13.7. Nucleophilic substitution of alcohols is not favored.

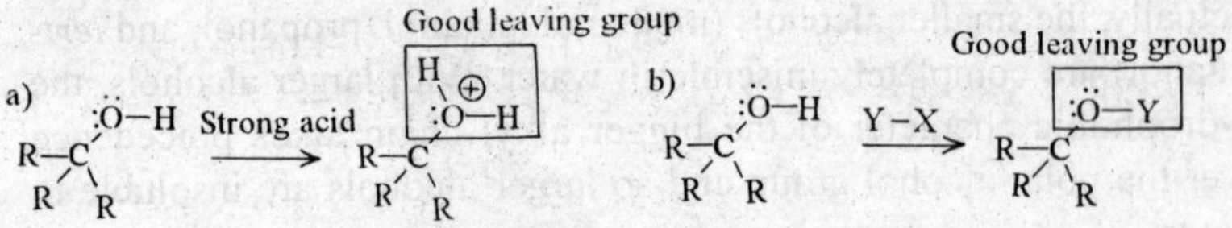

Fig. 13.8. Activation of an alcohol.

Phenols

Phenols are compounds that have an OH group directly attached to an aromatic ring. Therefore, the oxygen is sp^3 hybridized and the aryl carbon is sp^2 hybridized. Although phenols share some characteristics with alcohols, they have distinct properties and reactions that set them apart from that functional group.

Phenols can participate in intermolecular hydrogen bonding that means that they have a moderate water solubility and have higher boiling points than aromatic compounds lacking the phenolic group. Phenols are weakly acidic, and in aqueous solution an equilibrium exists between the phenol and the phenoxide ion (Fig. 13.9a). When treated with a base, the phenol gets converted to the phenoxide ion (Fig. 13.9b).

The phenoxide ion is stabilized by resonance and delocalization of the negative charge into the ring, therefore phenoxide ions are weaker bases than alkoxide ions. This means that phenols are more acidic than alcohols, but less acidic than carboxylic acids. The pK_a

useful reagents in organic synthesis. However, they cannot be used if water is the solvent since the alkoxide ion would act as a base and abstract a proton from water to regenerate the alcohol. Therefore, an alcohol would have to be used as solvent instead of water.

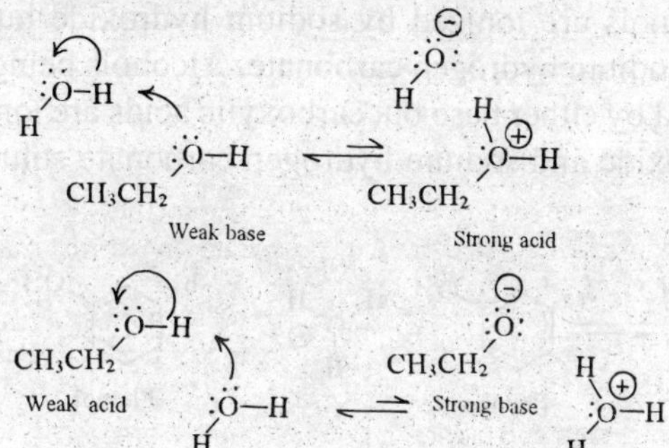

Fig. 13.5. Acid-base properties of alcohols.

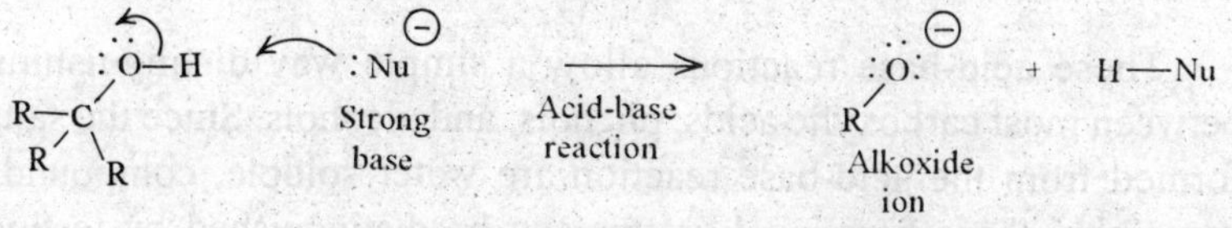

Strong base

Acid-base reaction

Alkoxide ion

+ H—Nu

Fig. 13.6. Formation of an alkoxide ion.

Nucleophiles that are also strong bases react with the electrophilic hydrogen of an alcohol rather than the electrophilic carbon. Nucleophilic attack at carbon would need the loss of a hydroxide ion in a nucleophilic substitution reaction. However, this is not favored as the hydroxide ion is a strong base and a poor leaving group (Fig. 13.7). However, reactions which involve the cleavage of an alcohol's C–O bond are possible if the alcohol is first 'activated' such that the hydroxyl group is converted into a better leaving group (Fig. 13.8). One method is to react the alcohol under acidic conditions such that the hydroxyl group is protonated before the nucleophile makes its attack (Fig. 13.8). Cleavage of the C–O bond would then be more likely because the leaving group would be a neutral water molecule that is a much better leaving group. Alternatively, the alcohol can be treated with an electrophilic reagent to convert the OH group into a different group (OY) that can then act as a better leaving group (Fig. 13.6). In both cases, the alcohol must first act as a nucleophile, with values of most phenols is in the order of 11, compared to 18 for alcohols and 4.74 for acetic acid. This means the phenols can be ionized with weaker bases than those needed to ionize alcohols, but need stronger bases than those needed to ionize carboxylic acids. For example, phenols are ionized by sodium hydroxide but not by the weaker base sodium hydrogen carbonate. Alcohols beings less acidic are not ionized by either base but carboxylic acids are ionized by both sodium hydroxide and sodium hydrogen carbonate solutions.

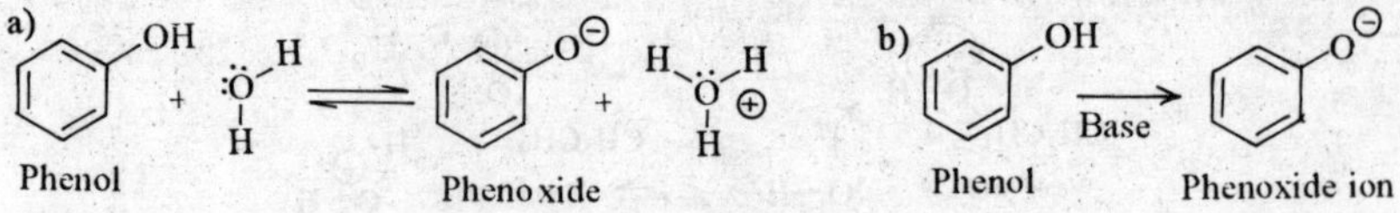

a) Phenol + H_2O ⇌ Phenoxide + H_3O^+ b) Phenol $\xrightarrow{\text{Base}}$ Phenoxide ion

Fig. 13.9. Acidic reactions of phenol.

These acid-base reactions allow a simple way distinguishing between most carboxylic acids, phenols, and alcohols. Since the salts formed from the acid-base reaction are water soluble, compounds containing these functional groups can be distinguished by testing their solubilities in sodium hydrogen carbonate and sodium hydroxide solutions. This solubility test is not valid for low molecular weight structures like methanol or ethanol since these are water soluble and dissolve in basic solution because of their water solubility rather than their ability to form salts.

13.4 Reactions of Alcohols

Acid-base Reactions

Alcohols are slightly weaker acids than water and thus the conjugate base generated from an alcohol (like **alkoxide** ion) is a stronger base than the conjugate base of water (the hydroxide ion). Due to this, it is not possible to generate an alkoxide ion using sodium hydroxide as base. Alcohols do not react with sodium bicarbonate or amines, and a stronger base like sodium hydride or sodium amide is needed to generate the alkoxide ion (Fig. 13.10). Alcohols can also be converted to alkoxide ions on treatment with potassium, sodium, or lithium metal. Some organic reagents can also act as strong bases, e.g. Grignard reagents and organolithium reagents.

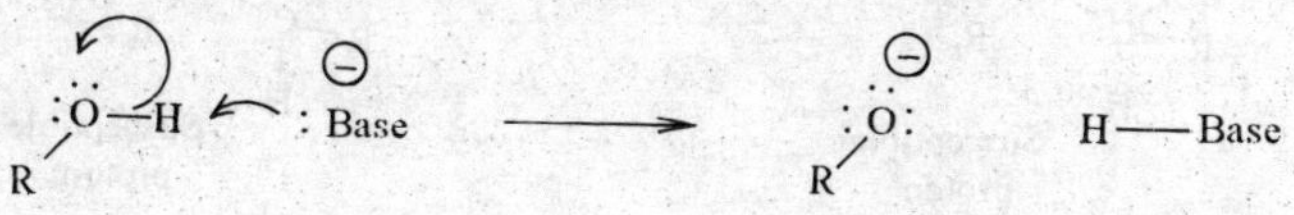

$$R-\ddot{O}-H + :Base^{\ominus} \longrightarrow R-\ddot{O}:^{\ominus} \quad H-Base$$

Fig. 13.10. Generation of alkoxide ion.

Alkoxide ions are neutralized in water and so reactions involving these reagents should be accomplished in the alcohol from which they were derived, that is reactions involving sodium ethoxide are best carried out in ethanol. Alcohols have a typical pK_a of 15.5–18.0 compared to pK_a values of 25 for ehtyne, 38 for ammonia and 50 for ethane.

Elimination

Alcohols, like alkyl halides, can undergo elimination reactions to form alkenes (Fig. 13.11). Since water is eliminated, the reaction is also called a dehydration.

$$HO-C(CH_3)_3 \xrightarrow{20\%\ H_2SO_4} (H_3C)_2C=CH_2 + H_2O$$

Fig. 13.11. Elimination of an alcohol.

Like alkyl halides, the elimination reaction of an alcohol needs the presence of a susceptible proton at the β-position (Fig. 13.12).

The elimination of alkyl halides is done under basic conditions, the elimination of alcohols is done under acid conditions. Under basic conditions, an E2 elimination would require the loss of a hydroxide ion as a leaving group. Since the hydroxide ion is a strong base, it is not a good leaving group and so the elimination of alcohols under basic conditions is difficult to achieve.

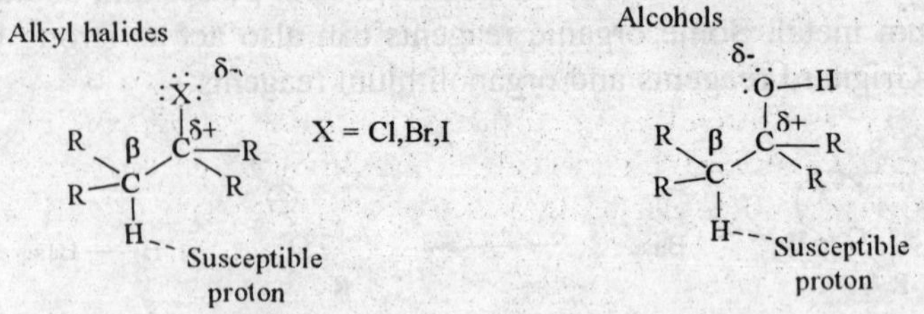

Fig. 13.12. Susceptible β-protons in an alkyl halide and an alcohol.

Elimination under acidic conditions is more successful because the hydroxyl group is first protonated and then it departs the molecule as a neutral water molecule (**dehydration**) that is a much better leaving group. If *different isomeric alkenes are possible, the most substituted alkene will be favored* (Fig. 13.13). The reaction occurs best with tertiary alcohols as the elimination proceeds by the E1 mechanism.

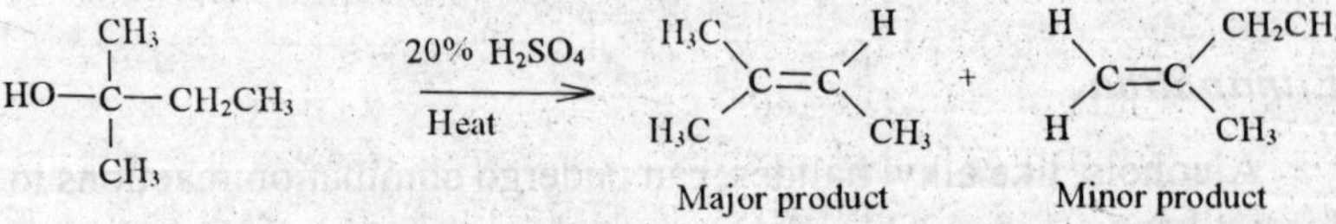

Fig. 13.13. Elimination of alcohols obeys Zaitsev's rule.

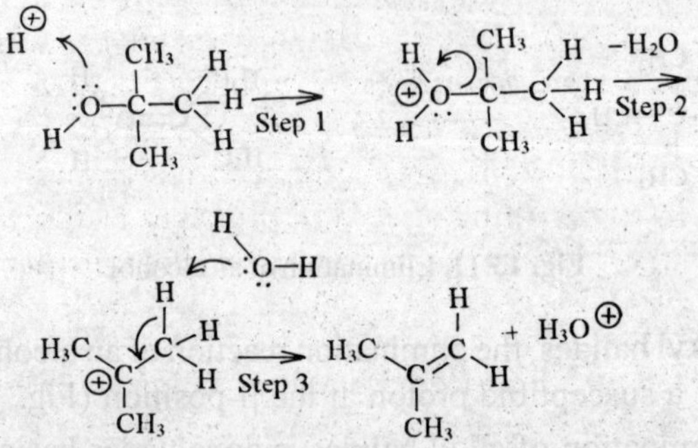

Fig. 13.14. E1 Elimination mechanism for alcohols.

The mechanism shown (Fig. 13.14) involves the nucleophilic oxygen of the alcohol making use of one of its lone pairs of electrons to form a bond to a proton to yield a charged intermediate (Step 1). When the oxygen gets protonated, the molecule has a much better leaving group because water can be ejected as a neutral molecule. The E1 mechanism can now proceed as normal. Water is lost and a carbocation is formed (Step 2). Water then acts as a base in the second step, making use of one of its lone pairs of electrons to form a bond to the β-proton of the carbocation. The C–H bond is broken and both the electrons in that bond are used to form a π bond between the two carbons. Because this is an E1 reaction, tertiary alcohols react better than primary or secondary alcohols.

The E1 reaction is not ideal for the dehydration of primary or secondary alcohols since vigorous heating is needed to force the reaction and this can result in rearrangement reactions. In alternative methods which are useful, reagents like phosphorus oxychloride ($POCl_3$) dehydrate secondary and tertiary alcohols under mild basic conditions using pyridine as solvent (Fig. 13.15). The phosphorus oxychloride serves to activate the alcohol, converting the hydroxyl function into a better leaving group. The mechanism involves the alcohol acting as a nucleophile in the first step. Oxygen uses a lone pair of electrons to form a bond to the electrophilic phosphorus of $POCl_3$ and a chloride ion is lost (Step 1). Pyridine then removes a proton from the structure to form a dichlorophosphate intermediate (Step 2). The dichlorophosphate group is a much better leaving group than the hydroxide ion and so a normal E2 reaction can occur. Pyridine acts as a base to remove a β-proton and as this is happening, the electrons from the old C–H bond are used to form a π bond and eject the leaving group (Step 3).

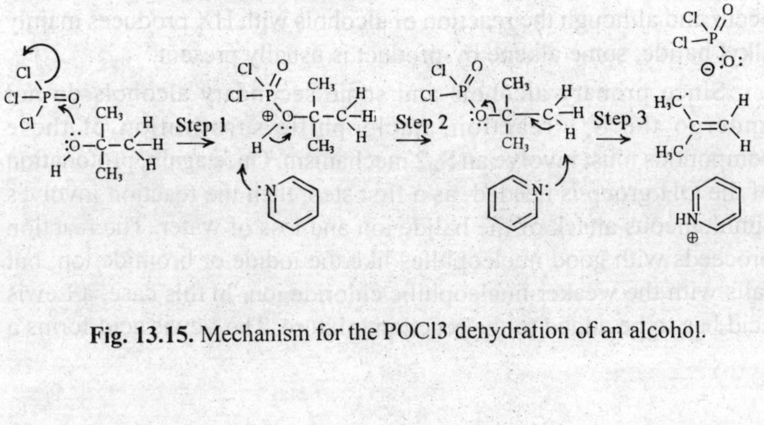

Step 1 Step 2 Step 3

Fig. 13.15. Mechanism for the POCl3 dehydration of an alcohol.

Synthesis of Alkyl Halides

Tertiary alcohols may undergo the S_N1 reaction to produce tertiary alkyl halides (Fig. 13.16). Since the reaction needs the loss of the hydroxide ion (a poor leaving group), so to convert the hydroxyl moiety into a better leaving group acidic conditions are achieved with the use of HCl or HBr. The acid serves to protonate the hydroxyl moiety as the first step and then a normal S_N1 mechanism occurs where water is lost from the molecule to form an intermediate carbocation. A halide ion then forms a bond to the carbocation center in the third step.

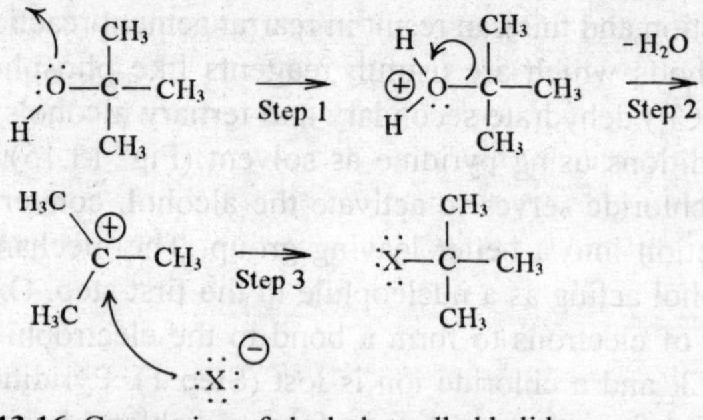

Fig. 13.16. Conversion of alcohols to alkyl halides.

The first two steps of this mechanism are the same as the elimination reaction. Both reactions are carried out under acidic conditions. The difference is that halide ion serve as good nucleophiles and are present in high concentration. The elimination reaction is carried out using concentrated sulfuric acid and only weak nucleophiles are present (i.e. water) in low concentration. Thus, some elimination may occur and although the reaction of alcohols with HX produces mainly alkyl halide, some alkene by-product is usually present.

Since primary alcohols and some secondary alcohols do not undergo the S_N1 reaction, nucleophilic substitution of these compounds must involve an S_N2 mechanism. Once again, protonation of the OH group is needed as a first step, then the reaction involves simultaneous attack of the halide ion and loss of water. The reaction proceeds with good nucleophiles like the iodide or bromide ion, but fails with the weaker nucleophilic chloride ion. In this case, a Lewis acid has to be added to the reaction mixture. The Lewis acid forms a complex with the oxygen of the alcohol group, which results in a much better leaving group for the subsequent S_N2 reaction.

However, the reaction of primary and secondary alcohols with hydrogen halides can generally be a problem since unwanted rearrangement reactions generally occurs.

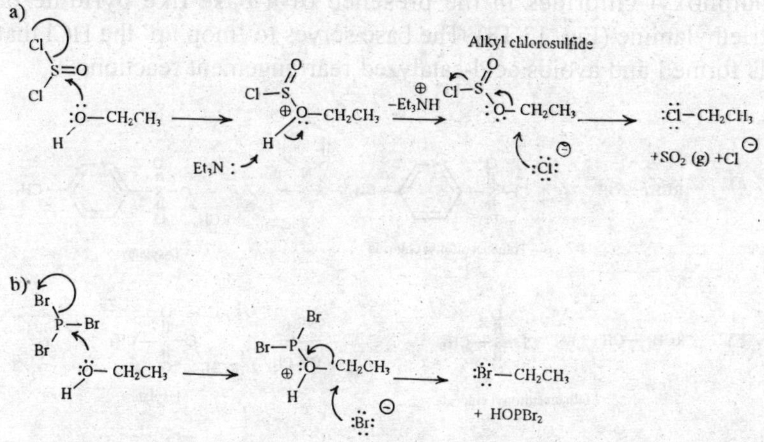

a)

Alkyl chlorosulfide

Et_3N :

$-Et_3NH$

$:\ddot{Cl}-CH_2CH_3$ $+SO_2$ (g) $+Cl^{\ominus}$

b)

$:\ddot{Br}-CH_2CH_3$

+ $HOPBr_2$

Fig. 13.17. Conversion of an alcohol to an alkyl halide using (a) thionyl chloride; (b) phosphorus tribromide.

To avoid this, the reaction is carried out under milder basic conditions using reagents like thionyl chloride or phosphorus tribromide (Fig. 13.17). These reagents act as electrophiles and react with the alcoholic oxygen to form an intermediate where the OH moiety gets converted into a better leaving group. A halide ion is released from the reagent in this process, and this can act as the nucleophile in the subsequent S_N2 reaction.

In the reaction with thionyl chloride, triethylamine is present to mop up the HCl formed during the reaction. The reaction is also helped by presence of one of the products (SO_2) as a gas which gets expelled thus driving the reaction to completion.

Phosphorus tribromide has three bromine atoms present and each PBr_3 molecule can react with three alcohol molecules to form three molecules of alkyl bromide.

Synthesis of Mesylates and Tosylates

Sometimes it is convenient to synthesize an activated alcohol that can be used in nucleophic substitution reactions like an alkyl halide. Mesylates and tosylates are such sulfonate compounds which serve this purpose. They can be synthesized by action of alcohols with sulphoxyl chlorides in the presence of a base like pyridine or triethylamine (Fig. 13.18). The base serves to 'mop up' the HCl that is formed and avoids acid-catalyzed rearrangement reactions.

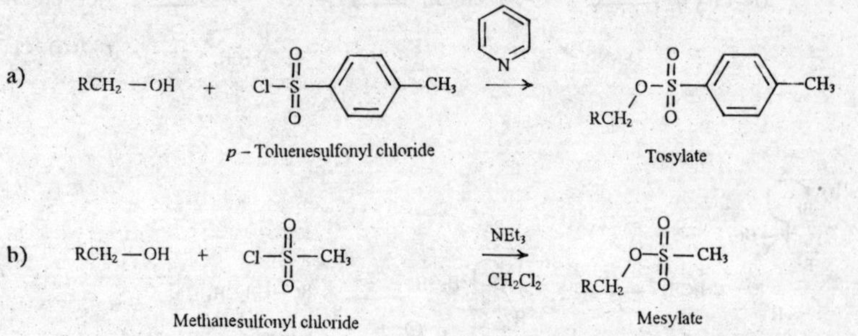

a) $RCH_2-OH + Cl-SO_2-C_6H_4-CH_3 \xrightarrow{\text{pyridine}} RCH_2-O-SO_2-C_6H_4-CH_3$

p – Toluenesulfonyl chloride → Tosylate

b) $RCH_2-OH + Cl-SO_2-CH_3 \xrightarrow[CH_2Cl_2]{NEt_3} RCH_2-O-SO_2-CH_3$

Methanesulfonyl chloride → Mesylate

Fig. 13.18. Synthesis of (a) tosylate and (b) mesylate.

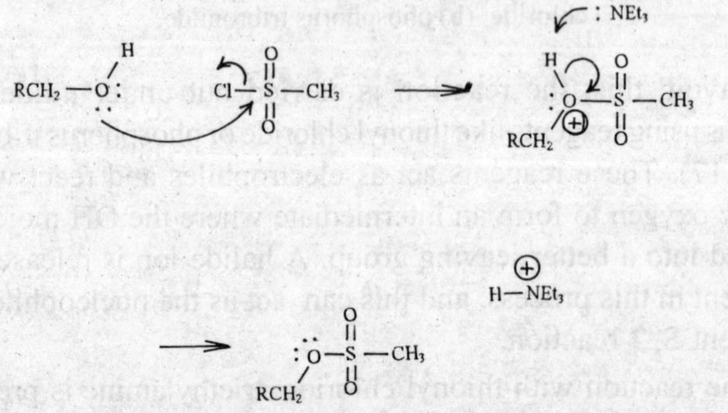

$RCH_2-\ddot{O}(H): + Cl-SO_2-CH_3 \longrightarrow RCH_2-\overset{\oplus}{O}(H)-SO_2-CH_3 \;\; (:NEt_3)$

$\longrightarrow \; :\ddot{O}(RCH_2)-SO_2-CH_3 \;\; + \;\; H-\overset{\oplus}{N}Et_3$

Fig. 13.19. Mechanism for the formation of a mesylate.

The reaction mechanism (Fig. 13.19) involves the alcohol oxygen acting as a nucleophilic center and substituting the chloride ion from the sulfonate. The base then removes a proton from the intermediate to give the sulfonate product. *Neither of these steps affects the stereochemistry of the alcohol carbon and so the stereochemistry o, chiral alcohols is retained.*

The mesylate and tosylate groups are excellent leaving groups and can be considered as the equivalent of a halide. Therefore mesylates and tosylates can undergo the S_N2 reaction in the same way as alkyl halides (Fig. 13.20).

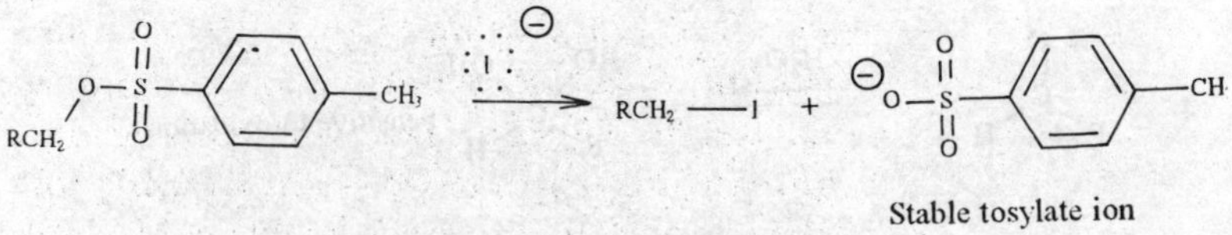

Fig. 13.20. Nucleophilic substitution of atosylate.

Oxidation

The oxidation of alcohols is quite an important reaction in organic synthesis. Primary alcohols can be oxidized to aldehydes, but the reaction is difficult because there is the danger of over-oxidation to carboxylic acids. In case of volatile aldehydes, the aldehydes can be distilled from the reaction solution as they are formed. However, this is not possible for less volatile aldehydes. To overcome this problem by a mild oxidizing agent like pyridinium chlorochromate (PCC; Fig. 13.21a) is used. If a stronger oxidizing agent is used in aqueous conditions (e.g. CrO_3 in aqueous sulfuric acid), primary alcohols are oxidized to carboxylic acids (Fig. 13.21b), while secondary alcohols are oxidized to ketones (Fig. 13.21c).

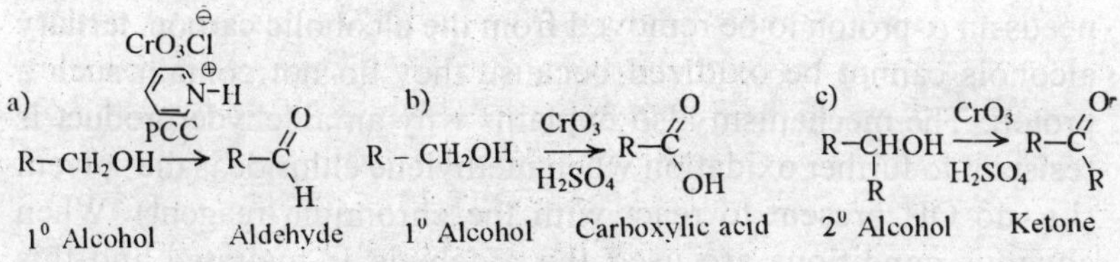

Fig. 13.21. Oxidation of alcohols.

The success of the PCC oxidation in stopping at the aldehyde stage is solvent dependent. The reaction is done in methylene chloride, whereas oxidation with CrO_3 is done in aqueous acid. Under aqueous conditions, the aldehyde that is formed by oxidation of the alcohol is hydrated and this structure is more sensitive to oxidation than the aldehyde itself (Fig. 13.22). In methylene chloride, hydration cannot take place and the aldehyde is more resistant to oxidation.

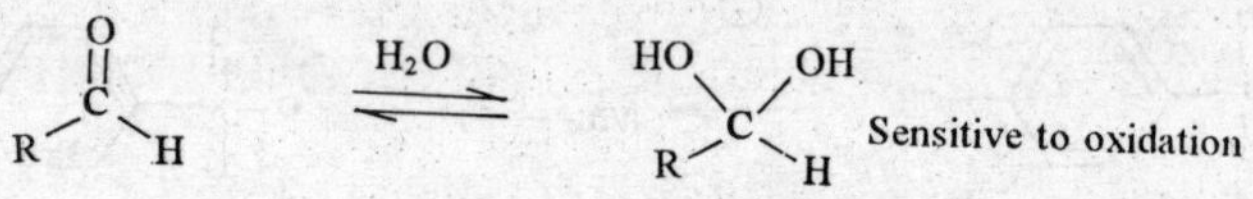

Fig. 13.22. Hydration of an aldehyde.

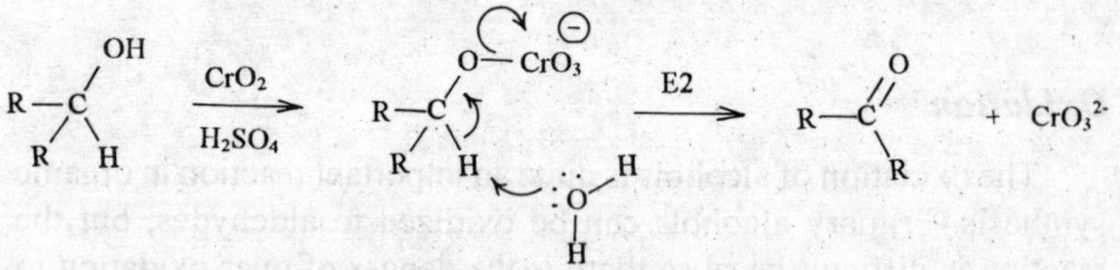

Fig. 13.23. Mechanism of oxidation of a secondary alcohol with CrO_3.

The mechanism of oxidation for a secondary alcohol with CrO_3 (Fig. 13.23) involves the nucleophilic oxygen reacting with the oxidizing agent to produce a charged chromium intermediate. Elimination then takes place where an α-proton is lost along with the chromium moiety to produce the carbonyl group. The mechanism can be considered as an E2 mechanism, the difference being that different bonds are being created and broken. As the mechanism needs an α-proton to be removed from the alcoholic carbon, tertiary alcohols cannot be oxidized because they do not contain such a proton. The mechanism also explains why an aldehyde product is resistant to further oxidation when methylene chloride is the solvent (i.e. no OH present to react with the chromium reagent). When aqueous conditions are used the aldehyde is hydrated and this generates two OH groups that are available to bond to the chromium reagent and result in further oxidation.

13.5 Reactions of Phenols

Acid-base Reactions

Phenols are stronger acids than alcohols. They react with bases like sodium hydroxide to form phenoxide ions. However, *they are weaker acids than carboxylic acids and do not react with sodium hydrogen carbonate.*

Phenols are acidic because the oxygen's lone pair of electrons can participate in a resonance mechanism involving the adjacent aromatic ring (Fig. 13.24). Three resonance structures are possible in which the oxygen gains a positive charge and the ring gains a negative charge. The net result is a slightly positive charge on the oxygen that accounts for the acidity of its proton. There are also three aromatic carbons with slightly negative charges.

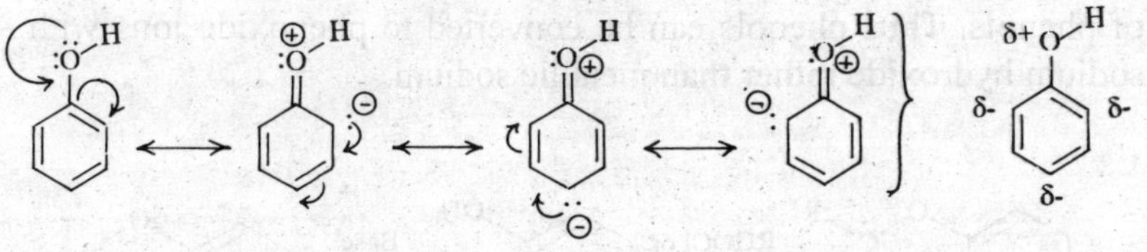

Fig. 13.24. Resonance structures for phenol.

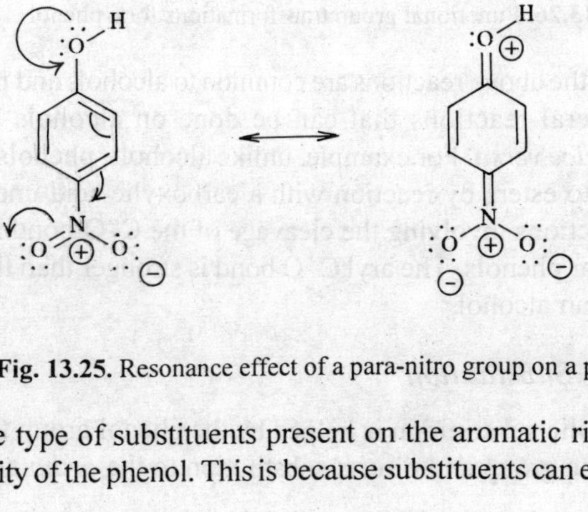

Fig. 13.25. Resonance effect of a para-nitro group on a phenol.

The type of substituents present on the aromatic ring can effect the acidity of the phenol. This is because substituents can either stabilize or destabilize the partial negative charge on the ring. The better the partial charge is stabilized, the more effective the resonance will be and the more acidic the phenol will be. Electron-withdrawing groups like a nitro substituent increase the acidity of the phenol since they stabilize the negative charge by an inductive effect. Nitro groups that are *ortho* or *para* to the phenolic group have an even greater effect. This is because fourth resonance structure is possible that delocalizes the partial charge even further (Fig. 13.25).

Electron-donating substituents (e.g. alkyl-groups) have the opposite effect and *decrease the acidity of phenols*.

Functional Group Transformations

Phenols can be converted into esters by reaction with acid chlorides or acid anhydrides and into ethers by reaction with alkyl halides in the presence of base (Fig. 13.26). These reactions can be done under milder conditions than those used for alcohols due to the greater acidity of phenols. Thus phenols can be converted to phenoxide ions with sodium hydroxide rather than metallic sodium.

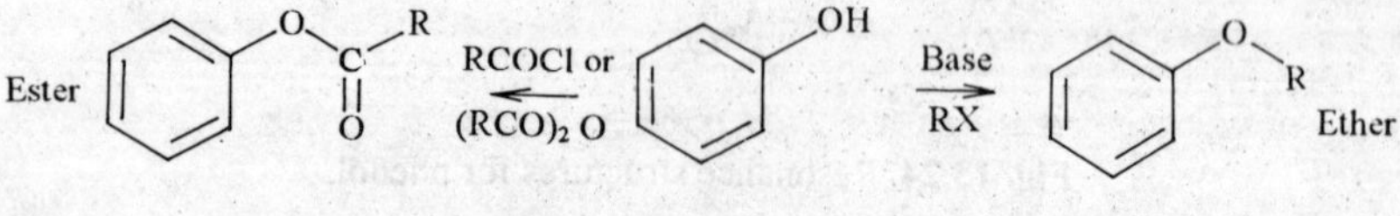

Ester ← RCOCl or $(RCO)_2O$ — phenol (OH) — Base, RX → Ether

Fig. 13.26. Functional group transformations for a phenol.

Although the above reactions are common to alcohols and phenols, there are several reactions that can be done on alcohols but not phenols, and *vice versa*. For example, unlike alcohols, phenols cannot be converted to esters by reaction with a carboxylic acid under acid catalysis. Reactions involving the cleavage of the C–O bond are also not possible for phenols. The aryl C–O bond is stronger than the alkyl C–O bond of an alcohol.

Electrophilic Substitution

Electrophilic substitution is helped by the phenol group that acts as an activating group and directs substitution to the *ortho* and *para* positions. Sulfonation and nitration of phenols are both possible to give *ortho* and *para* substitution products. Sometimes, the phenolic groups can be too powerful an activating group and it is difficult to control the reaction to one substitution. e.g., the bromination of phenol leads to 2,4,6-tribromophenol even in the absence of a Lewis acid (Fig. 13.27).

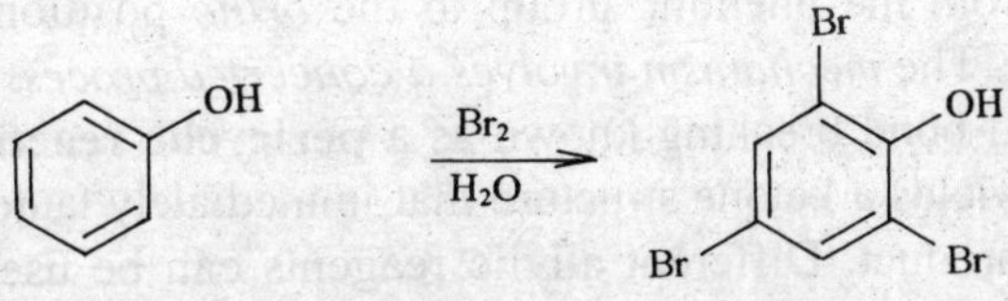

$$\text{Phenol} \xrightarrow[H_2O]{Br_2} \text{2,4,6-tribromophenol}$$

Fig. 13.27. Bromination of phenol.

The activating power of the phenolic group can be decreased by converting the phenol to an ester that can be removed by hydrolysis once the electrophilic substitution reaction had been carried put (Fig. 13.28). Since the ester is a weaker activating group, substitution takes place only once. Moreover, since the ester is a bulkier group than the phenol, *para* substitution is favored over *ortho* substitution.

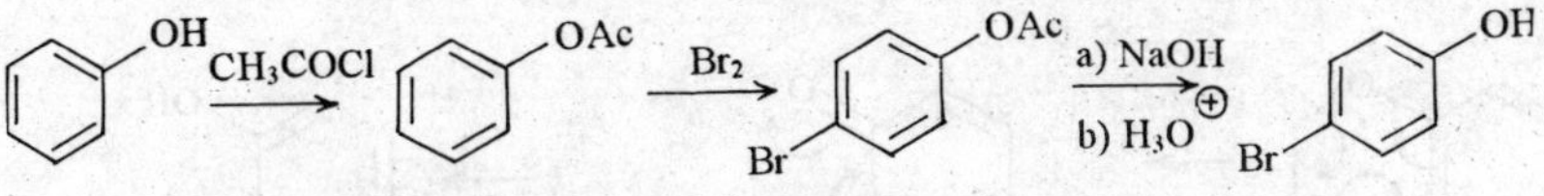

$$\text{PhOH} \xrightarrow{CH_3COCl} \text{PhOAc} \xrightarrow{Br_2} p\text{-Br-C}_6H_4\text{-OAc} \xrightarrow[b)\ H_3O^{\oplus}]{a)\ NaOH} p\text{-Br-C}_6H_4\text{-OH}$$

Fig. 13.28. Synthesis of para-bromophenol.

Oxidation

Phenols are susceptible to oxidation to quinones (Fig. 13.29).

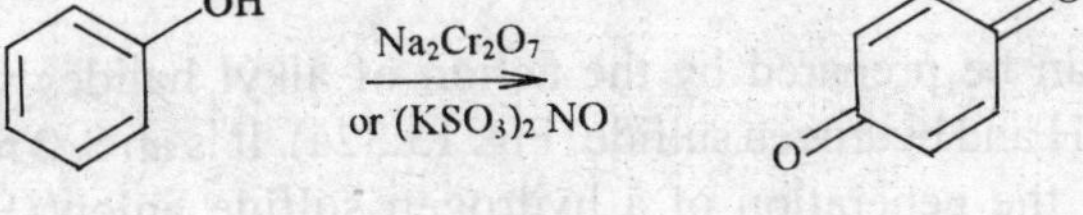

$$\text{PhOH} \xrightarrow[\text{or } (KSO_3)_2\,NO]{Na_2Cr_2O_7} \text{p-benzoquinone}$$

Fig. 13.29. Oxidation of phenol.

Claisen Rearrangement

It is a useful method of introducing an alkyl substituent to the *ortho* position of a phenol (Fig. 13.30). The phenol gets converted to the phenoxide ion, then treated with 3-bromopropene (an alkyl bromide) to form an ether. On heating, the allyl group ($–CH_2–CH=CH_2$) is transferred from the phenolic group to the *ortho* position of the aromatic ring. The *mechanism involves a concerted process* of bond formation and bond breaking known as a **pericyclic reaction** (Fig. 13.31). This yields a ketone structure that immediately tautomerizes to the final product. Different allylic reagents can be used in the reaction and the double bond in the final product can be reduced to form alkane substituent without affecting the aromatic ring.

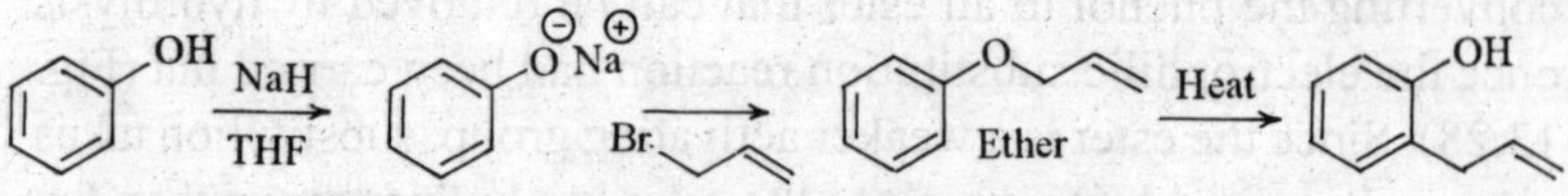

Fig. 13.30. Claisen rearrangement.

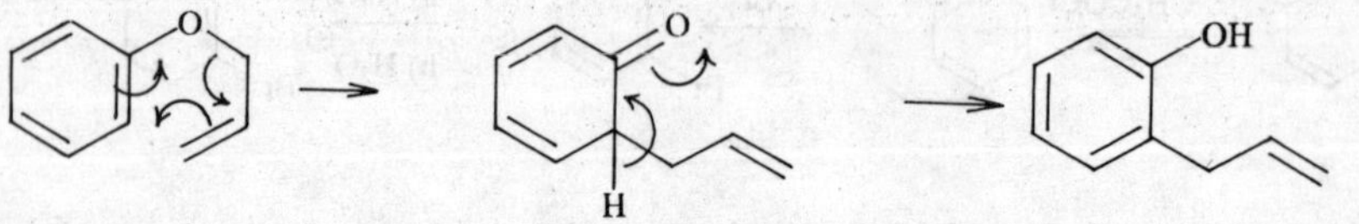

Fig. 13.31. Mechanism for the Claisen rearrangement.

13.6 Chemistry of Thiols

Preparation

Thiols can be prepared by the action of alkyl halides with an excess of KOH and hydrogen sulfide (Fig. 13.32a). It is an S_N2 reaction and involves the generation of a hydrogen sulfide anion (HS^-) as nucleophile. In this reaction there is the possibility of the product being ionized and reacting with a second molecule of alkyl halide to produce a thioether (RSR) as a byproduct. An excess of hydrogen sulfide is normally used to avoid this problem.

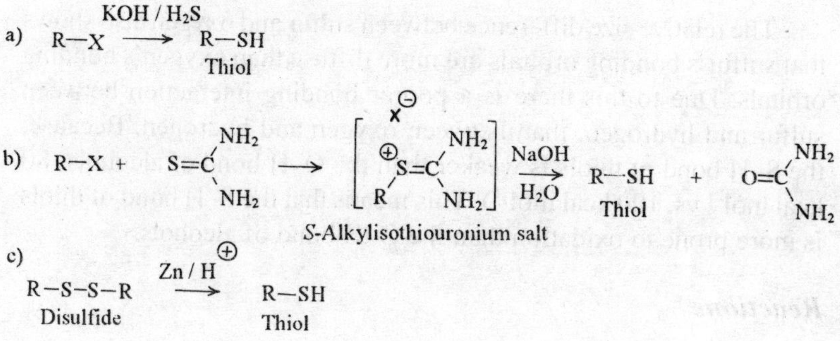

a) $R{-}X \xrightarrow{KOH / H_2S} R{-}SH$ (Thiol)

b) $R{-}X + S{=}C(NH_2)_2 \longrightarrow \left[R{-}\overset{\oplus}{S}{=}C(NH_2)_2 \right] X^{\ominus} \xrightarrow[H_2O]{NaOH} R{-}SH + O{=}C(NH_2)_2$

S-Alkylisothiouronium salt; Thiol

c) $R{-}S{-}S{-}R \xrightarrow{Zn / H^{\oplus}} R{-}SH$

Disulfide; Thiol

Fig. 13.32. Synthesis of thiols.

The formation of thioether can also be avoided by using an alternative procedure that involves thiourea (Fig. 13.32b). The thiourea acts as the nucleophile in an S_N2 reaction to produce an *S*-alkylisothiouronium salt that is then hydrolyzed with aqueous base to give the thiol.

Thiols can also be obtained by reducing disulfides with zinc in the presence of acid (Fig. 13.32).

Properties

Thiols form extremely weak hydrogen bonds–much weaker than alcohols – and so thiols have boiling points that are similar to comparable thioethers and which are lower than comparable alcohols. e.g. ethanethiol boils at 37°C whereas ethanol boils at 78°C.

Low molecular weight thiols are process disagreeable odours.

Reactivity

Thiols are the sulfur equivalent of alcohols (RSH). The sulfur atom is larger and more polarizable than oxygen which means that *sulfur compounds as a whole are more powerful nucleophiles than the corresponding oxygen compounds.* Thiolate ions (e.g. $CH_3CH_2S^-$) are stronger nucleophiles and weaker bases than corresponding alkoxides ($CH_3CH_2O^-$). Conversely, *thiols are stronger acids than corresponding alcohols.*

The relative size difference between sulfur and oxygen also shows that sulfur's bonding orbitals are more diffuse than oxygen's bonding orbitals. Due to this there is a poorer bonding interaction between sulfur and hydrogen, than between oxygen and hydrogen. Because, the S–H bond of thiols is weaker than the O–H bond of alcohols (80 kcal mol^{-1} vs. 100 kcal mol^{-1}). This means that the S–H bond of thiols is more prone to oxidation than the O–H bond of alcohols.

Reactions

Thiols can be easily oxidized by mild oxidizing agents like bromine or iodine to give disulfides (Fig. 13.33).

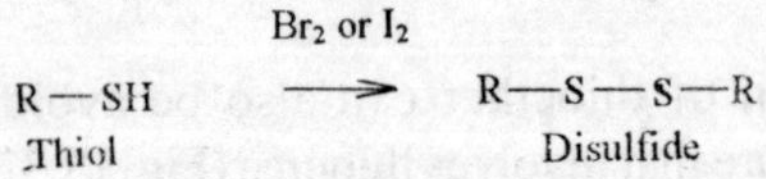

$$\underset{\text{Thiol}}{R—SH} \xrightarrow{Br_2 \text{ or } I_2} \underset{\text{Disulfide}}{R—S—S—R}$$

Fig. 13.33. Oxidation of thiols.

Thiols react with base to form thilate ions which can act as powerful nucleophiles (Fig. 13.34).

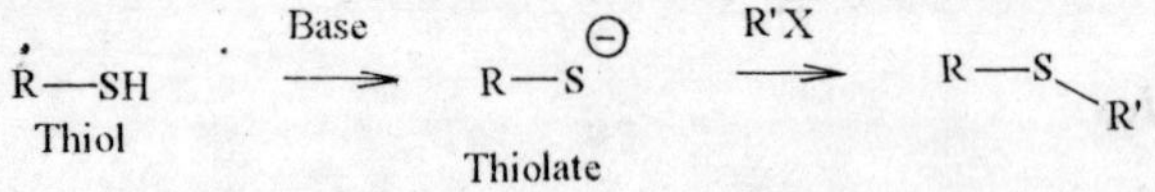

$$\underset{\text{Thiol}}{R—SH} \xrightarrow{\text{Base}} \underset{\text{Thiolate}}{R—S^{\ominus}} \xrightarrow{R'X} R—S—R'$$

Fig. 13.34. Formation of thiolate ions.

14

Ethers, Epoxides and Thioethers

14.1 Preparation of Ethers, Epoxides, and Thioethers

Ethers

For the synthesis of ether the Williamson ether synthesis is considered as the best method (Fig. 14.1a). It involves the S_N2 reaction between a metal alkoxide and a primary alkyl halide or tosylate. The alkoxide needed for the reaction is obtained by treating an alcohol with a strong base like sodium hydride. An alternative procedure is to treat the alcohol directly with the alkyl halide in the presence of silver oxide, thus avoiding the need to prepare the alkoxide beforehand (Fig. 14.1b).

$$\text{a)}\quad R-X \xrightarrow{NaOR''} \underset{\text{Ether}}{R-OR''} \qquad \text{b)}\quad R-X \xrightarrow[Ag_2O]{HOR''} \underset{\text{Ether}}{R-OR''}$$

Fig. 14.1. Synthesis of ethers.

For synthesis of an unsymmetrical ether, the most hindered alkoxide should be reacted with the simplest alkyl halide, rather than the other way round (Fig. 14.2). As this is an S_N2 reaction, primary alkyl halides react better then secondary or tertiary alkyl halides.

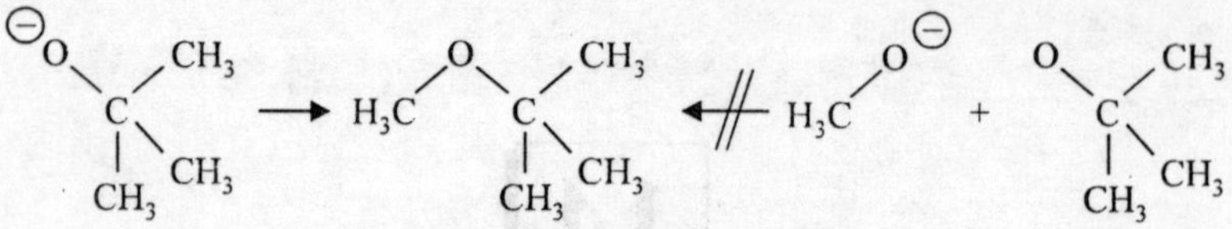

Fig. 14.2. Choice of synthetic routes to an unsymmetrical ether.

Alkenes can be converted to ethers by the electrophilic addition of mercuric trifluoroacetate, followed by addition of an alcohol. An organomercuric intermediate is obtained that can be reduced with sodium borohydride to yield the ether (Fig. 14.3).

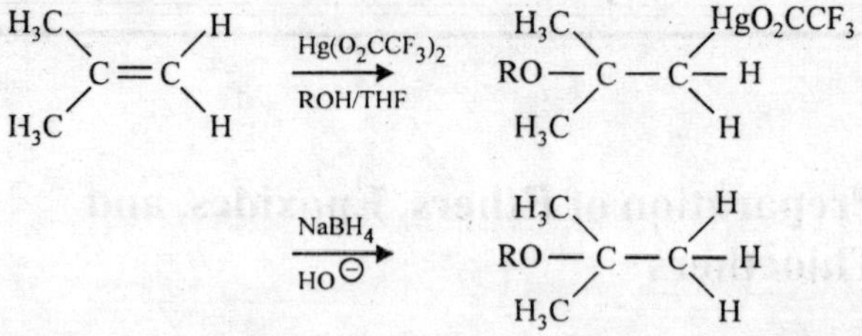

Fig. 14.3. Synthesis of an ether from an alkene and an alcohol.

Epoxides

Epoxides can be synthesized by the action of aldehydes or ketones with sulfur-ylides. They can also be prepared from alkenes by reaction with *m*-chloroperoxybenzoic acid. They can also be obtained from alkenes in a two-step process (Fig. 14.4). The first step involves electrophilic addition of a halogen in aqueous solution to form a halohydrin. Treatment of the halohydrin with base then ionizes the alcohol group, that can then act as a nucleophile (Fig. 14.5). The oxygen uses a lone pair of electrons to form a bond to the neighboring electrophilic carbon, thus displacing the halogen by an intramolecular S_N2 reaction.

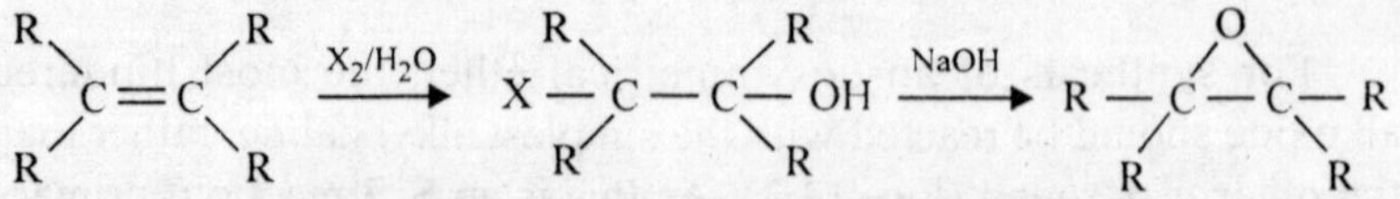

Fig. 14.4. Synthesis of an epoxide via a halohydrin.

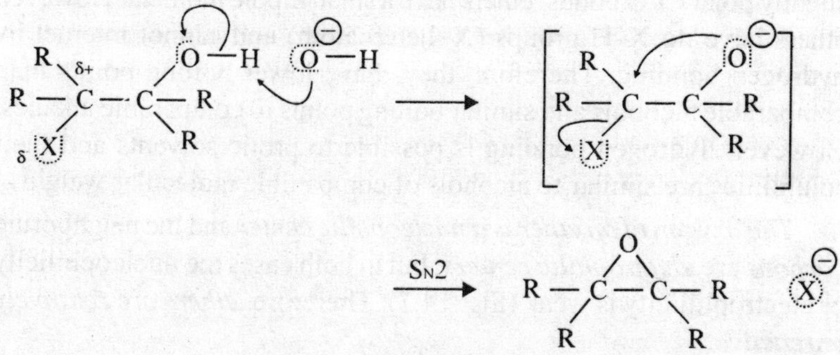

Fig. 14.5. Mechanism of epoxide formation from a halohydrin.

Thioethers

Thioethers (or sulfides) can be prepared by the S_N2 reaction of primary or secondary alkyl halides with a thiolate anion (RS^-), (Fig. 14.6). *The reaction is similar to the* **Williamson ether synthesis.**

$$R - X \xrightarrow{NaSR''} R - SR''$$

Fig. 14.6. Synthesis of a disulfide from an alkyl halide.

Symmetrical thioethers can be prepared by treating an alkyl halide with KOH and an equivalent of hydrogen sulfide. The reaction produces a thiol which is ionized again by KOH and reacts with another molecule of alkyl halide.

14.2 Properties of Ethers, Epoxides and Thioethers

Ethers

Ethers are made up an oxygen linked to two carbon atoms by σ bonds. In aliphatic ethers (ROR), the three atoms involved are sp^3 hybridized and have a bond angle of 112°. In Aryl ethers the oxygen is linked to one or two aromatic rings (ArOR or ArOAr) and in such a case the attached carbon(s) is sp^2 hybridized.

The C–O bonds are polarized in such a way that the oxygen is slightly negative and the carbons are slightly positive. Because of the slightly polar C–O bonds, ethers have a small dipole moment. However, ethers have no X–H groups (X=heteroatom) and cannot interact by hydrogen bonding. Therefore, they have lower boiling points than comparable alcohols and similar boiling points to comparable alkanes. However, hydrogen bonding is possible to protic solvents and their solubilities are similar to alcohols of comparable molecular weight.

The oxygen of an ether is a nucleophilic center and the neighboring *carbons* are *electrophilic centers*, but in both cases the nucleophilicity or electrophilicity is weak (Fig. 14.7). Therefore, *ethers are relatively unreactive*.

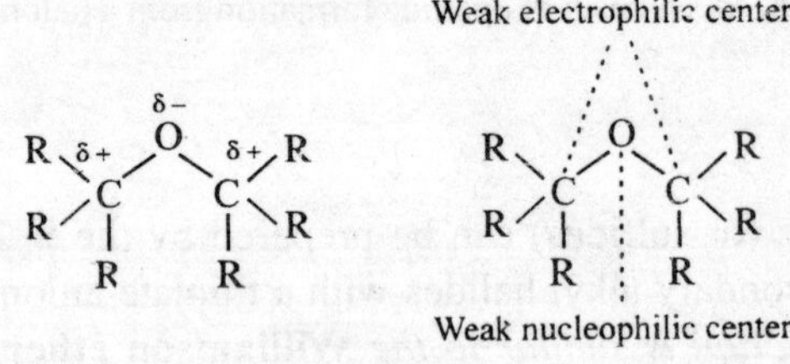

Fig. 14.7. Properties of ethers.

Epoxides

Epoxides (or *oxiranes*) are three-membered cyclic ethers and differ from other cyclic and acyclic ethers in that they are reactive with different reagents. The reason for this difference in reactivity is the strained three-membered ring. Reactions with nucleophiles can result in ring opening and relief of strain. Nucleophiles will attack either of the electrophilic carbons present in an epoxide by an S_N2 reaction (Fig. 14.8).

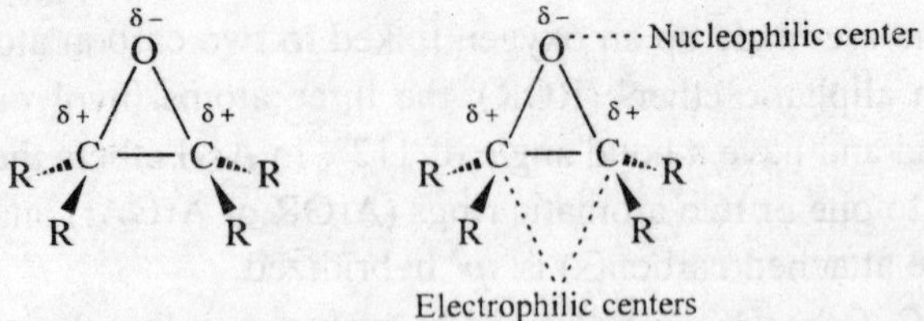

Fig. 14.8. Properties of an epoxide.

Thioethers

Thioethers (or **sulfides;** RSR) are the sulfur equivalents of ethers (ROR). Because the sulfur atmos are polarizable, they can stabilize a negative charge on an adjacent carbon atom. Thus hydrogens on this carbon are more acidic than those on comparable ethers.

14.3 Reactions of Ethers, Epoxides and Thioethers

Ethers

Ethers are generally unreactive functional groups and the only useful reaction that they undergo is cleavage by strong acids like HI and HBr to produce an alkyl halide and an alcohol (Fig.14.9). The ether is first protonated by the acid, then nucleophilic substitution occurs where the halide ion acts as the nucleophile. Primary and secondary ethers react by the S_N2 mechanism and the halide reacts at the least substituted carbon atom to produce an alkyl halide and an alcohol (Fig. 14.10). The initial protonation is a must because it converts a poor leaving group (an alkoxide ion) into a good leaving group (the alcohol).

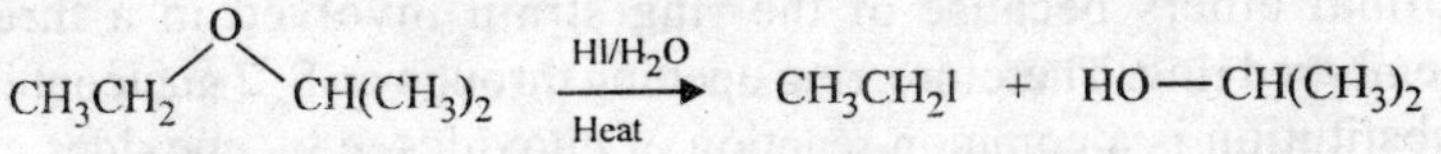

$$CH_3CH_2-O-CH(CH_3)_2 \xrightarrow[\text{Heat}]{HI/H_2O} CH_3CH_2I + HO-CH(CH_3)_2$$

Fig. 14.9. Cleavage of an ether to an alkyl halide and an alcohol.

$$CH_3CH_2-\ddot{O}-CH(CH_3)_2 \xrightarrow{H^{\oplus}} CH_3CH_2-\overset{\oplus}{O}(H)-CH(CH_3)_2 \xrightarrow{:\ddot{I}:^{-}} CH_3CH_2I + HO-CH(CH_3)_2$$

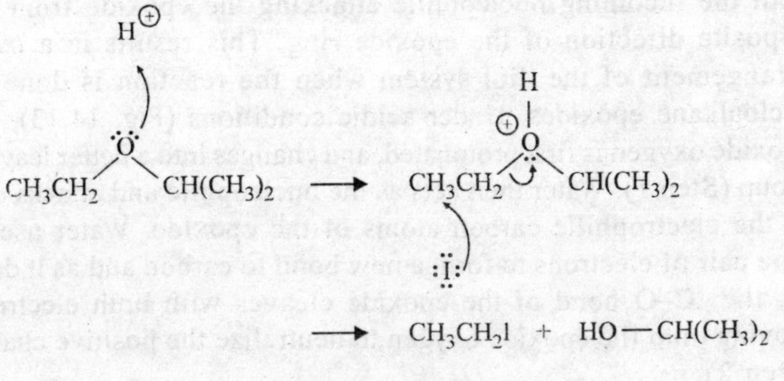

Fig. 14.10. Mechanism for the cleavage of an ether.

Primary alcohols formed from this reaction can be converted further to an alkyl halide. Tertiary ethers react by the S_N1 mechanism to produce the alcohol. However, an alkene may also be formed due to E1 elimination and this may be the major product (Fig. 14.11).

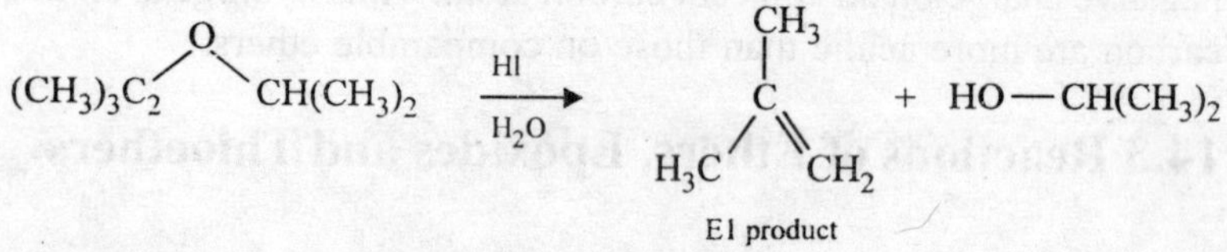

$$(CH_3)_3C-O-CH(CH_3)_2 \xrightarrow[H_2O]{HI} H_3C-C(CH_3)=CH_2 + HO-CH(CH_3)_2$$

E1 product

Fig. 14.11. Cleavage of a tertiary ether.

Trifluoroacetic acid can be used in place of HX, resulting in an E1 reaction and production of the alcohol and the alkene.

Most of the ethers undergo slow reaction with atmospheric oxygen by a radical process to form **hydroperoxides** (ROOH) and **peroxides** (ROOR). These products can prove to be explosive if old solvents are concentrated to dryness.

Epoxides

Epoxides are *cyclic ethers*, but they are more reactive than normal ethers because of the ring strain involved in a three-membered ring. Therefore, ring opening through an S_N2 nucleophilic substitution is a common reaction of epoxides. e.g., epoxides can be ring-opened under acidic or basic conditions to yield 1,2-diol (Fig. 14.12). In both cases, the reaction involves an S_N2 mechanism with the incoming nucleophile attacking the epoxide from the opposite direction of the epoxide ring. This results in a *trans* arrangement of the diol system when the reaction is done on cycloalkane epoxides. Under acidic conditions (Fig. 14.13), the epoxide oxygen is first protonated, and changes into a better leaving group (Step 1). Water then acts as the nucleophile and attacks one of the electrophilic carbon atoms of the epoxide. Water uses a lone pair of electrons to form a new bond to carbon and as it does so, the C–O bond of the epoxide cleaves with both electrons moving onto the epoxide oxygen to neutralize the positive charge (Step 2).

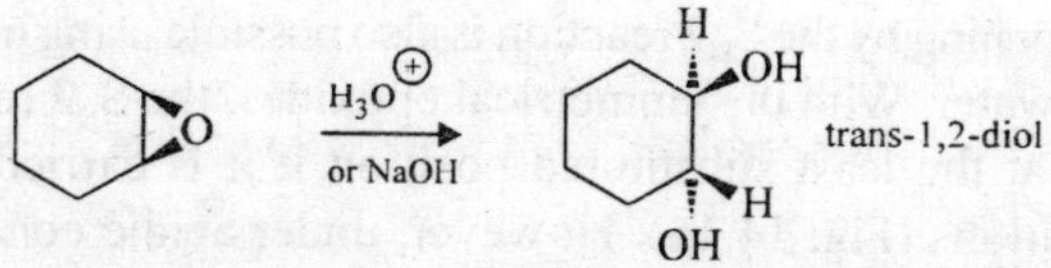

$H_3O^{\oplus}$ or NaOH → trans-1,2-diol

Fig. 14.12. Ring openeing of an epoxide under acidic or basic conditions.

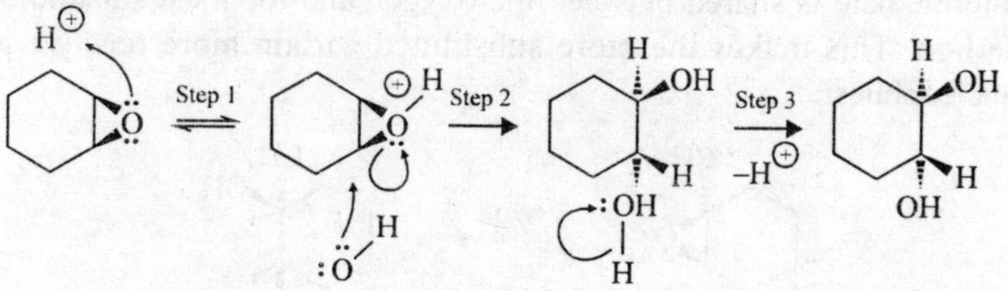

Step 1 — Step 2 — Step 3 $-H^{\oplus}$

Fig. 14.13. Mechanism for the ring opening of an epoxide under acidic conditions.

Although water is a poor nucleophile, the reaction is favored because of the neutralization of the positive charge on oxygen and the relief of ring strain once the epoxides opened up. Unlike other S_N2 reactions of course, the leaving group is still tethered to the molecule.

Ring opening under basic conditions is also possible with heating, but needs the loss of a negatively charged oxygen (Fig. 14.14). This is a poor leaving group and would not occur with normal ethers. It is only possible here because the reaction opens up the epoxide ring and relieves ring strain.

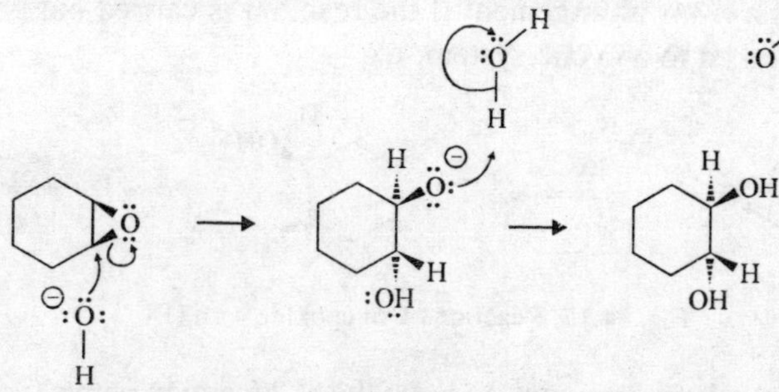

Fig. 14.14. Mechanism for the ring opening of an epoxide under basic conditions.

Ring opening by the S_N2 reaction is also possible using nucleophiles other than water. With unsymmetrical epoxides, the S_N2 reaction will take place at the least substituted position if it is carried out under basic conditions (Fig. 14.15). However, under acidic conditions, the nucleophile will generally attack the most substituted position (Fig. 14.16). This is because the positive charge in the protonated intermediate is shared between the oxygen and the most substituted carbon. This makes the more substituted carbon more reactive to nucleophiles.

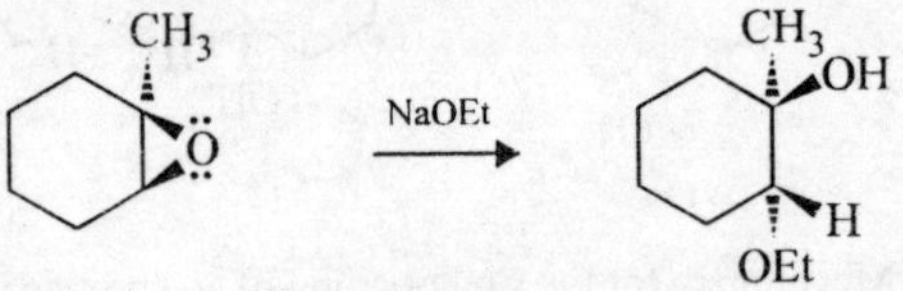

CH3 NaOEt CH3 OH H OEt

Fig. 14.15. Ring opening of an epoxide with the ethoxide ion.

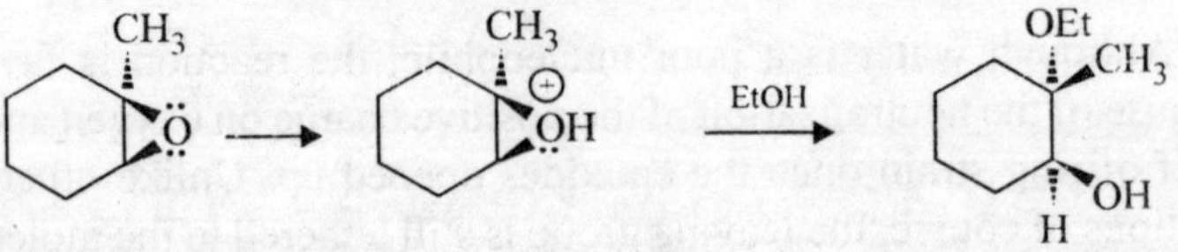

CH3 CH3 OH EtOH OEt CH3 OH H

Fig. 14.16. Ring opening of an epoxide with ethanol under acidic conditions.

The reaction of epoxides with hydrogen halides (Fig. 14.17) is analogous to the reaction of normal ethers with HX. Protonation of the epoxide with acid is followed by nucleophilic attack by a halide ion resulting in 1,2-halohydrins. The halogen and alcohol groups will also be in a *trans* arrangement if the reaction is carried out with an epoxide linked to a cyclic system, e.g.

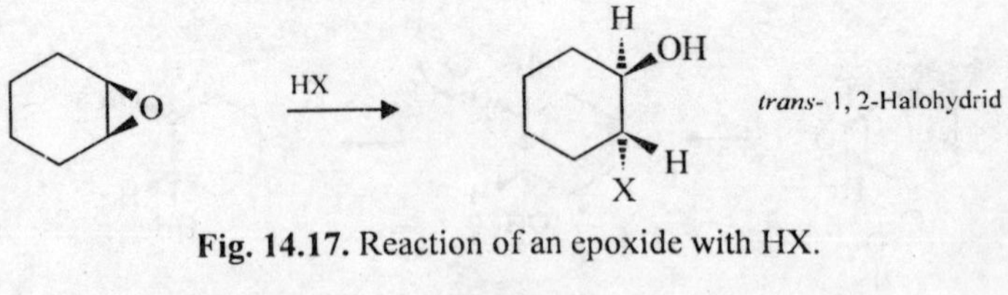

O HX H OH H X *trans*- 1, 2-Halohydrid

Fig. 14.17. Reaction of an epoxide with HX.

Unlike ethers, epoxides undergo the S_N2 reaction with a Grignard reagent (Fig. 14.18).

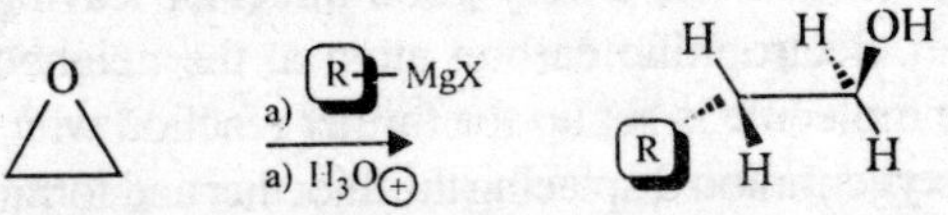

Fig. 14.18. Grignard reaction with an epoxide.

Thioethers

Unlike ethers, thioethers are good nucleophiles because of the sulfur atom. This is because the sulfur atom has its valence electrons further away from the nucleus, and these electrons experience less attraction from the nucleus, making them more polarizable and more nucleophilic. As they are good nucleophiles, thioethers can react with alkyl halides to form trialkylsulfonium salts (R_3S^+;Fig.14.19). This reaction is impossible for normal ethers. Sulfur is also able to stabilize a negative charge on a neighboring carbon atom, particularly when the sulfur itself is positively charged. This makes the protons on neighboring carbons acidic, permitting them to be removed with base to form sulfur-ylides.

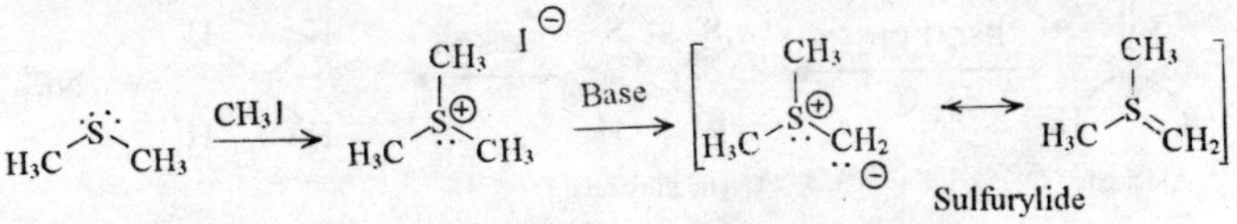

Fig. 14.19. Formation of a sulphur-ylide.

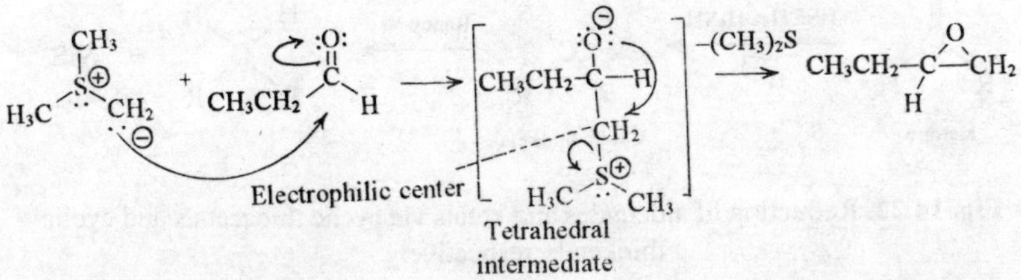

Fig. 14.20. Reaction of an aldehyde with a sulfur-ylide to produce an epoxide.

Sulfur-ylides are useful in the synthesis of epoxides from aldehydes or ketones (Fig. 14.20). They undergo a typical nucleophilic addition with the carbonyl group to yield the expected tetrahedral intermediate.

This intermediate now has a very good thioether leaving group which also creates an electrophilic carbon atom at the neighboring position. Therefore, the molecule is set up for further reaction which involves the nucleophilic oxygen anion displacing the thioether and forming an epoxide.

Thioethers can also be oxidized with hydrogen peroxide a yield a sulfoxide (R_2SO) that on oxidation with a peroxyacid, yields a sulfone (R_2SO_2; Fig.14.21).

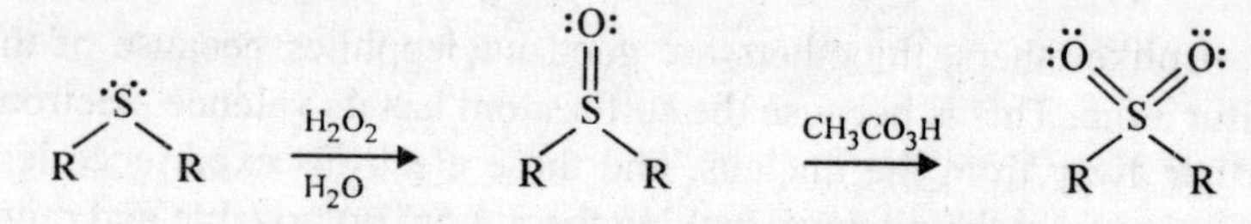

$$R_2S \xrightarrow[H_2O]{H_2O_2} R_2S{=}O \xrightarrow{CH_3CO_3H} R_2SO_2$$

Fig. 14.21. Synthesis of sulfoxides and sulfones.

Thioethers can be reduced using **Raney nickel** (a catalyst which has hydrogen gas adsorbed onto the nickel surface) (Fig. 14.22). This reaction is specially useful for reducing thioacetals or thioketals as this provides a means of converting aldehydes or ketones to alkanes.

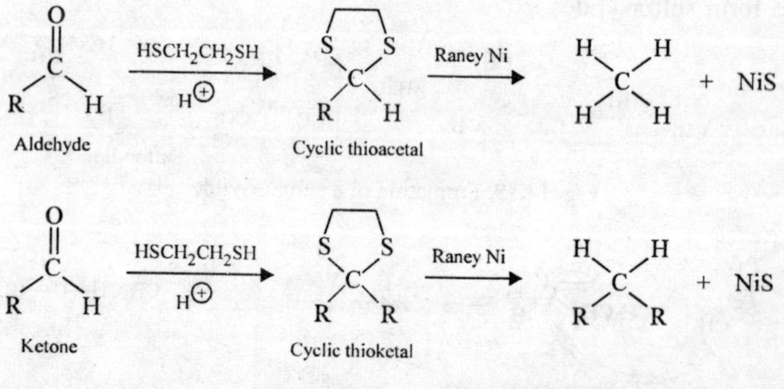

$$\underset{\text{Aldehyde}}{RCHO} \xrightarrow[H^{\oplus}]{HSCH_2CH_2SH} \underset{\text{Cyclic thioacetal}}{\text{cyclic } RCH(SCH_2CH_2S)} \xrightarrow{\text{Raney Ni}} CH_4 + NiS$$

$$\underset{\text{Ketone}}{RCHO} \xrightarrow[H^{\oplus}]{HSCH_2CH_2SH} \underset{\text{Cyclic thioketal}}{\text{cyclic } R_2C(SCH_2CH_2S)} \xrightarrow{\text{Raney Ni}} R_2CH_2 + NiS$$

Fig. 14.22. Reduction of aldehydes and ketals via cyclic thioacetals and cyclic thioketals respectively.

15

Amines and Nitriles

15.1 Preparation of Amines

Reduction

Nitriles and amides can be easily reduced to alkylamines using lithium aluminum hydride ($LiAlH_4$). In the case of a nitrile, a primary amine is the only possible product. Primary, secondary, and tertiary amines can be prepared from primary, secondary and tertiary amides, respectively.

Substitution with NH_2

Primary alkyl halides and some secondary alkyl halides can undergo S_N2 nucleophilic substitution with an azide ion (N_3^-) to yield an alkyl azide. The azide can then be reduced with $LiALH_4$ to give a primary amine (Fig. 15.1).

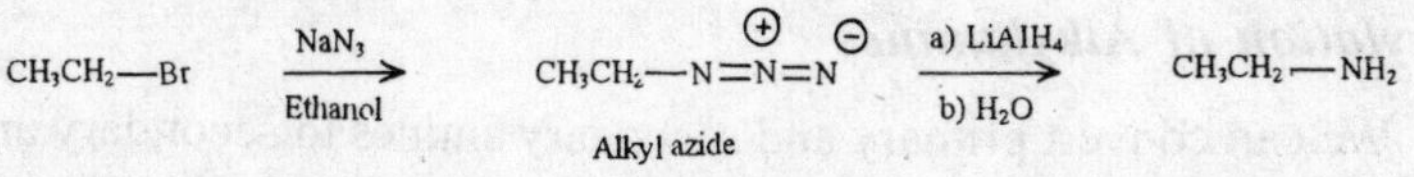

$$CH_3CH_2—Br \xrightarrow[\text{Ethanol}]{NaN_3} \underset{\text{Alkyl azide}}{CH_3CH_2—\overset{\oplus}{N}{=}N{=}\overset{\ominus}{N}} \xrightarrow[\text{b) } H_2O]{\text{a) } LiAlH_4} CH_3CH_2—NH_2$$

Fig. 15.1. Synthesis of a primary amine from an alkyl halide via an alkyl azide.

The overall reaction involves replacing the halogen atom of the alkyl halide with an NH_2 unit. Another method is the **Gabriel synthesis** of amines. This involves treating phthalimide with KOH to abstract the N–H proton (Fig. 15.2). The N–H proton of phthalimide is more acidic (pK_a 9) than the N–H proton of an amide since the anion formed can be stabilized by resonance with both neighboring carbonyl groups. The phthalimide ion can then be alkylated by treating it with an alkyl halide in nucleophilic substitution.

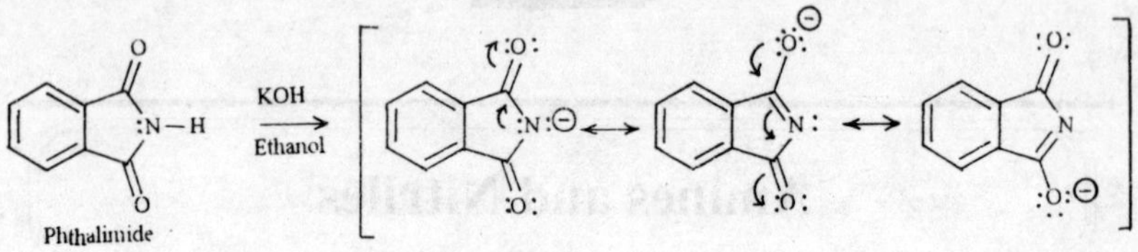

Fig. 15.2. Ionization of phthalimide.

Subsequent hydrolysis releases a primary amine (Fig. 15.3). Still other possible method is to react an alkyl halide with ammonia, but this is less satisfactory because over-alkylation is possible. The reaction of an aldehyde with ammonia by reductive amination is another method of obtaining primary amines.

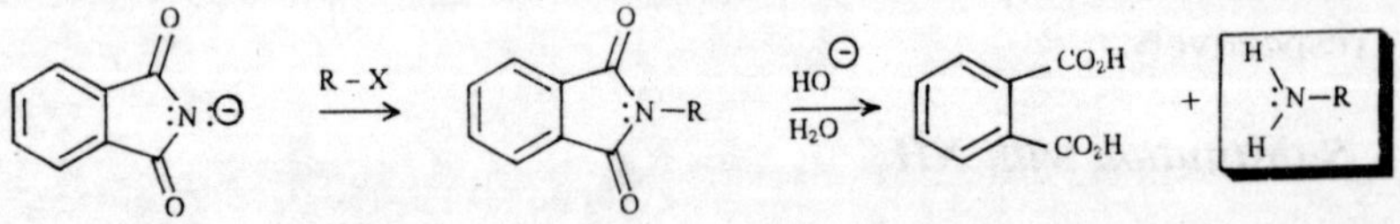

Fig. 15.3. Gabriel synthesis of primary amines.

Alkylation of Alkylamines

We can convert primary and secondary amines to secondary and tertiary amines respectively, by alkylation with alkyl halides by the S_n2 reaction. However, over-alkylation may occur and so better methods of amine synthesis which are available are used.

Reductive amination these are:— It is a more controlled method of adding an extra alkyl group to an alkylamine (Fig. 15.4). Primary and secondary alkylamines can be treated with a ketone or an aldehyde in the presence of a reducing agent known as **sodium cyanoborohydride.** The alkylamine reacts with the carbonyl compound by nucleophilic addition followed by elimination to give an imine or an iminium ion which is immediately reduced by sodium cyanoborohydride to yield the final amine. This is the equivalent of adding one extra alkyl group to the amine. Therefore, primary amines get converted to secondary amines and secondary amines are converted to tertiary amine. The reaction is suitable for the synthesis of primary amines if ammonia is used instead of an alkylamine. The reaction goes through an imine intermediate if ammonia or a primary amine is used (Fig. 15.4a). When a secondary amine is used, an iminium ion intermediate is involved (Fig. 15.4b).

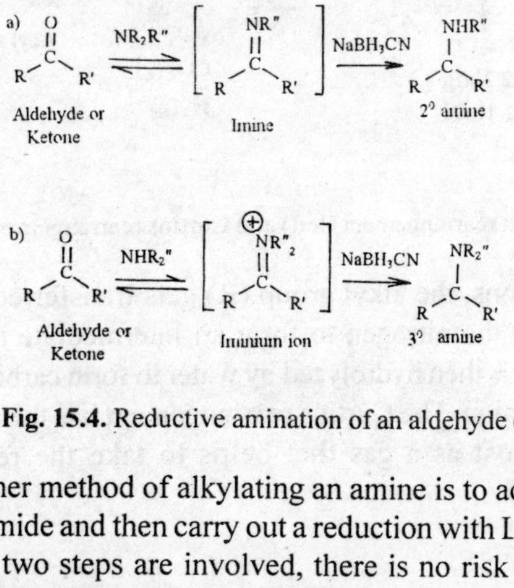

a) $R(C{=}O)R'$ $\underset{}{\overset{NR_2R''}{\rightleftharpoons}}$ $[R(C{=}NR'')R']$ $\xrightarrow{NaBH_3CN}$ $R(CH\text{-}NHR'')R'$

Aldehyde or Ketone — Imine — 2^0 amine

b) $R(C{=}O)R'$ $\overset{NHR_2''}{\rightleftharpoons}$ $[R(C{=}\overset{\oplus}{N}R''_2)R']$ $\xrightarrow{NaBH_3CN}$ $R(CH\text{-}NR_2'')R'$

Aldehyde or Ketone — Iminium ion — 3^0 amine

Fig. 15.4. Reductive amination of an aldehyde or ketone.

Another method of alkylating an amine is to acylate the amine to yield an amide and then carry out a reduction with $LiALH_4$ (Fig. 15.5). Although two steps are involved, there is no risk of over-alkylation since acylation can only occur once.

$$R{-}\ddot{N}H_2 \xrightarrow[\text{Pyridine}]{CH_3COCl} R{-}\ddot{N}H{-}C({=}O){-}CH_3 \xrightarrow[\text{b) } H_2O]{\text{a) } LiAlH_4} R{-}\ddot{N}H{-}CH_2{-}CH_3$$

Fig. 15.5. Alkylation of an amide via an amine.

Rearrangements

The following two rearrangement reactions can be used to convert carboxylic acid derivatives into primary amines in which the carbon chain in the product has been shortened by one carbon unit (Fig. 15.6). These are called the **Hofmann and the Curtius rearrangements.** The **Hofmann rearrangement** involves the treatment of a primary amide with bromine under basic conditions, while the **Curtius rearrangement** involves heating an acyl azide. In both cases we get a primary amine with loss of the original carbonyl group.

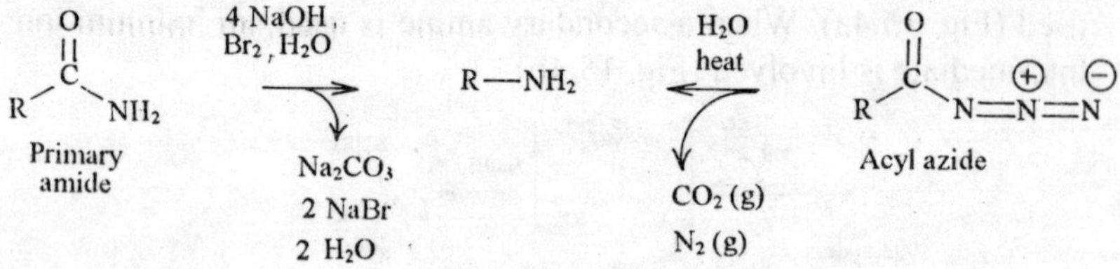

$$\underset{\text{Primary amide}}{R-C(=O)-NH_2} \xrightarrow[-\,Na_2CO_3,\ 2\,NaBr,\ 2\,H_2O]{4\,NaOH,\ Br_2,\ H_2O} R-NH_2 \xleftarrow[-\,CO_2(g),\ N_2(g)]{H_2O,\ heat} \underset{\text{Acyl azide}}{R-C(=O)-N=\overset{\oplus}{N}=\overset{\ominus}{N}}$$

Fig. 15.6. Hofmann rearrangement (left) and Curtius rearrangement (right).

In both reactions, the alkyl group (R) gets transferred from the carbonyl group to the nitrogen to form an intermediate isocyanate (O=C=N–R). This is then hydrolyzed by water to form carbon dioxide and the primary amine. The Curtius rearrangement has the advantage that nitrogen is lost as a gas that helps to take the reaction to completion.

Arylamines

The direct introduction of an amino group to an aromatic ring is not possible. But nitro groups can be added directly by electrophilic substitution and then reduced to the amine (Fig. 15.7). The reduction is done under acidic conditions yielding an arylaminium ion as product. The free base can be isolated by basifying the solution with sodium hydroxide to precipitate the arylamine.

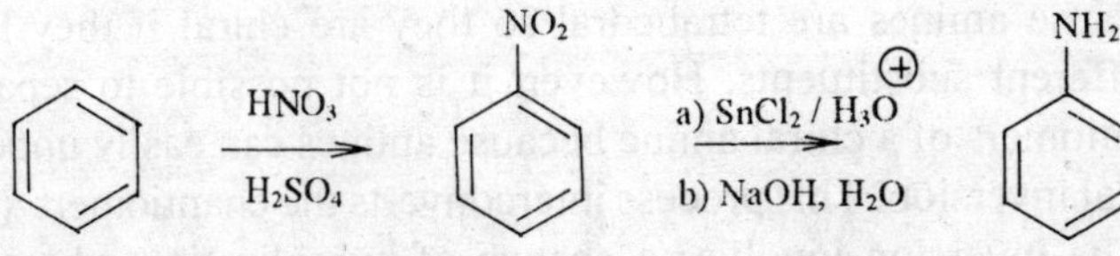

$$\text{C}_6\text{H}_6 \xrightarrow[\text{H}_2\text{SO}_4]{\text{HNO}_3} \text{C}_6\text{H}_5\text{NO}_2 \xrightarrow[\text{b) NaOH, H}_2\text{O}]{\text{a) SnCl}_2\text{ / H}_3\text{O}^{\oplus}} \text{C}_6\text{H}_5\text{NH}_2$$

Fig. 15.7. Introduction of an amine to an aromatic ring.

Once an amino group has been introduced to an aromatic ring, it can be alkylated with an alkyl halide, acylated with an acid chloride or converted to a higher amine by reductive amination as already described for an alkylamine.

15.2 Properties of Amines

Structure

Amines are made up of an sp^3 hybridized nitrogen linked to three substituents by three σ bonds. The substuents can be hydrogen, alkyl or aryl groups, but at least one of the substituents must be an alkyl or aryl group. If only one such group is present, the amine is known as primary. If two groups are present, the amine is secondary. If three groups are present, the amine is tertiary. If the substituents are all alkyl groups, the amine is referred as being an alkylamine. If there is at least one aryl group directly attached to the nitrogen, then the amine is known as an arylamine.

The nitrogen atom has four sp^3 Hybridized orbitals pointing to the corners of a tetrahedron in the same way as an sp^3 hybridized carbon atom. However, one of the sp^3 orbitals is occupied by the nitrogen's lone pair of electrons.Therefore the atoms in an amine functional group are pyramidal in shape. The C–N–C bond angles are approximately 109° which is consistent with a tetrahedral nitrogen. However, the bond angle is slightly less than 109° since the lone pair of electrons demands a slightly greater amount of space than a σ bond.

Pyramidal Inversion

Because amines are tetrahedral so they are chiral if they have three different substituents. However, it is not possible to separate the enantiomers of a chiral amine because amines can easily undergo pyramidal inversion. This process interconverts the enantiomers (Fig. 15.8). The inversion involves a change of hybridization where the nitrogen becomes sp^2 hybridized rather than sp^3 hybridized. Because of this, the molecule becomes planar and the lone pair of electrons occupy a p orbital. Once the hybridization reverts back to sp^3, the molecule can either revert back to its original shape or invert.

Although the enantiomers of chiral amines cannot be separated, such amines can be alkylated to form quaternary ammomium salts where the enantiomers can be separated. Once the lone pair of electrons is locked up in a σ bond, pyramidal inversion becomes impossible and the enantiomers can no longer interconvert.

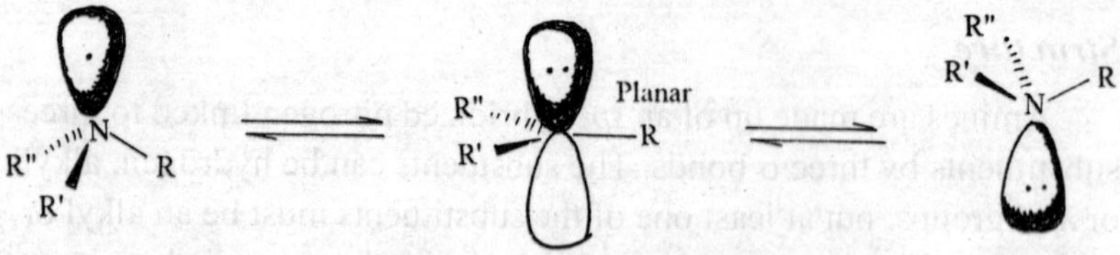

Fig. 15.8. Pyramidal inversion.

Physical Properties

Amines are polar compounds and intermolecular hydrogen bonding is possible for primary and secondary amines. Therefore, primary and secondary amines have higher boiling points than alkanes of similar molecular weight. Tertiary amines also have higher boiling points than comparable alkanes, but have slightly lower boiling points than comparable primary or secondary amines as they cannot participate in intermolecular hydrogen bonding.

However, all amines can participate in hydrogen bonding with protic solvents, so amines have similar water solubilities to comparable alcohols. Low molecular weight amines are freely miscible with water. Low molecular weight amines have an offensive fish-like smell.

Basicity

Amines are weak bases but they are more basic than alcohols, ethers, or water. Due to this, amines act as bases when they are dissolved in water and an equilibrium is set up between the ionized form (the **ammonium** ion) and the unionized form (the free base; Fig. 15.9).

The basic strength of an amine can be measured by its pK_b value (typically 3–4). The lower the value of pK_b, the stronger the base. The pK_b for ammonia is 4.74, which compares with pK_b values for methylamine, ethylamine, and propylamine of 3.36, 3.25 and 3.33, respectively. This shows that larger alkyl groups increase base strength. This is an inductive effect by which the ion is stabilized by dispersing some of the positive charge over the alkyl group (Fig. 15.10). This shifts the equilibrium of the acid base reaction towards the ion, which means that the amine is more basic. The larger the alkyl group, the more significant this effect.

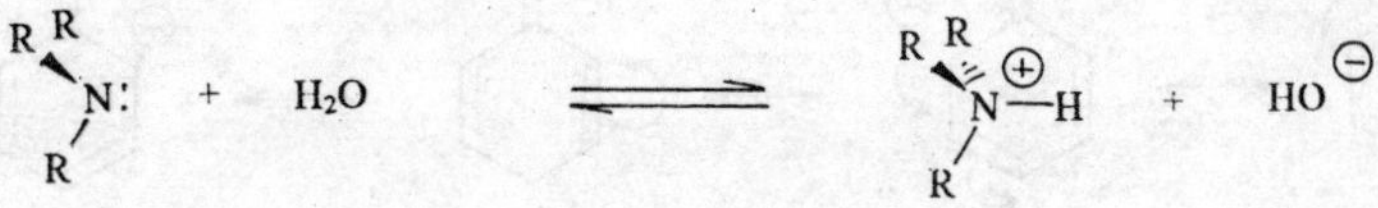

$$R_3N: + H_2O \rightleftharpoons R_3N^{\oplus}\text{—}H + HO^{\ominus}$$

Fig. 15.9. Acid-base reaction between an amine and water.

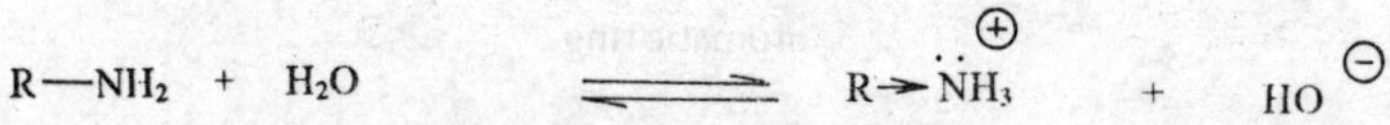

$$R\text{—}NH_2 + H_2O \rightleftharpoons R \rightarrow \overset{\oplus}{N}H_3 + HO^{\ominus}$$

Fig. 15.10. Inductive effect of an alkyl group on an alkylammonium ion.

Moreover alkyl substituents should have an even greater inductive effect and we can expect secondary and tertiary amines to be stronger bases than primary amines. This is not necessarily the case and there is no direct relationship between basicity and the number of alkyl groups attached to nitrogen. The inductive effect of more alkyl groups is counterbalanced by a **salvation effect**.

Once the ammonium ion is formed, it is solvated by water molecules – a stabilizing factor that involves hydrogen bonding between the oxygen atom of water and any N–<u>H</u> group present in the ammonium ion (Fig. 15.11). The more hydrogen bonds that are possible, the greater the stabilization. Due to this, solvation and solvent stabilization is stronger for alkylaminium ions formed from primary amines than for those formed from tertiary amines. *The solvent effect tends to be more important than the inductive effect as far as tertiary amines are concerned and so tertiary amines are generally weaker bases than primary or secondary amines.*

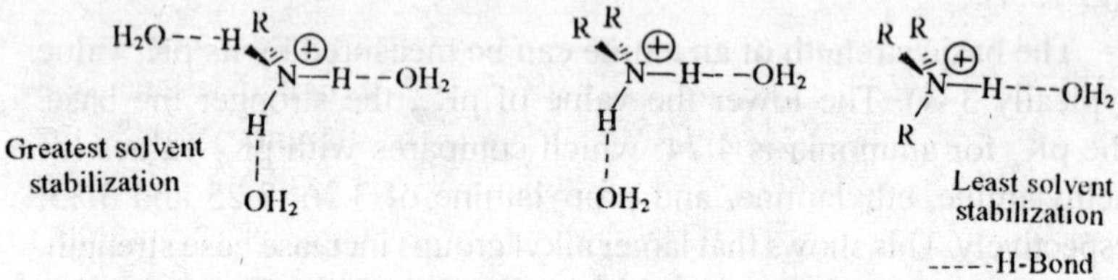

Greatest solvent stabilization

Least solvent stabilization

-----H-Bond

Fig. 15.11. Solvent effect on the stabilization of alkylammonium ions.

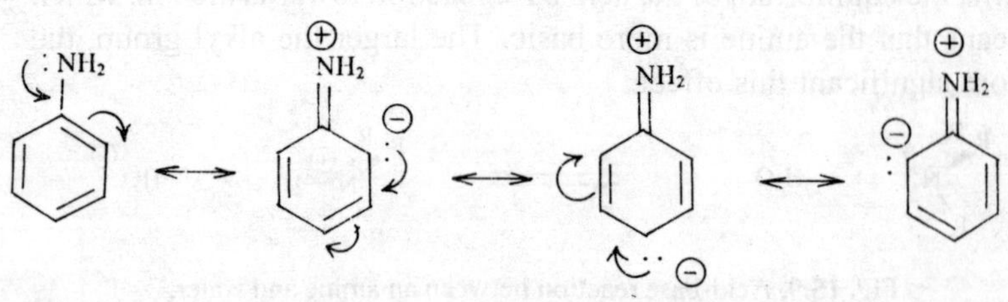

Fig. 15.12. Resonance interaction between nitrogen's lone pair and the aromatic ring.

Aromatic amines (arylamines) are weaker bases than alkylamines as the orbital containing nitrogen's lone pair of electrons overlaps with the π system of the aromatic ring. In terms of resonance, the lone pair of electrons can be used to form a double bond to the aromatic ring, resulting in the possibility of three **zwitterionic** resonance structures (Fig. 15.12). (A zwitterion is a molecule containing a positive and a negative charge.) Since nitrogen's lone pair of electrons is involved in this interaction. It is less available to form a bond to a proton and so the amine is less basic.

The nature of aromatic substituent also affects the basicity of aromatic amines. Substituents that deactivate aromatic rings (e.g. NO_2, Cl, or CN) lower electron density in the ring, which means that the ring will have an electron-withdrawing effect on the neighboring ammonium ion. Because of this the charge will be destabilized and the amine will be a weaker base. Substituents that activate the aromatic ring enhance electron density in the ring and the ring will have an electron-donating effect on the neighboring charge. This has a stabilizing effect and so the amine will be a stronger base. The relative position of aromatic substituents can be important if resonance is possible between the aromatic ring and the substituent. In such cases, the substituent will have a greater effect if it is at the *ortho* or *para* position, e.g., *para*-nitroaniline is a weaker base than *meta*-nitroaniline. This is because one of the possible resonance structures for the *para* isomer is highly disfavored since it places a positive charge immediately next to the ammonium ion (Fig. 15.13). Therefore, the number of feasible resonance structures for the *para* isomer is limited to three, compared to four for the *meta* isomer. Due to this the *para* isomer experience less stabilization and so the amine will be less basic.

If an activating substituent is present that is capable of interacting with the ring by resonance, the opposite holds true and the *para* isomer will be a stronger base than the *meta* isomer. This is because the crucial resonance structure mentioned above would have a negative charge immediately next to the ammonium ion and this would have a stabilizing effect.

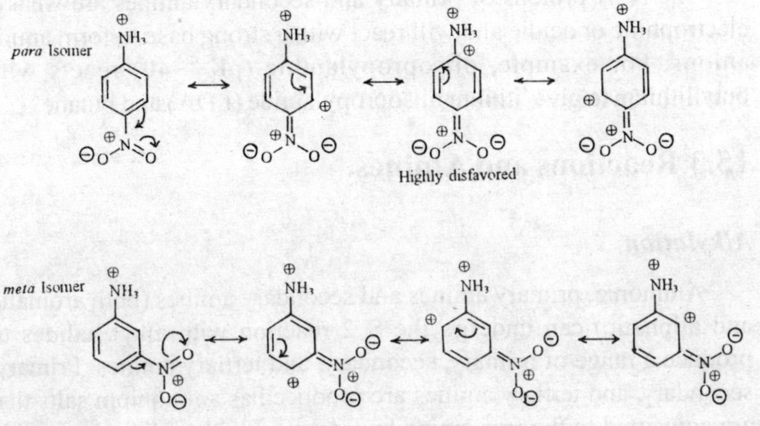

Fig. 15.13. Resonance structures for para-nitroaniline and meta-nitroaniline.

Reactivity

Amines react as nucleophiles or bases, since the nitrogen atom has a readily available lone pair of electrons that can participate in bonding (Fig. 15.14). Because of this the amines react with acids to form water soluble salts. This permits the easy separation of amines from other compounds. A crude reaction mixture can be extracted with dilute hydrochloric acid such that any amines present are protonated and dissolve into the aqueous phase as water-soluble salts. The free amine can be recovered by adding sodium hydroxide to the aqueous solution such that the free amine precipitates out as a solid are as an oil.

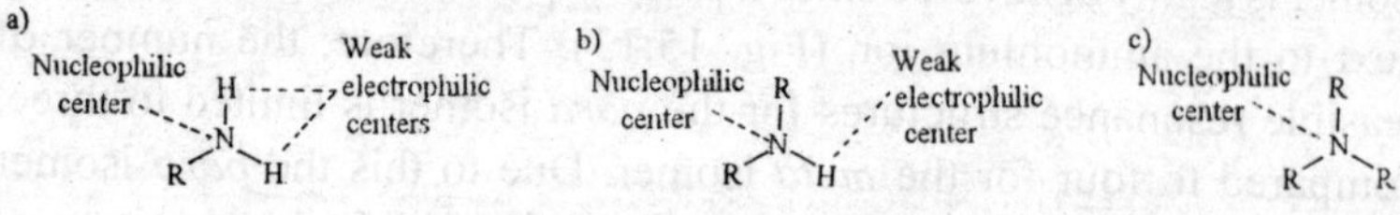

Fig. 15.14. Nucleophilic and electrophilic centers in (a) primary, (b) secondary, and (c) tertiary amines.

Amines will also react as nucleophiles with a wide range of electrophiles including alkyl halides, aldehydes, ketones, and acid chlorides.

The N–H protons of primary and secondary amines are weakly electrophilic or acidic and will react with a strong base to form amide anions. For example, diisopropylamine (pK_a ~40) reacts with butyllithium to give lithium diisopropylamide (LDA) and butane.

15.3 Reactions and Amines

Alkylation

Ammonia, primary amines and secondary amines (both aromatic and aliphatic) can undergo the S_N2 reaction with alkyl halides to produce a range of primary, secondary, and tertiary amines. Primary, secondary, and tertiary amines are produced as ammonium salts that are converted to the free amine by treatment with sodium hydroxide (Fig. 15.15a).

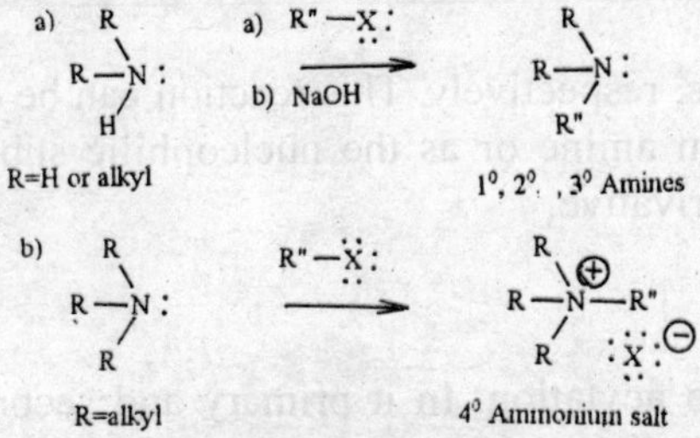

a) $R_2N(H)$: + a) $R''-X$; b) NaOH → R_2NR'' (R=H or alkyl; 1°, 2°, 3° Amines)

b) R_3N: + $R''-X$ → $R_3N^{+}R''\ X^{-}$ (R=alkyl; 4° Ammonium salt)

Fig. 15.15. Alkylation of amines.

Theoretically, it should be possible to synthesize primary amines from ammonia, secondary amines from primary amines, and tertiary amines from secondary amines. Generally, over-alkylation occurs e.g. reaction of ammonia with methyl iodide yields a mixture of primary, secondary, and tertiary amines along with a small quantity of the quaternary ammonium salt (Fig. 15.16).

Alkylation of tertiary amines by this method is a good method of obtaining quaternary ammonium salts (Fig. 15.15b) as no other products are possible. However, alkylation of lower order amines is not so satisfactory.

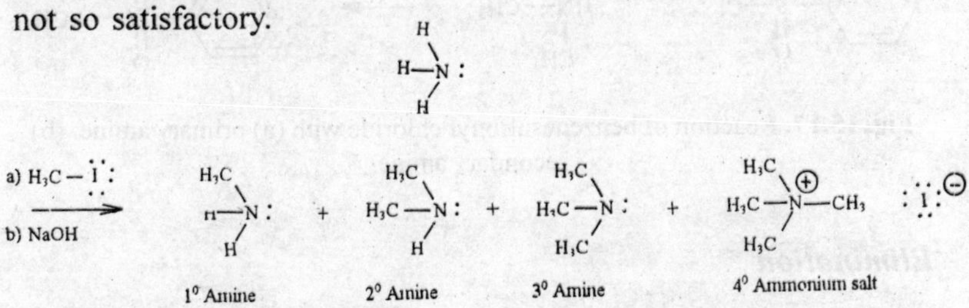

NH_3 + a) H_3C-I ; b) NaOH → CH_3NH_2 (1° Amine) + $(CH_3)_2NH$ (2° Amine) + $(CH_3)_3N$ (3° Amine) + $(CH_3)_4N^{+}\ I^{-}$ (4° Ammonium salt)

Fig. 15.16. Alkylation of ammonia with methyl iodide.

A better method of alkylating a primary or secondary amine is to treat the amine with a ketone or an aldehyde in the presence of a reducing agent such as *sodium cyanoborohydride*. This reaction is called a **reductive amination**. Over-alkylation cannot occur by this method.

Acylation

Primary and secondary amines (both aromatic and aliphatic) can be acylated with an acid chloride or acid anhydride to yield secondary and tertiary amides, respectively. This reaction can be considered as the acylation of an amine or as the nucleophilic substitution of a carboxylic acid derivative.

Sulfonylation

It is similar to acylation. In it primary and secondary amines (both aromatic and aliphatic) can be treated with a sulfonyl chloride to yield a sulfonamide (Fig. 15.17). Tertiary amines do not give a stable product and are recovered unchanged.

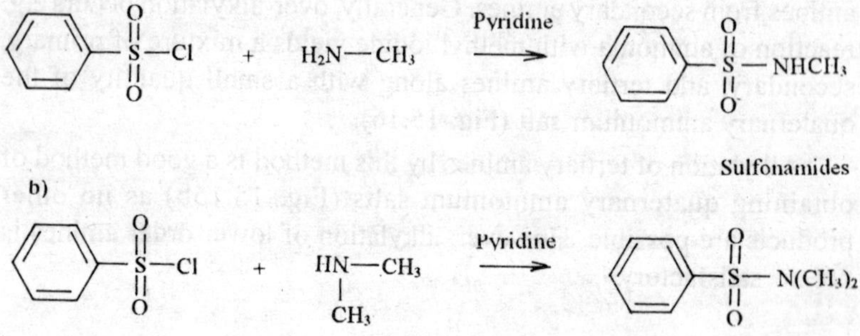

$$C_6H_5SO_2Cl + H_2N{-}CH_3 \xrightarrow{\text{Pyridine}} C_6H_5SO_2{-}NHCH_3$$

Sulfonamides

b)

$$C_6H_5SO_2Cl + HN(CH_3){-}CH_3 \xrightarrow{\text{Pyridine}} C_6H_5SO_2{-}N(CH_3)_2$$

Fig. 15.17. Reaction of benzenesulfonyl chloride with (a) primary amine; (b) secondary amine.

Elimination

Primary amines can be converted to alkenes if we can eliminate ammonia from the molecule. However, the direct elimination of ammonia is not possible. A less direct method of achieving the same is to exhaustively methylate the amine by the S_N2 reaction to obtain a quaternary ammonium salt. From this salt, it is possible to eliminate triethylamine in the presence of silver oxide and to form the desired alkene. The reaction is known as the **Hofmann elimination** (Fig. 15.18). The silver oxide provides a hydroxyl ion that acts as the base for an E2 elimination (Fig. 15.19). However, unlike most E2 eliminations, the less substituted alkene is preferred if a choice is available (Fig. 15.20). The reason for this preference is not fully understood, but may have something to do with the large bulk of the triethylamine leaving group hindering the approach of the hydroxide ion such that it approaches the least hindered β-carbon.

Secondary and tertiary amines can also be exhaustively methylated then treated with silver oxide. However, mixture of different alkenes may be obtained if the *N*-substituents are different alkyl groups (Fig. 15.21).

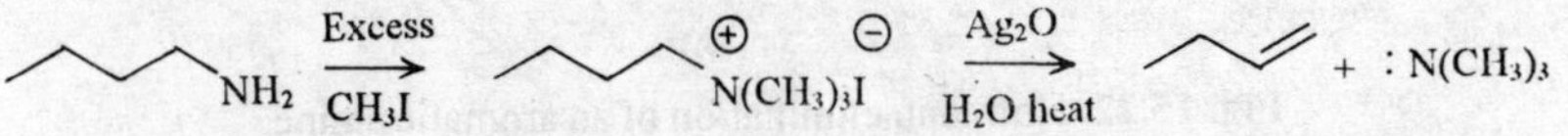

NH_2 Excess CH_3I → $N(CH_3)_3I$ Ag_2O H_2O heat → + : $N(CH_3)_3$

Fig, 15.18. Hofmann elimination.

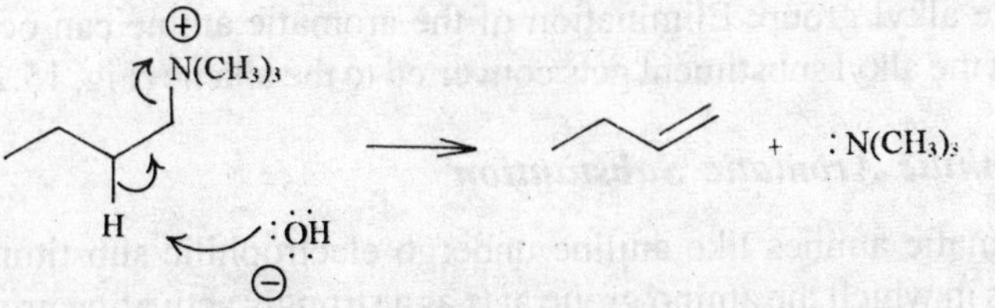

$N(CH_3)_3$ H :ÖH → + : $N(CH_3)_3$

Fig. 15.19. Mechanism of the Hofmann elimination.

NH_2 Excess CH_3I → $N(CH_3)_3I$ β Least β binder

Ag_2O H_2O heat → Minor alkene + Major alkine + : $N(CH_3)_3$

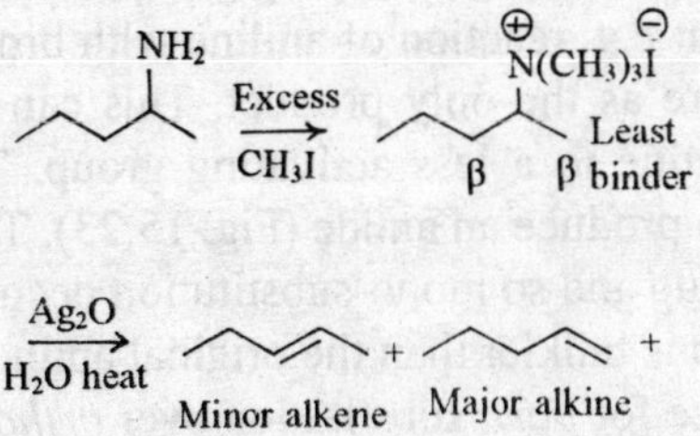

Fig. 15.20. The less substituted alkene is preferred in the Hofmann elimination,

NH Excess CH_3I → β $N(CH_3)_2I$ β Ag_2O H_2O heat → +

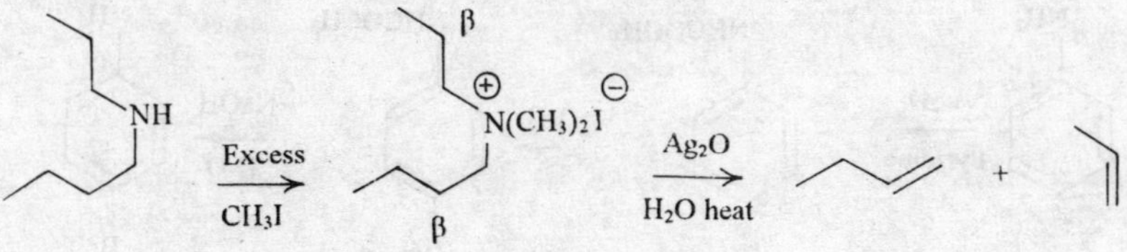

Fig. 15.21. Hofmann elimination of a secondary amine.

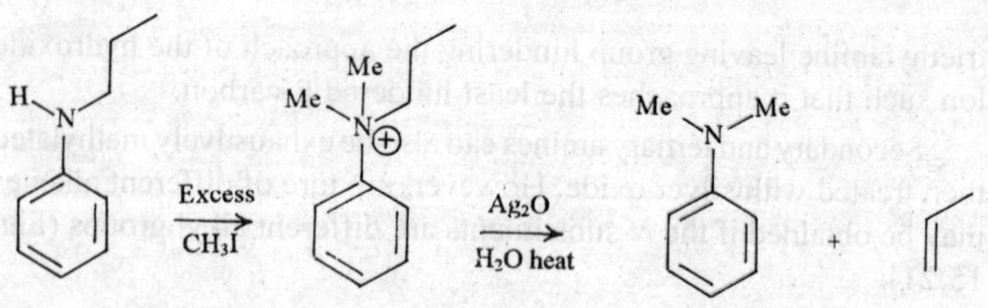

Fig. 15.22. Hofmann elimination of an aromatic amine.

The *Hofmann elimination is not possible with primary arylamines*, but secondary and tertiary arylamines will react of the substituents is a suitable alkyl group. Elimination of the aromatic amine can occur such that the alkyl substituent gets converted to the alkene (Fig. 15.22).

Electrophilic Aromatic Substitution

Aromatic amines like aniline undergo electrophilic substitution reactions in which the amino group acts as a strongly activating group, directing substitution to the *ortho* and *para* positions. Like phenols, the amino group is such a strong activating group that more than one substitution may occur e.g. reaction of aniline with bromine yields a tribrominated structure as the only product. This can be overcome by converting the amine to a less activating group. This involves acylating the group to produce an amide (Fig. 15.23). This group is a weaker activating group and so mono-substitution occurs. Moreover, since the amide group is bulkier than the original amino group, there is more of a preference for *para* substitution over *ortho* substitution. Once the reaction has been carried out the amide can be hydrolyzed back to the amino group.

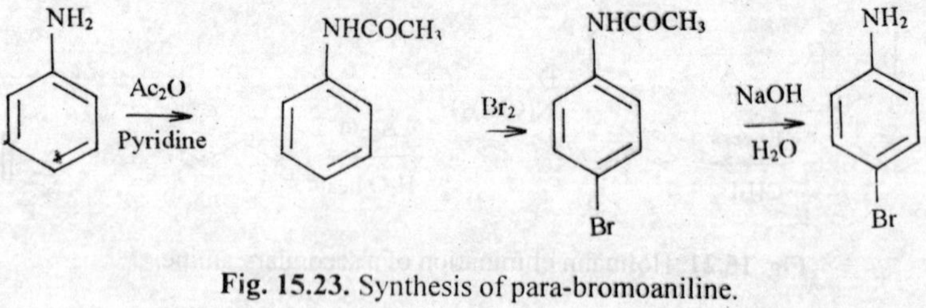

Fig. 15.23. Synthesis of para-bromoaniline.

Anilines can be sulfonated and nitrated, but the Friedel-Crafts alkylation and acylation are not possible because the amino group forms an acid base complex with the Lewis acid needed for this reaction. To overcome we convert the aniline to the amide before carrying out the reaction.

Diazonium Salts

Primary arylamines or anilines can be converted to diazonium salts, that in turn can be converted to a large variety of substituents (Fig. 15.24.)

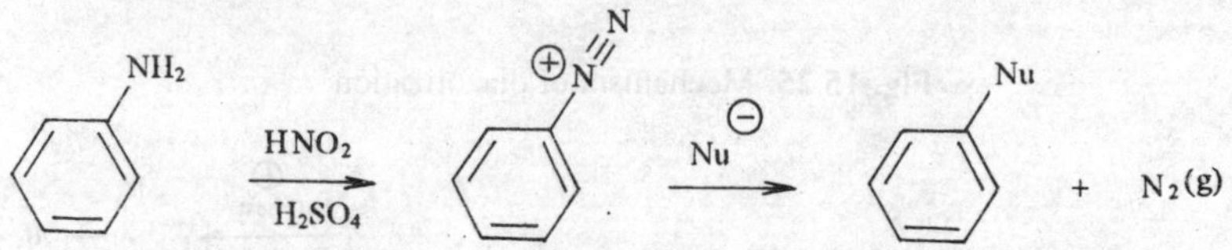

$$C_6H_5NH_2 \xrightarrow[H_2SO_4]{HNO_2} C_6H_5N^{+}\equiv N \xrightarrow{Nu^{-}} C_6H_5Nu + N_2(g)$$

Fig. 15.24. Synthesis and reactions of diazonium salts.

Reaction of an aniline with nitrous acid forms the stable diazonium salt and the process is called **diazotization** (Fig. 15.25). In the strong acid conditions, the nitrous acid dissociates to form an ^{+}NO ion that can then act as an electrophile. The aromatic acid uses its lone pair of electrons to form a bond to this ^{+}NO ion. Loss of a proton from the intermediate formed, followed by a proton shift leads to the formation of a diazohydroxide. The hydroxide group is then protonated turning it into a good leaving group, and a lone pair from the aryl nitrogen forms a second π bond between the two nitrogen atoms and expels water.

Once the diazonium salts has been formed, it can be treated with various nucleophiles such as Br^{-}, Cl^{-}, I^{-}, ^{-}CN and ^{-}OH (Fig. 15.26). The nucleophile displaces nitrogen from the aromatic ring and the nitrogen which is formed is lost from the reaction mixture as a gas, and helps to take the reaction to completion. Those reactions involving Cu(I) are also called as the **Sandmeyer reaction**.

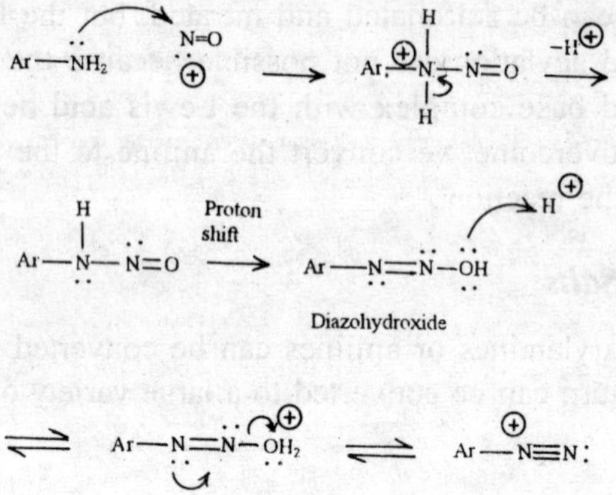

Fig. 15.25. Mechanism of diazotization.

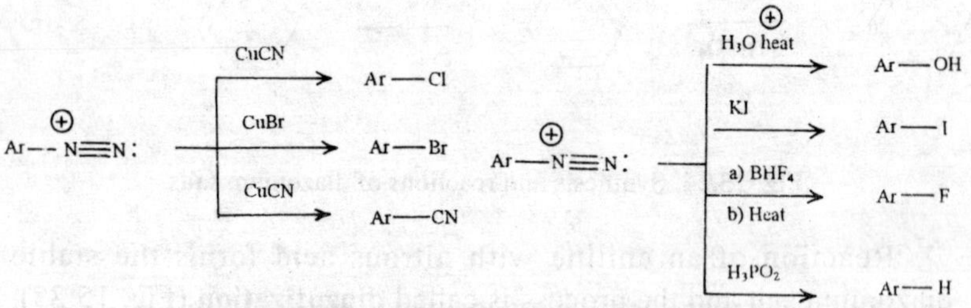

Fig. 15.26. Reactions of diazonium salts.

Diazonium salts are also used in a reaction called **diazonium coupling** where the diazonium salt is coupled to the *para* position of a phenol or an arylamine (Fig. 15.27). The azo products obtained have an extended conjugated system that includes both aromatic rings and the N=N link. Due to this, these compounds are generally colored and are used as dyes.

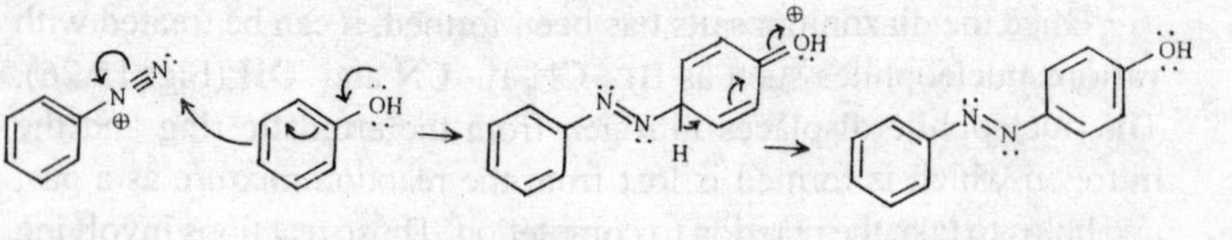

Fig. 15.27. Diazonium coupling.

The above coupling is more efficient if the reaction is done under slightly alkaline conditions (NaOH) such that the phenol is ionized to a phenoxide ion (ArO^-). Phenoxide ions are more reactive to electrophilic addition than phenols themselves. Strong alkaline conditions cannot be used since the hydroxide ion adds to the diazonium salt and prevents coupling. If the *para* position of the phenol is already occupied, diazo coupling can occur at the *ortho* position instead.

Aliphatic amines, as well as secondary and tertiary aromatic amines, react with nitrous acid, but these reactions are less useful in organic synthesis.

15.4 Chemistry of Nitriles

Preparation

Nitriles can be prepared by the S_N2 reaction of a cyanide ion with a primary alkyl halide. However, this limits the nitriles that can be synthesized to those having the following general formula RCH_2CN. A more general synthesis of nitriles involves the dehydration of primary amides with reagents such as thionyl chloride or phosphorus pentoxide (Fig. 15.28).

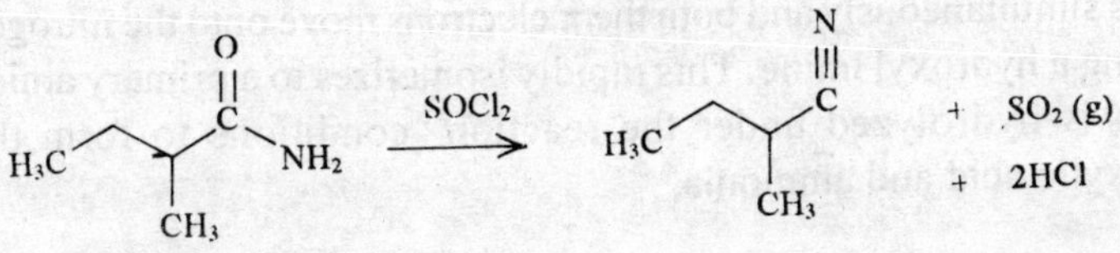

$$H_3C\text{-}CH_2\text{-}CH(CH_3)\text{-}C(=O)NH_2 \xrightarrow{SOCl_2} H_3C\text{-}CH_2\text{-}CH(CH_3)\text{-}C{\equiv}N + SO_2\,(g) + 2HCl$$

Fig. 15.28. Dehydration of primary amides with thionyl chloride.

Properties

The nitrile group (CN) is linear in shape with both the carbon and the nitrogen atoms being *sp* hybridized. The triple bond linking the two atoms consists of one σ bond and two π bonds. Nitriles are strongly polarized. The nitrogen is a nucleophilic center and the carbon is an electrophilic center. Nucleophiles react with nitriles at the electrophilic carbon (Fig. 15.29). Generally, the nucleophile will form a bond to the electrophilic carbon resulting in simultaneous breaking of one of the π bonds. The π electrons end up on the nitrogen to form an sp^2 hybridized imine anion which then react further to give different products depending on the reaction conditions used.

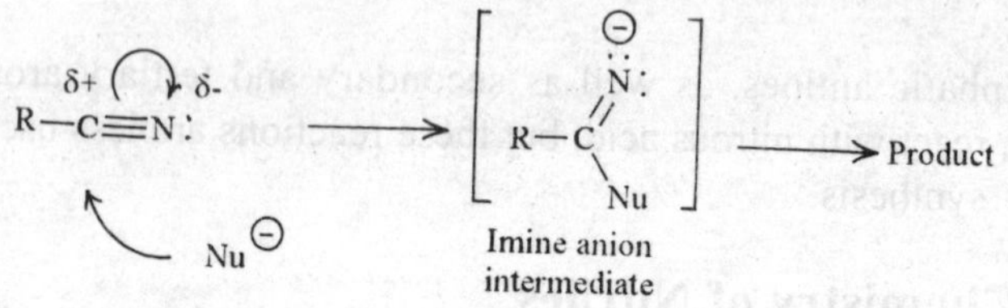

Imine anion intermediate

Fig. 15.29. Reaction between nucleophile and nitriles.

Reactions

Nitriles (RCN) get hydrolyzed to carboxylic acids (RCO_2H) in acidic or basic aqueous solutions. The mechanism of the acid-catalyzed hydrolysis (Fig. 15.30) involves initial protonation of the nitrile's nitrogen atom. This activates the nitrile group towards nucleophilic attack by water at the electrophilic carbon. One of the nitrile π bonds breaks simultaneously and both the π electrons move onto the nitrogen yielding a hydroxyl imine. This rapidly isomerizes to a primary amide which is hydrolyzed under the reaction conditions to form the carboxylic acid and ammonia.

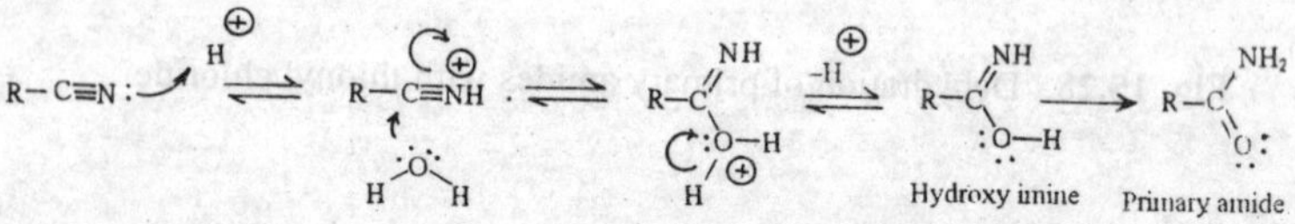

Hydroxy imine Primary amide

Fig. 15.30. Acid-catalyzed hydrolysis of nitrile to carboxylic acid.

Nitriles (RCN) can be reduced to primary amines (RCH_2HN_2) with lithium aluminum hydride that provides the equivalent of a nucleophilic hydride ion. The reaction can be explained by the nucleophilic attack of two hydride ions (Fig. 15.31).

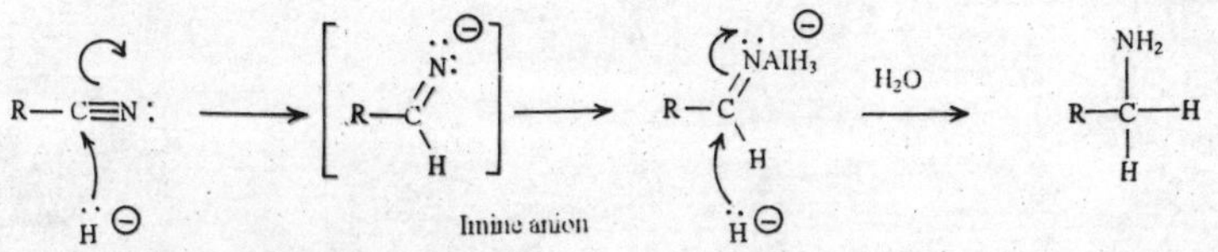

$$R{-}C{\equiv}N: \xrightarrow{H^{\ominus}} \left[R{-}C(H){=}\ddot{N}:^{\ominus} \right] \xrightarrow{H^{\ominus}} R{-}C(H){=}\ddot{N}AlH_3^{\ominus} \xrightarrow{H_2O} R{-}C(NH_2)(H){-}H$$

Imine anion

Fig. 15.31. Reduction of nitriles to form primary amines.

With a milder reducing agent like DIBAH (diisobutylaluminum hydride), the reaction stops after the addition of one hydride ion, and an aldehyde is obtained instead (RCHO).

Grignard Reaction

Nitriles can be treated with Grignard reagents or organolithium reagents to give ketones (Fig. 15.32).

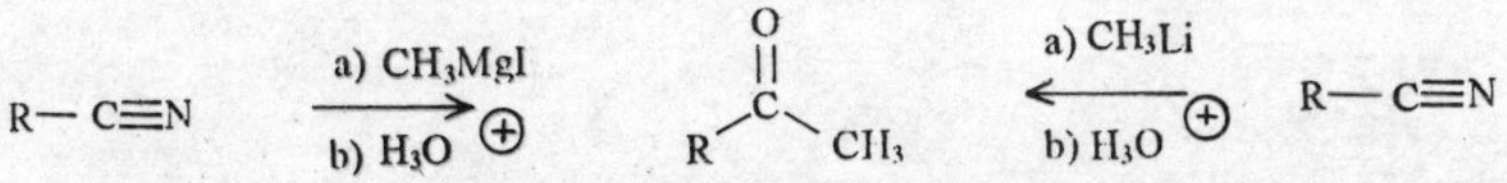

$$R{-}C{\equiv}N \xrightarrow[\text{b) } H_3O^{\oplus}]{\text{a) } CH_3MgI} R{-}C(=O){-}CH_3 \xleftarrow[\text{b) } H_3O^{\oplus}]{\text{a) } CH_3Li} R{-}C{\equiv}N$$

Fig. 15.32. Nitriles react with Grignard reagent or organolithium reagents to produce ketones.

Grignard reagents provide the equivalent of a nucleophilic carbanion which can attack the electrophilic carbon of a nitrile group (Fig. 15.33). One of the π bonds is broken simultaneously forming an intermediate imine anion that is converted to a ketone on treatment with aqueous acid.

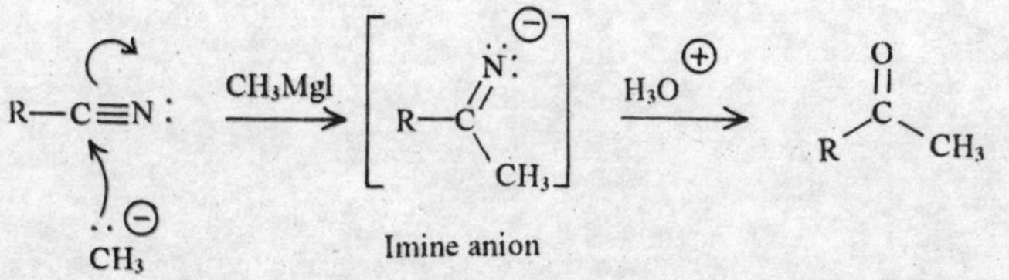

$$R{-}C{\equiv}N: \xrightarrow{CH_3MgI\ (\ddot{C}H_3^{\ominus})} \left[R{-}C(CH_3){=}\ddot{N}:^{\ominus} \right] \xrightarrow{H_3O^{\oplus}} R{-}C(=O){-}CH_3$$

Imine anion

Fig. 15.33. Mechanism of the Grignard reaction on a nitrile group.